Routledge Contemporary Human
Geography

Geographies of Globalization

Warwick E. Murray

Routledge
Taylor & Francis Group

LONDON AND NEW YORK

First published 2006
by Routledge
2 Park Square, Milton Park, Abingdon, Oxon OX14 4RN

Simultaneously published in the USA and Canada
by Routledge
270 Madison Ave, New York, NY 10016

Routledge is an imprint of the Taylor & Francis Group

© 2006 Warwick E. Murray

Typeset in Times Franklin Gothic by
Keystroke, Jacaranda Lodge, Wolverhampton
Printed and bound in Great Britain by
MPG Books Ltd, Bodmin

British Library Cataloguing in Publication Data
A catalogue record for this book is available from the British Library

Library of Congress Cataloging in Publication Data
Murray, Warwick E.
    Geographies of globalization / Warwick E. Murray.
      p. cm. – (Routledge contemporary human geography series)
    Includes bibliographical references and index.
    ISBN 0–415–31799–1 (hardcover : alk. paper) – ISBN 0–415–31800–9
    (softcover : alk. paper) 1. Human geography–Economic aspects.
    2. Human geography–Political aspects. 3. Globalization. I. Title.
    II. Series.
    GF50.M87 2005
    303.48'2–dc22
    2005006891

ISBN10: 0–415–31799–1 ISBN13: 9–78–0–415–31799–3 (hbk)
ISBN10: 0–415–31800–9 ISBN13: 9–78–0–415–31800–6 (pbk)

Geographies of Globalization

Geographies of Globalization offers a lively exploration of the
geographical impacts of globalization and the distinctive contribution
of human geography to studies and debates in this field. Fully up-to-date
and engaging, this work:

- Critically appraises the concept and processes of globalization from
  a geographical perspective
- Debates the historical evolution of globalized society
- Illustrates how the core principles of human geography – such as
  space and scale – lead to a better understanding of the phenomenon
- Analyses the interconnected economic, political and cultural
  geographies of globalization
- Examines the impact of global transformations 'on the ground' using
  examples from six continents
- Discusses the challenges for the environment and the Third World
  created by globalizing processes
- Articulates a human geographical framework for progressive
  globalization

Boxed sections highlight key concepts and innovative work by
geographers. The book is also illustrated with a wide range of figures,
photographs, and maps. *Geographies of Globalization* is indispensable
for human geography undergraduates and postgraduates studying the
phenomenon in both dedicated and linked courses.

Warwick E. Murray is Senior Lecturer in Human Geography and
Development Studies at Victoria University of Wellington, New Zealand.

revolutionary
evolutionary

Routledge Contemporary Human Geography Series

Series Editors:
David Bell, Manchester Metropolitan University
Stephen Wyn Williams, Staffordshire University

This series of texts offers stimulating introductions to the core subdisciplines of human geography. Building between 'traditional' approaches to subdisciplinary studies and contemporary treatments of these same issues, these concise introductions respond paticularly to the new demands of modular courses. Uniformly designed, with a focus on student-friendly features, these books will form a coherent series which is up-to-date and reliable.

Existing Titles:

Urban Geography, Second edition

Cultural Geography

Tourism Geography

Development Geography

Political Geography

Geographies of Globalization

Para Gabriela, Francisca, y Ximena
and
For Ronald Stokes

Contents

Plates

Figures

Tables

Cartoons

Maps

Boxes

Acknowledgements

In another book in this series, *Urban Geography*, Tim Hall wrote that writing a book is like winning an Oscar – it involves a lot of people. The last person from New Zealand to win an Oscar, Peter Jackson, was accused of thanking the whole of New Zealand by his hosts. Please don't worry, I am not going to do that. However, as this is my first book I do have a relatively long list of people I would like to acknowledge for personal and professional reasons. First of all, I would like to thank Rob Imrie and five other anonymous reviewers for their comments at various stages of this book. The patience of the editorial staff at Routledge, especially Andrew Mould and Zoe Kruze, has been saintly. For opening my eyes to the wonders of our subject, I am particularly indebted to Dorothy Kennedy and Martin Ganderton for their inspiring early lessons. As an undergraduate and postgraduate at Birmingham University, UK, I was lucky to be taught by Anne Gellately, Brian Giles, Peter Daniels, Michael Bradshaw and James Sidaway. The latter two, in particular, have continued to support me professionally and intellectually, for which I am very grateful. I would like to acknowledge the friendship and stimulation of colleagues at the University of the South Pacific, Victoria University of Wellington and beyond. In no particular order, these include: Cristóbel Kay, Patrick Nunn, Randy Thaman, John Soulsby, Philip Morrison, Vijay Naidu, Tim Vowles, Richard Willis, Jonathan Barton, Juan Salazar, Alfredo Apey and Sean Weaver. A special thank you to Sara Kindon for helping me to broaden the way I look at things (a bit, but I am not yet a post-structuralist!). In the New Zealand academic community Richard LeHeron, John Overton and Donovan Storey convinced me to come here in the first place and have made me feel welcome, especially considering that I am a pom. Thank you to the latter for 'lending' me some of his best photos for the book. The grants from the School of Earth Sciences to help me complete this work are gratefully acknowledged. My students have also played a much bigger role than they probably realize in shaping the way I think. Edward Challies, Tony Chandler and Anton Griffith have been especially stimulating and fun to have around, and the latter drew and compiled many of the maps and diagrams in the book. My greatest intellectual debt, however, is to Robert Gwynne, who has been an invaluable mentor and friend over the past fourteen years – *gracias Tio Bob*.

I would like to give thanks to Louis Hellman for his permission to reprint cartoon 1.1, Kirk Anderson for cartoon 5.1 and 7.1, Routledge for figures 3.3, 5.1, 5.2, 8.1, 8.2 and 8.3, Getty for plates 1.1, 5.1 and 5.2, Donovan Storey for plates 1.2, 1.3, 4.5 and 7.2, Robert N. Gwynne for plate 4.6, The Potteries Museum and Art Gallery Stoke on Trent for plate 4.1, Still Pictures for plate 4.2, April Henderson for plate 6.9 and Asia Pacific Viewpoint (Blackwell/Victoria University of Wellington) for plate 8.1. If any unintentonal use of copyright material has been made in this book, the author would be grateful if the copyright owners could contact him via the publisher. Every effort has been made to trace the owners of copyright material.

For personal reasons I would like to express gratitude to Mark Thomas and Aaron Nonoa of *Strange Brew*, as well as Craig Nicholson, each of whom has helped me practise the globalization of music until very late at night at least once a week for the past three years. My greatest debt is to my family, however. Thank you to Carmen Espina, Enrique Garcia and Jochen Hess for supporting my Chilean research efforts. My sincere thanks to Ray and Jean Wright, Irene Stokes and – most of all – Stafford Murray and Lynda Stokes for encouraging me to study, and putting up with my being half-way across the world most of the time. Ronald Stokes, who was my first geography teacher, has a special place in everything I do. Finally, I would like to thank Gabriela and Francisca for not crying when I close the study door, for sharing their toys and for being truly wonderful. Ximena, thanks for your unconditional support over the years, but, most importantly, thank you for completing a loving home.

Warwick E. Murray, Wellington, New Zealand

Abbreviations

AFTA	ASEAN Free Trade Area
AGM	Anti-globalization Movement
ANCOM	Andean Community
APEC	Asia-Pacific Economic Cooperation
ARENA	Action, Research, Education Network Aotearoa
ASEAN	Association of Southeast Asian Nations
AUSAID	Australian Agency for International Development
BMG	Bertelsmann Music Group
CARCOM	Caribbean Common Market
CENTO	Central Treaty Organization
CET	Common external tariff
CJD	Creutzfeldt-Jakob Disease
DIRECON	Office of Regional Trade Agreements, Chile
ECSC	European Coal and Steel Community
EFTA	European Free Trade Area
EMI	Electric and Musical Industries Ltd
ETA	Euskadi Ta Askatasuna (Basque Fatherland and Liberty) – Basque Separatists
EU	European Union
EURATOM	European Atomic Energy Commission
EZLN	Ejercito Zapatista Liberacíon Nacional
FAO	Food and Agriculture Organization
FDI	Foreign direct investment
FTAA	Free Trade Area of the Americas
G3	USA, Japan and the EU
G7	Canada, France, Germany, Italy, Japan, United Kingdom, USA
G8	G7 + Russia
GATT	General Agreement on Tariffs and Trade

GCC	Global commodity chain
GDP	Gross domestic product
GM	Genetic modification
GNP	Gross national product
GUUAM	Georgia, Ukraine, Uzbekistan, Azerbaijan and Moldova
HDI	Human Development Index
HYV	High-yielding variety
IBRD	International Bank for Reconstruction and Development (World Bank)
ILO	International Labour Office
IMF	International Monetary Fund
INGO	International Non-government Organization
IPCC	International Panel on Climate Change
ISI	Import substitution industrialization
KFC	Kentucky Fried Chicken
LDC	Less developed country
MAD	Mutually assured destruction
MERCOSUR	Mercado Común del Sur (Common Market of the Southern Cone)
MIRAB	Migration, remittances, aid and bureaucracy
MNC	Multi-national company
MTV	Music television
NAFTA	North American Free Trade Agreement
NATO	North Atlantic Treaty Organization
NGO	Non-government organization
NIC	Newly industrializing countries
NIDL	New international division of labour
NSM	New social movement
NTAX	Non-traditional agricultural exports
NZAID	New Zealand Agency for International Development
OECD	Organization for Economic Cooperation and Development
OPEC	Organization of Petroleum Exporting Countries
PGA	People's Global Action
PIC	Pacific Island country
RIAA	Recording Industry Association of America
R&D	Research and development
RTA	Regional Trade Agreement
SAP	Structural Adjustment Programme
SEATO	Southeast Asia Treaty Organization
TNC	Transnational corporation/company

TRIAD	USA, Japan, EU
UN	United Nations
UNCTAD	United Nations Conference on Trade and Development
UNDP	United National Development Programme
UNEP	United Nations Environment Programme
UNESCO	United Nations Educational, Scientific and Cultural Organization
UNFCCC	United Nations Framework Convention on Climate Change
USSR	United Soviet Socialist Republics
WB	World Bank (International Bank for Reconstruction and Development)
WEF	World Economic Forum
WHO	World Health Organization
WST	World Systems Theory
WTO	World Trade Organization

Part I
Transformed geographies

Geography is dead?
The rise of globalization

- Globalization and the end of geography
- Anti-globalization – what is all the fuss about?
- Competing discourses of globalization
- Defining globalization
- Geography and the study of globalization
- New globalizing spaces?
- This book

Globalization and the end of geography

We live in a world that many of our grandparents could never have dreamed of, and that it is an increasingly 'small world' is commonly heard in conversation. Much of this is tied up with the twentieth-century revolution in transport and communications technology that has captured the popular imagination. It is possible, for example, to journey around the world in just over a day on commercial airlines. Less than fifty years ago, the journey from England to Australia took about a week. In 1870, it would take seventy days for surface mail to travel from London to New Zealand. With the advent of the telephone, faxes, and more recently e-mail and video conferencing, such communication has become almost instant. We live, if the hyperbole is to be believed, in a 'global village' and the term that is most commonly used to refer to this apparent shrinking is 'globalization' (see Figure 1.1).

Increasingly used to rationalize a wide range of economic and political policies, and to explain a plethora of cultural, social and economic processes and outcomes, the concept of 'globalization' has assumed enormous power. Despite this, it is not always well defined or critically appraised in either popular or even academic usage. A common image of globalization is one of a process that unfolds like a blanket across the globe, homogenizing the world's economies, societies and cultures as it falls. Everywhere becomes the same, boundaries don't matter and distance disappears.

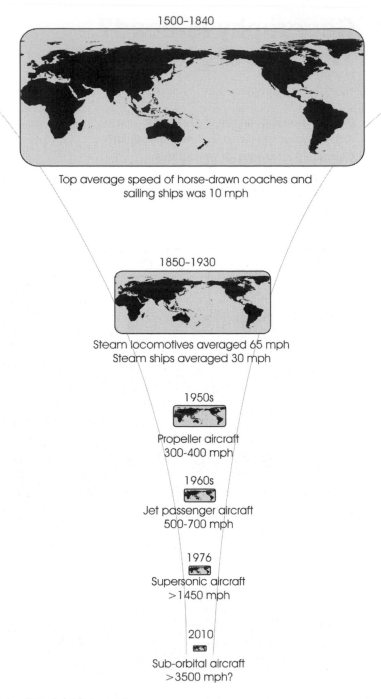

Figure 1.1 *The shrinking world*

Source: Adapted from Dicken (1998)

The end of geography?

People have been predicting the death of geography for nearly four decades on the basis of the trends discussed above. Toffler (1970) led this perspective in *Future Shock*, arguing that the evolution of transport and communications technologies and the intensified flows of people that have resulted imply that 'place is no longer a primary source of diversity' (quoted in Leyshon, 1995, p. 35). More recently, the subtitle of O'Brien's *Global Financial Integration* proclaimed 'the end of geography' – a situation involving 'a state of economic development where geographical location no longer matters' (O'Brien, 1992, p. 1) (see Cartoon 1.1).

Cartoon 1.1 *Homogenized destinations*
Source: Louis Hellman

There can be little doubt that, relative to the past, people and processes in 'far-off' places can have almost instant 'local' impacts. Geographers and other social scientists have used a number of other terms for this process including 'annihilation of space by time', 'time–space convergence' and 'time–space compression' (see Figure 1.1) (Harvey, 1989). Economic *contagion* has become a new buzz-word in financial sectors, for example. The Asian 'crisis' of 1997 quickly spread around the Pacific Asia region from Thailand, to South Korea, and on to Japan, and the ramifications of the collapse had a rapid and tangible impact on economies all across the world to varying degrees. But this 'shrinking' appears to be permeating all spheres of human activity. Twenty-four-hour news providers, such as CNN and the BBC, broadcast unfolding events, such as the aftermath of the 2004 Indian Ocean tsunami, in real time to people's homes and offices across the planet and this prompts an instant outpouring of relief donations from around the globe. Carbon dioxide emissions in the USA or China have a direct impact on small island states as sea level rises and cyclone events become more common in localities such as Grenada in the Caribbean or Niue in the Pacific. Feasibly, then, we could talk of increased cultural, social, environmental and political *contagion* given the new fluidity of global flows. What is for sure is that the way that many of us experience the world is shifting in sometimes dizzying ways and that a revolution in technologies of interaction has played a central role in this (see Table 1.1).

Popular notions of what globalization is often misunderstand what geography is both as an entity and an academic discipline, and fail to appreciate how contemporary geographers define central components of their analyses such as space, place, scale and location. The core aim of this book is to illustrate that as the processes of globalization fundamentally alter the way people, commodities and information flow and interact, this creates new and complex geographies. There are two notes of caution that need to be sounded about the 'shrinking world' concept immediately. First, *the relative distance between some places and people has become greater*. For example, the income gap between the poorest and richest countries and peoples has increased over the past fifty years (Potter *et al.*, 2004). Those who are hooked up to the internet may enjoy rapid communications with distant friends who can appear to be round the corner, but the majority who do not have access to such technologies have become *relatively* more isolated. The fact that 'shrinking' technology can drive places apart is illustrated in Map 1.1,

Table 1.1 *Transport and communication milestones*

Date	Event
c. 60,000	*Homo sapiens'* colonization of Australia implies the construction of seaworthy boats
c. 2900	Egyptian hieroglyphic script fully developed
c. 2500	Chariots in Mesopotamia, skis in Scandinavia
c. 1200	Phoenician alphabet of 22 letters invented
c. 220	Three-masted ships in Greece
271	Magnetic compass invented in China
c. 650	Chinese scholars develop technique for printing texts from engraved wooden blocks
1403	Moveable metal type first used for printing in Korea
1522	Spanish ships returning from Magellan's voyage complete the first circumnavigation of the world
1807	First steamboat service begins operation on Hudson river, USA
1825	Stockton–Darlington railway, Britain, opens with locomotive designed by George Stephenson
1843	*SS Great Britain* is first propeller-driven ship to cross the Atlantic
1866	First transatlantic cable is successfully laid
1876	Telephone is invented by US scientist Alexander Graham Bell
1878	First electric railway is demonstrated in Germany by Werner von Siemens
1886	Four-wheeled petrol-powered automobile is built by German engineer Gottlieb Daimler
1901	Marconi transmits radio signals across the Atlantic
1903	Wright brothers make first sustained flight of powered aircraft
1908	Mass production of cars begins with Model T Ford in USA
1927	First public broadcast of television is transmitted in Britain
1941	First flight of jet-powered plane, with engine designed by British engineer Frank Whittle, takes place
1951	US engineers build UNIVAC I, the first commercial computer
1957	Soviet Union launches an artificial satellite – *Sputnik I*
1962	US communications satellite Telstar is launched
1964	Japanese railways begin running high-speed 'bullet' trains
1970	Boeing 747 enters service
1975	A personal computer (PC) in kit form goes on sale in the USA
1984	A cellphone network is launched in Chicago, USA
1994	World Wide Web is created
1997	Mobile internet technologies emerge (WAP)

which shows the simultaneous time–space convergence and divergence of the world measured in terms of the cost of a minute-long phone call from the USA in 2000. In short, rather than smothering the earth like a blanket, globalization has cast a net across it and this has increased spatial differentiation accordingly.

A second point is that it is tempting to run away with the idea that technology itself is *driving* the processes of spatial shrinking and that therefore globalization is technologically determined. Underpinning technological change are social, cultural, economic and political processes and human agendas. Globalization processes are *facilitated* by technological change but are driven by much more fundamental forces. The study of globalization has political overtones precisely because it widens gaps, creates unevenness and creates winners and losers. This is why globalization can inspire fear and loathing among some, while being seen as a saviour by others. In short, globalization is about both processes and political-economic agendas.

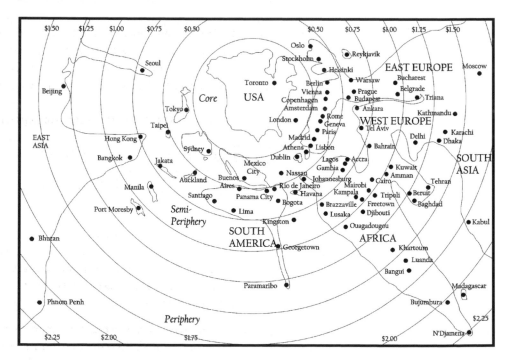

Map 1.1   *Time–space convergence and divergence*

Source: Adapted from Gwynne *et al.* (2003)

These initial qualifications to the 'shrinking world' concept expose some of the weaknesses of 'geography is dead' arguments. Contemporary research in human geography and some other social sciences points towards the differentiating impacts of 'global' processes as they interact with 'local' places, institutions and people. It is indeed true that the world's economies and cultures are increasingly interconnected, and that 'global' forces are penetrating even the most remote and peripheral regions and localities on Earth. However, as such processes are articulated and resisted in *specific* places, complete with particular histories, societies and environments, they create more uneven geographies. These geographies of globalization, although increasingly dynamic, sometimes transient and therefore harder to grasp, are none the less significant, and have real social, economic and political implications. Underpinning this book is the proposition that understanding globalization, and attempting to regulate and reform it, requires that geography be taken more seriously.

Anti-globalization – what is all the fuss about?

Every reader will be aware that globalization has been a site of enormous struggle over recent years. Images of 'anti-globalization' protests have been flashed across our television screens, beginning with Seattle in 1999. Predictably, mainstream media have often portrayed such protests as violent and anarchistic. There can be no doubt that huge passions have been stirred, but in general we have witnessed the evolution of a peaceful global protest movement (see Plate 1.1). Although this is complex, diverse and incoherent in many respects (see Chapter 5), there can be no doubt that the movement has fulfilled a major purpose in bringing the issue of globalization and its perceived regressive impacts to the attention of the global public. One of the great ironies, as is explored in Chapter 5, is that the movement has used the technologies of globalization, notably the internet, to spread its message. The movement has attracted some unlikely supporters, such as former World Bank chief economist and economics Nobel prize winner Joseph Stiglitz, who writes:

> [f]ormerly uneventful meetings of obscure technocrats discussing mundane subjects such as concessional loans and trade quotas have now become the scene of raging street battles and huge demonstrations. The protests at the Seattle meeting of the World Trade Organisation in 1999 were a shock. Since then, the movement has grown stronger and the fury

has spread. Virtually every major meeting of the international Monetary
Fund, the World Bank, and the World Trade Organisation is now the
scene of conflict and turmoil. The death of a protestor in Genoa in 2001
was just the beginning of what may be many more casualties in the war
against globalization. . . . It is clear to almost everybody that something
has gone horribly wrong.

(Stiglitz, 2002, pp. 3–4)

'Anti-globalization' did not start in Seattle 1999 however, as the
mainstream media might have us believe. There have been similarly
motivated protests and battles across the world in response to corporate
globalization's greatest facilitator, neoliberalism (see Chapter 4),
for decades. More generally, resistance to capitalism has a long and
turbulent history over two centuries long. Neither are protests always as
spectacular as those that have made the recent headlines; daily resistance
to globalization and neoliberalism continues perpetually in the lives of
many marginalized in both the rich and poor worlds (Routledge, 2002).
There can be little doubt however that the movement is unprecedented
in terms of its breadth and reach, and that globalization has become one
of the most contentious issues of our times.

What has all of this got to do with time–space compression and the
popular notions associated with globalization as a process? In general,
the 'anti-globalization movement' has a particular belief in terms of
what it believes globalization is – what might be termed 'corporate/
neoliberal globalization'. The movement's supporters, to the extent
that it is possible to represent their opinions in one idea, believe that there
is a powerful state-sponsored corporate agenda at work which seeks to
spread free-market capitalism and that 'shrinking' processes provide a
'spatial fix' which helps achieve that goal (Harvey, 1989). But this is only
one way of conceiving of the phenomenon – the globalization debate is
one of competing discourses about the nature of geographical
interactions.

Competing discourses of globalization

The reaction of the so-called anti-globalization movement illustrates
the contested nature of globalization – it is perceived and represented
in many ways, none of which are necessarily 'correct' (Rupert, 2000;
Schirato and Webb, 2003; Scholte, 2000). Before attempting more formal
definitions it is useful to try to boil down some of the broad contours of

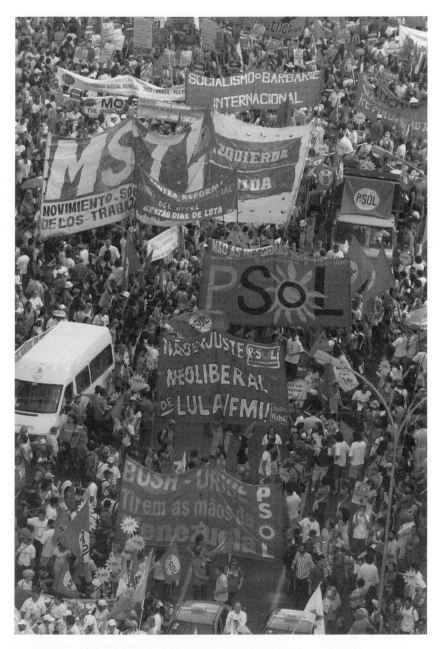

Plate 1.1 *Anti-globalization protests at the World Social Forum, Porto Alegre, Brazil* (AFP)

Source: Getty

the debate which are then taken up in depth through the remainder of this book. This debate is not just abstract and has real political ramifications. Although discourses overlap (see Chapter 2), it is useful to divide normative perspectives into three camps.

Pro-globalization

This perspective holds that the spread of market economics, competition, free trade and Western democracy – that is to say globalization as currently practised – are important *progressive* trends. Capitalism is seen as a moral good which promotes economic growth and efficiency and leads to global welfare gains overall. This is the type of view that is associated with the IMF and its sister institutions, the World Bank and the WTO. In general most national governments are of this opinion at the present time, especially those that are most powerful in the global capitalist economy such as the USA and the UK. The right-wing media also support and disseminate this view (Herman and Chomsky, 1988). Pro-globalists often conceive of globalization as a continuation of the logic of modernization and progress, and as something that is an inevitable part of human evolution. Underlying this view is usually the idea that technology is the driving force of progress. The political ramifications are 'there is no alternative' and citizens and nation-states will have to learn to participate for better or for worse. This view is closely associated with the *hyperglobalist* take on globalization (see Chapter 2).

Anti-globalization

As explored briefly above, this perspective holds that globalization is a threat to local society and environment in a way that echoes colonialism of the past. Some argue that those who bear the heaviest load are already marginalized groups, especially in the poor world. Under this view globalization is seen as increasing – not decreasing – the unevenness of development and perpetuating inequality in ways that are not necessarily reversible. Anti-globalists argue that pro-globalists make globalization *seem* inevitable because it is in the interests of to those who promote it – the business class, industrialists and rich countries. The political ramifications tend towards the revolutionary and against the institutions and corporations of global capitalism (Korten, 1995).

Alter-globalization

[handwritten margin notes: "evolutionary?", "driven by humans not technology?"]

This view posits that the nature of globalization is not predetermined and does not follow a given evolutionary path. Rather, it is a consequence of human actions and particular political choices. The political implications of this view are reformist; citizens and nation-states have a role to play in resisting and regulating it, and alternative and progressive globalizations are possible (see Chapter 9). As such, globalization can yield both positive and negative outcomes depending on how it is constructed. The major task then is changing the *nature* of globalization through human action – but not destroying it. A number of the groups lumped together under the 'anti-globalization' label might be more appropriately placed in this school of thought. This view is associated to a large degree with the *transformationalist* school of thought (see Chapter 2), and the majority of the work undertaken in human geography to date falls into this camp – although it has tended to be a little more radical than the weak or passive transformationalist view, as we will see.

Defining globalization

Given these multiple and overlapping discourses, how might globalization be defined? Not surprisingly, no single definition exists. As Taylor *et al.* (2002, p. 2) argue, '[G]lobalisation is a very fluid and flexible concept; indeed it is a very slippery customer to deal with.' Daniels *et al.* (2001, appendix) highlight the controversy over the use of the term in 'defining' it in the following way:

> A contested term relating to the transformation of spatial relations that involves a change in the relationship between space, economy and society.

Illustrating the competing views, the following two definitions represent a hyperglobalist view and a sceptical view respectively (see Chapter 2 for a full discussion of these terms):

> Information, capital, and innovation flow all over the world at top speed, enabled by technology and fuelled by consumers' desires for access to the best and least expensive products.
>
> (Ohmae, 1995, in Dicken, 1998, p. 4)

> Globalization seems to be as much an overstatement as it is an ideology and an analytical concept.
>
> (Ruigrok and van Tulder, 1995, p. 22)

In a leading book on globalization, *Global Transformations*, Held *et al.* (1999, p. 16) emphasize a transformationalist argument:

> [Globalization is] a process (or set of processes) which embodies a transformation in the spatial organisation of social relations and transactions – assessed in terms of their extensity, intensity, velocity and impact – generating transcontinental or interregional flows and networks of activity, interaction and the exercise of power.

As Waters (2001) argues, a useful exercise in seeking a definition is to consider what a completely globalized world would look like: single society, a homogenized global culture, a single global economy, and no nation-states? That is clearly not what we have. We are therefore living in a *globalizing* world – where processes do appear to be taking us closer to the above situation. Building on the work of Robertson which highlights the development of a 'global consciousness' (1992, p. 7; see also 2003), Waters offers a sociological definition:

> A process in which the constraints of geography on social and cultural arrangements recede and in which people become increasingly aware that they are receding.

Human geographical work has tended to build upon transformationalist and sociological work. A definition reflecting this perspective is found in Cloke *et al.*'s *Introducing Human Geographies* (1999, p. 337):

> The economic, political, social and cultural processes whereby: a) places across the globe are increasingly interconnected; b) social relations and economic transactions increasingly occur at the intercontinental scale; and c) the globe itself comes to be a recognizable geographic entity. As such, globalization does not mean that everywhere in the world becomes the same. Nor is it an entirely even process; different places are differently connected into the world and view that world from different perspectives. Globalization has been occurring for several hundred years, but in the contemporary world that scale and extent of social, political and economic interpenetration appears to be qualitatively different to international networks in the past.

There are, of course, many more definitions, and we will come across some of them later into the book, one of the most important being Harvey's (1989) concept of time–space compression (see Chapter 2). I have attempted to summarize my opinion on what globalization is below and to catch the spirit of a geographical perspective. There are two versions – a long one and a short one. The short one reads:

> Globalization is a collection of dialectical human agency driven processes which create local–local and person–person networks of inclusion/compression that increasingly transcend territorial/national borders and stretch to become global in proportion.

For the long version, take the short one and add the following:

> The processes are dialectical in that the relative social distance of those not on the net from those on the net widens as the intensity and extensity of the processes increase. Thus globalization simultaneously creates spaces of exclusion/marginalization leading to an increase in social, economic, political and cultural unevenness across space. For those on the net however, globalizing technologies – in transport and communications – give the impression that the world is one system that is becoming smaller. The rise of such processes is intimately tied up with the rise and expansion of capitalism and thus goes back to the first 'global' empires. Given the endless capitalist imperative of increasing profits, new technologies have been developed which have greatly increased the velocity of capital, information and ideas, especially over the past three decades. Individuals, social groups, governments and transnational bodies both respond to and create the processes of globalization in their struggles to create and protect viable and sustainable livelihoods and identities, and supranational organizations have been developed to attempt to regulate and direct such flows. From a political viewpoint globalization has often been recast as an agenda or a discourse where the self-interest of certain groups leads them to accord normative or moral status to the processes.

You must arrive at your own definition – and students are notoriously divided on the subject of what globalization means (see Box 1.1 and Map 1.2)! The above definition, however, is the one that is assumed for the remainder of this book and many of the terms within it will be elaborated upon. In short, this text takes the explicit view that globalization is real, inherently geographical in its nature, and creates unevenness and inequality.

Box 1.1

Student definitions of globalization

These definitions are reproduced from a class competition run in 2004 in a human geography course for third-year students at Victoria University, New Zealand. These definitions were presented at the start of the course – before any substantive lectures

Box 1.1
continued

took place – and students were given ten minutes to write them. Note how many students tend to define globalization in terms of 'loss of diversity' often with a political motive behind this:

> 'Globalization is seen, both positively and negatively, as the increased connectiveness of peoples/cultures. The primary issue is who is globalization for? And, who has the power at present?'

> 'The spread of ideas, theories and practices through the realms of commerce, culture, society. The ideas that are spread are those of the dominant cultures and societies and often result in the loss of diversity and imposition of ideas, values, practices, etc. upon others, not always best suited to those they are imposed upon.'

> 'The process of making any given part of the world indistinguishable from any other part.'

> 'Countries linked through world systems (e.g. economically, politically, etc.) yet incorporating these into their pre-existing cultures.'

> 'Globalization is the increased communication, transportation of goods, services, people and ideas around the world.'

> 'People need people, and people need adventure. These two ideas have led us to where we are in the world. This is the phenomenon known as globalization.'

'Globalization is everything interconnected working to make the rich richer and move the poor further into the depths of despair and deprivation. Globalization is a frenzy of mass media and booming industry. It is labour at its cheapest and fashion at a high price. It is stripping the Earth of its natural resources for every last drop of Mother Nature's juices.'

'The sharing of new technologies with the aim of making life easy and "better."'

'Is the gradual disappearance of the illusion of space created by physical distance, through various technologies created by the human race.'

'Globalization is the fact that words such as "globalization" with a z are used in New Zealand when the correct spelling is globalisation with an "s".'

'McDonald's and Starbucks.'

'I believe that globalization is the next stage in the growth of humanity that has been brought about through both modernization and curiosity.'

'Globalization is the gradual awareness of the human race that its existence is intrinsically interconnected to even the tiniest particle in the universe and our lives and actions are completely dependent on everything that surrounds us.'

'I think globalization is a very complex idea that affects people differently,

Box 1.1
continued

which creates different perceptions of globalization and its advantages and disadvantages – this makes it hard to define.'

'A constant and unstoppable process of gradual change which we must own and try to influence, not just be influenced by.'

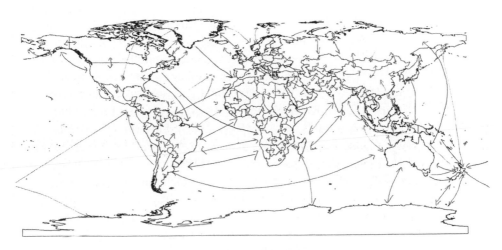

Map 1.2   *A student's spatial perspective on globalization*

Geography and the study of globalization

The term 'globalization' was first used in the mid-1960s and entered into popular usage in the mid-1990s. Few deny there is something going on 'out there' which is radically restructuring the world. There is, as already noted, disagreement on what drives it, how it should be theorized and what impacts it is having. Theoretical and empirical research on these issues has generated a vast and growing literature in the social sciences. The *International Bibliography of the Social Sciences* (IBSS) records articles published in the most prestigious academic journals. The first article with the term globali*ation (that is spelt with either an s or a z) in either the title or the keywords appeared in 1983. In 1990 only fifteen articles included the term. By 1998 this figure had risen to over 1000 and in 2003 stood at 2909. In the twenty years from 1983 12,859 articles dealing substantively with globalization issues were listed in the IBSS. In Figure 1.2 this trend is compared, on a log scale, to the number of articles over time with *both* the terms 'gobali*ation' and 'geograph*' in the title, keywords or journal title. In proportional terms there was a rapid rise in the number of such articles in the second half of the 1990s rising from one in 1993 to seventy-two in 2000. In 2002, 152 articles of this nature were listed. There are 725 articles with these two keywords listed in the IBSS. This measure gives us only a feel for the increasing importance of the topic, as many articles deal with issues surrounding globalization less explicitly.

Although much of human geography is intimately tied to globalization debates and the contribution of geographers to those debates has risen in absolute terms, geography books and articles on the subject are not referred to a great deal by academics in other disciplines. Some authors, such as Harvey (1989, 1995, 2000, 2003); Sassen (2000, 2001); Amin and Thrift (2004); Knox and Taylor (1995); Taylor and Flint (2000); Johnston *et al.* (2002); Massey (1984, 1995); LeHeron (1993); Cox (1997); Corbridge *et al.* (1994); Knox and Agnew (1998); Peet (1991); Smith (1984, 2003); Daniels and Lever (1996); Dicken (2003); Gwynne *et al.* (2003); and Swyngedouw (1997), to mention a few, have made significant interventions, some of which will be explored later in this book. In general, however, geographers' work on globalization is influenced by research from beyond the discipline far more than the other way around (see Dicken, 2004).

Figure 1.2   *Globalization in social science and geography journals 1983–2003*

Source: Calculated from the *International Bibliography of the Social Sciences*

So what does geography have to offer? What is intriguing for those
who study globalization, and possibly confusing for those who are new
to it, is that there appear to be 'globalizing' and 'localizing' forces
simultaneously at work in politics, culture and the economy. For example,
in Europe we have witnessed over the past fifty years the evolution of
an integration movement culminating in a political and economic union
of twenty-five countries (from 1 May 2004). In 2002, twelve of these
countries adopted a common currency, the Euro. And yet, at the same
time, Tony Blair's Labour government has devolved power to the Scottish
Parliament, the Welsh Assembly and, with less success, the Northern
Irish Assembly. In the economic sphere, we have witnessed the evolution
of a globalized financial sector coordinated through a number of world
cities across the planet such as London, New York, Tokyo and Singapore.
Through the operation of this sector, economic change in one region can
be instantly transmitted to another through currency and stock markets,
for example. Simultaneously, in countries such as Australia and the UK,
Local Economic Trading Schemes (LETS) which bypass these circuits of
capital entirely have evolved. Finally, in the cultural sphere, wherever you
go in the world and tune into commercial radio, it is likely at some point
that you will hear a song by Madonna, Robbie Williams or Eminem.
However, as global music 'brands' have risen to prominence on the
airwaves a 'world' music scene has evolved which markets 'local'
non-Western musical styles to global markets through retail outlets
such as HMV in the UK or Sounds in New Zealand (see Global music
industry study in Chapter 6).

How do we make sense of these apparently oppositional processes
at different scales? Human geography offers a distinct framework for
understanding the complexities of globalization in at least three ways:

1 *The discipline is inherently concerned with space, and in particular
   the interaction of processes, structures and agents at different scales
   of analysis.* The debate surrounding the relative importance of
   structure and agency in human geography has raged for many years.
   Most geographers now concede that, although the relative power of
   each varies across time and space and that this power is contingent on
   a range of political, economic and environmental factors, structure and
   agency play a mutually determining role in the local spatial outcomes
   of broad processes. Furthermore, the 'global' is constituted by 'local'
   – everything, ultimately, has a local expression, even if it is stretched
   to become 'global'. For example, in the cultural example above it is
   wrong to assume that Robbie Williams produces 'global' music. His

music combines particular elements emanating from specific places (rock and pop principally) and this hybrid is marketed through networks stretching across the globe that link together particular sites of consumption. This conception avoids the determinism of globalist economic discourses and escapes the idealized localism of some anthropological viewpoints.

2 Linked to the first point, *human geography 'peoples' discourses of globalization*. Globalization is constructed from neither 'above' nor 'below'; rather it is constantly refashioned by interaction between people and institutions on these various scales. Although there is no doubt that some are more powerful than others in shaping the processes it is nevertheless driven by humans. For example, the financial sector is not some kind of floating global force; it comprises individual decision-makers located at particular nodes in the network. Through the problematization of scale as an objective entity some human geographies have warned of the futility of attempting to separate the 'global' and the 'local'. These forces are not oppositional; they are part of the same *dialectical process*.

3 *Human geography is inherently eclectic*. Geographers deal with economic, cultural, political and environmental processes and the spatial impacts of these for, and due to, the actions of people. Globalization involves change in each of these spheres, which is one reason why it has proven so difficult for academics in their traditional disciplinary roles to understand it. For example, the evolution of the European Union is much more than an economic union. Its roots are political and cultural and its impacts wide-ranging. Human geography is in a strong position to reflect upon such change.

Much will be said to elaborate the above discussion in the remainder of this book. At this point it is useful to briefly take stock of the position and role of human geography to date. Taylor *et al.*, (2002) suggest four ways in which geography as a discipline and globalization are interrelated.

- Geography and globalization
- Geography in globalization
- Geographies of globalization
- Geographies for and against globalization

Geography and globalization

Geography has a long 'global tradition' and this was evident before
the rise of globalization as a concept in the social sciences. This puts
geographers in a strong position in terms of placing globalization in
historical perspective. Taylor *et al.*, (2002) trace the changing emphasis
placed on the 'global' and the 'local' through various stages of the
discipline's evolution. Geography was founded as a tool for exploration
of the globe, largely to support the needs of Western imperialism.
Following the Second World War regional geography eclipsed this
globalism but this in turn was eclipsed in the 1960s by a reinvigorated
concern for the global. This new focus linked three issues; first, the
growing concern with the global physical and ecological environment;
second, the work concerned with the 'international division of labour'
and the role of multinational companies; and third, the increasing concern
for global inequalities. With the rise of postmodernism in the 1980s
concern for the local came to the fore once again (see Crang, 1999;
Esteva and Prakash, 1998). As is evidenced in Figure 1.2 the recent
evolution of the discipline is characterized by a significant engagement
with global issues. The approach is different this time however, since
there is an explicit objective of linking global and local change, as
the following paragraph elaborates (see for example Larner and
LeHeron, 2002).

Geography in globalization

Geography has played an important role in the pursuit of meanings
and measurements of globalization. In particular, through its concern
with understanding scale, the subject has tended to 'avoid the notion of
the global as a stage, an inert space on which events inevitably unfold'
(Taylor *et al.*, 2002, p. 3). In work that sets out to define the 'global', the
authors identify two central points. First, scales are relational – in other
words no scale is static or pre-given and the effects of change at no one
scale predetermine the impacts at another. This is particularly important
in terms of the idea that the 'global' determines the 'local'. Two
constructs, 'glocalization' and 'local-global', are important responses
to this false dichotomy, implying that the global is constantly in a state
of flux and being reconstituted from below. In other words, things are
constantly being 'rescaled'. Thus, 'sometimes this involves processes
moving to institutions "above the state", other times to "below the state",
and through all of this the state itself is changing and adapting' (p. 7).

Second, scale is *created* through the flows and interactions that occur in the political economy and society, and thus there is nothing inevitable about the rise of 'global scale' phenomena which result from the action of agents. For example, as we will see in Chapter 3, colonial globalization was constructed based on notions of Western supremacy. Arguably, similar forces underpin the latest phase of globalization. In short, echoing an earlier point, the scale of the global and thus the rise of globalization are human constructs.

Another important way in which geographers have contributed is elucidating the distinction between 'spaces of flows' and 'spaces of places'. This idea was first touted by Manuel Castells in *The Rise of the Network Society* (1996), and we will take up these arguments in more detail in Chapter 2. Suffice to say here, the spaces of places (nations, territories, villages) have been challenged by new spaces of flows where knowledge and other forms of information are transmitted through electronically based webs and other networks. But 'spaces of flows' have not eradicated places – a 'world city', for example, is a node on a network but it is still a distinctive place (see Plate 1.2). But the change in relative importance of places and flows tends to collapse scale, making 'out there' more immediate, thereby increasing the velocity of transactions and their impacts on the ground.

Geographies of globalization

As previously mentioned, one of the abounding myths which accompanies certain discourses of globalization is the idea that it homogenizes the world. This is clearly not true, and one of geography's major contributions has been to continue to provide both empirical data and theoretical constructs which both illustrate and explain this. As we will see in Chapter 7 concerning development and inequality, the world is now more unequal than ever. Under colonial globalization, unevenness was created as part of the necessary process of fuelling of the imperial economies. Following the Second World War, the rise of the development project (McMichael, 2004) furthered the expansion of the capitalist nucleus, perpetuating marginalization and exclusion. The more recent phase (and we will delineate these waves more formally in Chapter 3) is also predicated on inequality. The nature of that inequality has changed however; what we have now is networks of privilege, with interstices (gaps in the net) of poverty and deprivation. This is of course only one

Plate 1.2 *Singapore: a unique place in the financial space of flows*
Source: Donovan Storey

reading of the phenomenon. However, given geography's preoccupation with differentiation it is the one that we as geographers are naturally predisposed to. Harvey (1995), for example, argues that globalization may be better described as 'uneven spatio-temporal development of capitalism' (cited in Gwynne *et al.*, 2003, p. 10). Geography's role in understanding 'spaces of flows' must also understand the 'spaces of places' that are *not* subject to those flows (see Plate 1.3). In this regard, Taylor *et al.* (2002, p. 9) argue:

> The rise of globalization has been marked by increased material polarisation between regions; for every global city in the network of global cities there is what Castell's calls a 'black hole' of marginalisation and exclusion form the global network society.

Plate 1.3 *Black holes in the net: a highway passes over a deprived community in Manila, Philippines*
Source: Donovan Storey

Geographies for and against globalization

As we know from our brief look at the rise of the anti-globalization movement, major challenges have been laid down. Globalization is resisted in many ways and has been since its inception. As also noted above, the recent wave of protests is not unprecedented in history; resistance to colonial rule in India, resistance to the new international division of labour in the form of the rise of unions in the USA, and the anti-debt agendas which have been evident in Latin America for many years, are but a few examples of this. Geography has played a role in documenting the rise of such resistance movements. As already noted, one of the more ironic outcomes of the rise of the technologies which have enabled globalization is that they have been mobilized to resist it. This leads us once more to the point that it is not necessarily globalization per se which is the problem – rather it is the way that it has been practised. The *moral* role of geography (D.M. Smith, 2000) in this respect is not just to document the rise of such resistance but also to suggest ways as to how it might alter globalization for the better – what is called for here is a 'grassroots globalization' or a 'globalization from below'. From geographical studies that have taken place to date however, it appears that global civil society is a long way from achieving the kind of resolve and consistency required to face-off the 'spaces of flows' emanating from the selected 'spaces of places' that dominate. What geography does at least show, given its inherent global/local focus, is that this is at least possible in theory.

Overall, then, human geography has played a role in elaborating the nature, roots and impacts of globalization – but it has a long way to go in order to suggest the kinds of solutions that might be workable. Taylor *et al.*, suggest that we need to 'understand globalization today, to make a better tomorrow, for all of humanity' (2002, p. 17) – but we also need to understand the globalization of yesterday in order to make historically grounded suggestions.

New globalizing spaces?

To conclude this introductory chapter, we return to the death-knells sounded for geography in some quarters. The two views on this important debate are outlined below. The essence of the argument in this book is that new globalized spaces are being forged which are changing the nature of geography without eroding its relevance.

Geography is dead

The argument that geography is dead, or least less relevant today, put forward by some hyperglobalist economists and other non-geographers may be characterized as follows. Time–space compression renders distance unimportant. Locality has less meaning, since we live in a 'global village'. Difference is declining and culture homogenizing. Thus, for example, under this argument TNCs are completely 'footloose' and spread identical products across the planet. In the cultural sphere, global brands such as Nike or Madonna or the BBC send out identical messages which are interpreted in the same way across world society. This idea sees globalization as deterministic and inevitable, subject to its own inexorable logic. Territories are made less salient by the transcendence of process across and above them.

Geography has new life

In contrast, the more realistic argument that geography has new life through globalization may go like this. Distance, as measured *in an absolute sense*, is indeed less important, but place, space, locality and the *relative* distance between these things are not. 'Global' processes are actually stretched 'local to local' processes, and they unfold in localities that have a unique history and character. In the economic sphere, TNCs choose to locate in particular places due to a complex mixture of local characteristics in the destination area and factors specific to the firm influenced by the nature of their source location. In culture, hybridity becomes the new normality as 'global' trends mix with local ones to create new cultures, for example, New Yorican salsa or Polynesian rock. Geography is indeed transforming, both as a concrete reality and as a discipline that seeks to analyse those changes, but it is not irrelevant. This new, highly volatile unevenness makes geography – 'the study of spatial differentiation' – more important than ever as the world fragments and recrystallizes along new political, economic, social and cultural lines.

Altered spaces?

It is in the area of technological change, especially the rise of the internet, where the death of geography is often most readily proclaimed. *The Economist* (15 March 2003) ran an article entitled the 'Revenge of geography' where it was argued that: 'it was naïve to imagine that the global reach of the internet would make geography irrelevant. Wireline and wireless technologies have bound the virtual and physical worlds closer than ever' (p. 13). The article goes on to explain that the process of using such technology is inherently bound by place and 'real' or territorial space. In particular it is claimed:

> Actually, geography is far from dead. Although it is often helpful to think of the internet as a parallel universe, or an omnipresent cloud, its users live in the real world where limitations of geography still apply. And these limitations extend online. Finding information relevant to a particular place, or the location associated with a specific piece of information is not always easy. This has caused a surge of innovation,

as new technologies have developed to link places on the internet with places in the real world – stitching together the supposedly separate virtual and physical worlds.

This argument is echoed in the groundbreaking work of geographers such as Kitchin and Dodge (2002) in their article 'The emerging geographies of cyberspace', which argues that cyberspace exhibits uneven and real geographies where 'online interactions are not divorced from those offline, but rather are contexted by them' (p. 353).

The concept of space is important here; those who believe that geography is rendered less important have misinterpreted the discipline and some of its core elements. Crucially confusion exists concerning the difference between *geographic* space and *absolute* space. Geography is seen as something innate, usually represented on a topographic map upon which processes unfold. This is, of course, a central component of geographical analysis, but it is not the whole story. A wide body of literature in geography has argued that space has many meanings that are transformed by contemporary globalization (see Chapter 2). Geographers need to work extra hard to keep up with the challenges of these new spaces and to convince others of their importance.

This book

Essentially this book is about exploring the various concepts, theories, processes and impacts of globalization from a geographical perspective. We are going to have to carefully define concepts such as 'global' and 'local' and to think deeply about the implications of increased interaction within and between these geographic scales. We must make two important points here: first, the study of globalization is inherently multi-disciplinary; there is geography in the work of non-geographic specialists and vice versa. Second, the literature that has been generated around globalization is awesome; it would take a life's work to summarize the main works and their implications, and we have only nine brief chapters to survey some of the main contours here. Some important issues are left out, although readers are pointed to where they can find further information. The purpose of this book is to flag the most important ideas concerning the processes and agendas of globalization and to provide signposts to further study. It is intended that straightforward English will be used to do this – often, and necessarily, new concepts and ideas in the social and natural sciences generate a new language, or at least *academilects*, of their own. Keep a good dictionary of human geography handy as you read!

In order to achieve the above objectives the book is separated into three sections that flow from the conceptual to the concrete, the general to the specific and the global to the local. Each of these sections contains individual chapters that expand upon specific aspects of the organizing themes. *Part I: Transformed geographies*, which includes this chapter, introduces and critically appraises the concept of globalization. It then goes on to discuss various 'theories' and perspectives (Chapter 2); drawing out the spatial implications of the various viewpoints and placing them in historical context (Chapter 3). *Part II: Shifting spheres*, breaks down and analyses the process and outcomes of globalization in overlapping spheres in three chapters – economic (Chapter 4), political (Chapter 5) and cultural (Chapter 6), emphasizing the links and breaks between change in these various areas. Case studies and examples from a wide range of countries are inserted throughout the text, sometimes in boxed form. In setting out these examples an explicit attempt has been made to draw on materials and research from across the globe – something to which a coherent geography of globalization should aspire. *Part III: Global challenges*, deals with two areas of special concern to human geographers – development and inequality (Chapter 7) and the environment (Chapter 8), tracing the challenges thrown up by globalization in these areas.

In the final chapter the theoretical, historical and empirical analyses are brought together in order to reflect on the changing nature of globalization and the implications of this for both geographies on the ground and the way they are studied and researched. Particular attention is paid to the possibility of transformative *progressive* globalization (as opposed to 'globalization as currently practised') and the role geography might play in helping to forge this. There are a number of questions which the books seeks to address, and these are returned to throughout and in the concluding chapter explicitly:

1 How can globalization be defined?
2 Is globalization new?
3 What drives globalization, now and in the past?
4 Is contemporary globalization different from the globalization of the past?
5 Is it homogenizing global society?
6 Is globalization a process or an agenda?

7 How does globalization alter our concepts of space, place and scale?

8 Is globalization the same thing as internationalization?

9 Is globalization eroding the power of the nation-state?

10 Is globalization bad for the environment?

11 Is globalization bad for the 'poor world'?

12 Does globalization create more winners than losers?

13 Can globalization be reformed?

14 If so, how should it be reformed?

15 What role can geography and geographers play in forging an alternative progressive globalization?

Running throughout the book, then, is an explicit engagement with the often-assumed deterministic power of globalization. Centrally, it is argued that human geographies of difference remain as important as they have ever been. This general conclusion, illustrated through appropriate examples, is likely to be of relevance to human geography students who never cease to be amazed by the increasing complexity and heterogeneity of their own localities despite dominant discourses of homogenization.

Further reading

The following readings relate specifically to the relationship between geography and globalization. Further substantive readings on globalization in general are referenced in the more detailed chapters that follow.

Dicken (2004): This article suggests that geographers are missing the boat in terms of studying globalization and sets a convincing agenda for how geographers might make a difference.

Leyshon (1995): This chapter is especially useful in terms of the arguments concerning the 'death of geography'.

Taylor *et al.* (2002): Introduce the collection *Geographies of Global Change* and discuss the relationship between geography and globalization.

Yeung (2002): This paper interprets economic globalization from human geographical perspective in an accessible way.

Globalization across space
– contesting theories

- Early theories – Marx to the global village
- Contemporary theses – hyperglobalists, sceptics and transformationalists
- Four outstanding controversies
- Geography and space
- Geography and scale
- 'Spaces of flows' versus 'spaces of places'?
- Conclusion – new geographies

Early theories – Marx to the global village

In order to illustrate the long heritage of the idea, Waters (2001) reminds us that 'curiously, globalization, or a concept very much like it, put in an early appearance in the development of social science' (p. 7). He traces the links between some of the early classical accounts of 'internationalism' of Saint-Simon, Comte and Durkheim among others. Durkheim, for example, spoke of the implications of industrialization in terms of weakening state structures and nation-state culture, leading to the 'dismantling of boundaries between societies' (Waters, 2001, p. 8). In another example, Weber talked of the spread of Calvinism and rational thinking in Europe and eventual dominance of western culture. Marx had a globalization theory of sorts too, which was more formally elaborated by Lenin. He argued that the bourgeoisie (industrial class), in their search for profits and extraction of surplus from the proletariat (working class), would expand their geographical horizons drawing peripheral countries into the system as providers of cheap labour and raw materials. His conclusion was that this would eventually lead to revolution – echoing the utopian internationalism of the earlier ideas of others including Saint-Simon. Two interesting points stand in Marx's account: first, he is explicit that the underlying logic for the expansion of capitalism is cultural as well as economic; second, Marx viewed this process as one of 'the interdependence of nations and recognises the continuing existence of the nation-state' (Waters, 2001, p. 9). This latter point in particular went on to influence Marxist dependency theory, world systems theory and modernization theory.

Writing in the 1970s, Wallerstein (1980) proposed *World Systems Theory* (WST), arguing that the capitalist mode of production would supersede others as it expanded across the globe. For Wallerstein, and for dependency theorists who employ similar perspectives, this would not necessarily lead to an eventual utopian outcome as Marx predicted and would continue indefinitely. Central to Wallerstein's ideas is the centrality of the nation-state and co-existence of *different* world systems and cultures of which capitalism is the dominant one.

According to Waters, theories of globalization up until the mid-1980s were 'uni-dimensional', focusing on one sphere of human experience – usually the economic. In the 1970s and 1980s through the work of Burton (1972) and Rosenau (1980) in particular, political aspects were elaborated. It was increasingly argued, for example, that transnational governance was emerging alongside national polities and that this was leading to a new political dualism. This idea was elaborated most explicitly with Rosenau's work concerning 'state-centric' and 'multi-centric' worlds (Rosenau, 1990) (and see Chapter 5). Subsequently, Held (1991) argued most explicitly for the erosion of state power in the context of the rise of global governance.

A global village?

Among the most influential ideas of the second half of the twentieth century with respect to globalization was that of Canadian literary critic Marshall McLuhan, namely the 'global village' (McLuhan, 1962, 1964). This concept has entered the popular imagination in ways that were probably not intended by the author and has been central in the evolution of a popular global consciousness (Robertson, 2003). After the Second World War critics noted that social relations were extending greatly over space due to the evolution of communications technology, some of which had been advanced during the war effort. In the 1960s, McLuhan captured the essence of this and coined the term, arguing that 'the world has become compressed and electrically contracted, so that the globe is no more than one village' (McLuhan, 1964, p. 4). Although often employed in a simplistic manner subsequently, McLuhan's original concept was complex. It referred principally to the evolution of social relations and the cultural infrastructures used to communicate these over time. Human history was divided into three phases based on the dominant mode of communication:

- Oral
- Writing/painting
- Electronic

Over time, argued McLuhan, each phase had been superseded by another mode and this had spatial ramifications. Oral communication, by definition, restricted social interaction to relatively short distances, for example, within villages. The advent of writing and painting made the transport of ideas and cultural markers across space a lot easier leading to an increase in the 'extensibility' of human relations. The arrival of the electronic age and the even greater stretching of social relations had the paradoxical effect of once again making oral communication more common. This mode echoed the first period but this time the 'village' was stretched globally. In the introduction to *Understanding Media*, McLuhan sums this up: '(t)oday, after more than a century of electric technology, we have extended our central nervous system in a global embrace, abolishing both space and time as far as our planet is concerned' (1964, p. 3).

The construct is often misunderstood as a utopian reflection of reality, and one which promotes the idea of homogenization. The author himself was often ambivalent about the outcomes of 'electric contraction'. In 1964, for example, he writes: 'the aspiration of our time for wholeness, empathy and depth of awareness is a natural adjunct of electric technology. . . . There is a deep faith to be found in this attitude – a faith that concerns the ultimate harmony of all being' (1964, p. 5). In other writings, however, McLuhan refers to the global village as a place of 'terror', 'uncertainty' and 'tribalism'. The 'global village' idea has been co-opted by politicians, businesspeople and academics of all shades to justify and illustrate their particular worldviews. Employed uncritically it fails to capture the unevenness of access to electronic technology that characterizes the world. There is no doubt, however, that by injecting notions of cultural interaction and 'stretched' space into the debate the idea influenced a number of subsequent theorizations, including those of Robertson (1992) and Giddens (1990), as we explore below. McLuhan was also influential in critical media studies and media ecology where his notions of stretched networks that transmit cultural markers are analysed.

Contemporary theses – hyperglobalists, sceptics and transformationalists

From around the mid-1980s explicit theorization about globalization appeared as the term entered into academic and eventually popular usage. Waters (2001) cites figures such as Beck (1992), Rosenau (1990), Lash and Urry (1994) and geographer Harvey (1989) as important contributors in this endeavour. It is Robertston (1992) and Giddens (1991) who are credited with the major contributions, however, with their ideas concerning the 'relativisation of the individual with respect to higher scales of analysis' and 'time–space distanciation' respectively. It is argued that:

> [t]he significant features of each of their [Robertson and Gidden's] proposals are first that they are multi-causal or multi-dimensional in their approach, and second, that they emphasise subjectivity and culture as central factors in the current acceleration of the globalization process.
>
> (Waters, 2001, p. 14)

Through the work of these scholars Waters claims that a paradigm of globalization theory is evolving where common themes can now be identified. It could be argued that this is over-optimistic and that there is in fact a multiplicity of ways that globalization is being viewed and contested. A useful way of reviewing the diversity of ideas is provided by Held et al., (1999) in their three 'theses' of globalization schema. These authors do not see an evolving consensus, or even paradigm, arguing that:

> beyond a general acknowledgement of a real or perceived intensification of global interconnectedness there is substantial disagreement as to how globalization is best conceptualised, how one should think about its causal dynamics, and how one should characterise its structural consequences, if any.
>
> (Held et al., 1999, p. 2)

What is particularly noticeable about Held et al.,'s three-part division is that traditional ideological lines are blurred. In the hyperglobalist camp, for example, we find neo-Marxists such as Greider (1997), together with neoliberals such as Ohmae (1990, 1995). Similarly in the sceptical group both conservative and radical views exist. Nevertheless, as we see below, each school conforms around a central theme in terms of the nature and consequences of globalization (see also Chapter 3).

The hyperglobalists

According to hyperglobalists a new age of history is upon us, one that is entirely unprecedented. Ohmae, for example, argues that 'traditional nation-states have become unnatural, even impossible, units in the global economy' (Ohmae, 1995, p. 5). In this argument, the world economy is borderless and characterized by a single global market which functions through transnational networks of production, trade and finance (Khan, 1996). Therefore, emergent forms of governance, which sit above and below the scale of the nation-state, become increasingly important. The divergence of ideologies within this group is highlighted by the fact that both neoliberals and some neo-Marxists form part of the school. The former, epitomized by Ohmae (1995), view the erosion of state regulatory power as a positive factor which signals the victory of capitalism over socialism. It is notable that such rhetoric was written, and taken very seriously in business schools in particular, immediately following the collapse of the USSR in 1989. The second group, the radicals, also argue that global capitalism has triumphed. In contrast, they view global capitalism as both oppressive and regressive (Peet and Watts, 1993; Petras, 1999). Both perspectives posit that economic forces are dominant, a truly integrated global economy actually exists, and that governments are wedged uncomfortably in between local, regional and supranational bodies of governance, and are thus increasingly irrelevant.

The hyperglobalist thesis argues that new transnational class allegiances have evolved, creating a new global elite that shares an ideological attachment to neoliberalism and consumerism. Some see this as evidence of the first truly 'global civilization', facilitating the spread not only of neoliberal consumerism, but also of liberal democracy and new forms of global governance (Gill, 1995; Luard, 1990; Strange, 1996). Radical hyperglobalists do not celebrate this but do acknowledge the arrival of the 'global market civilisation' (Perlmutter, 1991; Greider, 1997). In sum, 'Whether issuing from a liberal or radical/socialist perspective, the hyperglobalist thesis represents globalization as embodying nothing less than a fundamental reconfiguration of the "framework of human action"' (Held et al., 1999, p. 5). Globalization then is reified as an unstoppable and all-determining juggernaut (Klak, 1998).

Anti-globalization?

The sceptics

At the other end of the spectrum are the sceptics who question the very existence of 'globalization' (Rugman, 2000). Although less vocal, they provide an important counterpoint to the accounts that claim globalization to be deterministic. The leading proponents of this view are Hirst and Thompson (1999) who use a broad range of empirical data to argue that the world economy is no more interconnected today than in the past (Boyer and Drache, 1996). Indeed, they argue that during the era of the Gold Standard, at the end of the nineteenth century, the world economy was *more* integrated (see Chapter 3 for a fuller analysis of this issue). Flows of labour, investment and trade would – if the hyperglobalist thesis were true – be unrestricted and free. This, they argue, is very far from reality and labour remains particularly immobile. For sceptics, national governments remain the central actors in constructing and regulating the global economy. What we have then is an *internationalized* political economy which is fragmenting into powerful economic blocs, the most powerful being Europe, East Asia and the USA. The process of regionalism, according to the sceptics, runs in opposition to globalization, whereas for other schools it is seen as a precursor to it (see Chapter 4 on regionalism).

Sceptics charge that hyperglobalist accounts of globalization are smoke-screens set up to obfuscate the real intentions of the powerful capitalist elite. Capitalist national governments, principally those of the G8, are the primary designers of the new global economy. This select group of core countries insist upon free-market reform in poorer countries – claiming that this is necessary for survival in the context of inevitable globalization. Simultaneously, these vociferous proponents of free-market reform are themselves the most protectionist. In this way, sceptics view globalization as an 'agenda' (Firth, 2000) or a 'project' (McMichael, 2004) which seeks actively to wedge open spaces of opportunity for capital emanating from the advanced capitalist economies (Hirst, 1997). This agenda came about as a result of US-led reconstruction of the global economic order following the Second World War which created a new agenda of liberalization – designed largely in the interests of the dominant powers at the time (Gilpin, 2001) (see Chapter 4). Neo-Marxists have seen recent 'globalization' as another phase in the expansion of Western imperialism where national governments act as the agents and conduits for monopoly capital (Firth, 2000).

Given the above, the sceptics question the rise of a global 'civilization' or a new global 'civil society'. Rather, deepening inequality heightens neocolonial tensions between civilizational blocs and ethic groupings (Huntington, 1996). The discourse of the 'homogenization' of global culture is seen as part of the West's historic project of domination initiated under colonialism. There are links here, which are just beginning to be explored, with postcolonial scholarship which tends to view globalization discourse as an agenda designed to extend the subordination of postcolonial subjects (see Blunt and McEwan, 2003; Gregory, 2004; Sidaway, 2000; Sidaway et al., 2003). Writers such as Chomsky (2001) see the recent 'war on terror' as part of the same project, complementing discourses of development and globalization as a strategy for continued dominance by the West. The principal outcome of the sceptical thesis is the exploding of the myth of the erosion of state power. Nation-states are seen as the principal disseminators of the agenda as well as, potentially at least, the principal sites of resistance to it.

The transformationalists

Lying at various points on the continuum between extremes is the transformationalist thesis (see Table 2.1). Scholars in this group argue, often in very different ways, that globalization is real and is restructuring society profoundly (Castells, 1996; Giddens, 1990). This change extends from historical transitions which contrast with the hyperglobalist claim that globalization is an entirely new condition. Tranformationalists argue that globalization is historically and geographically contingent, constructed by human action, and therefore has no predetermined outcome (Peet, 1991). The role of nation-states is altered, though not necessarily eroded. Novel policy responses are required given that traditional binaries such as international/domestic, internal/external and global/local are collapsed (Rosenau, 1990; Sassen, 2000). In this context, Rosenau (1990) talks of the rise of 'intermestic' affairs.

Although building explicitly on historical processes of change (see Chapter 3), contemporary globalization is none the less different to that of the past. In the economic sphere, for example, we see trade, financial and productive organizations transcending national borders so that the fate of distant groups is interlinked in ways that are qualitatively different from the past (Dicken, 2003) (and see Chapter 4). Cultural exchange is more extensive leading to the formation of new hybrids. In the political sphere, the links between the sovereign power of nation-states and the territory of those states is disrupted. National polities are no longer governed solely by national governments but also by supranational and international institutions. The formation of transnational social movements, based on new identities that link sub-national groups together across boundaries, challenges the state from below (Routledge, 2002). In this respect, 'sovereignty today is less a territorially defined

Table 2.1 *Three theses of globalization – a schema*

	Hyperglobalist	*Sceptical*	*Transformationalist*
What is happening?	The global era	Increased regionalism	Unprecedented interconnectedness
Central features	Global civilization based on global capitalism and governance	Core-led regionalism makes globe less interconnected than in late nineteenth century	'Thick globalization'. High intensity, extensity and velocity of globalization
Driving processes	Technology, capitalism and human Ingenuity	Nation-states and the market	'Modern' forces in unison
Patterns of differentiation	Collapsing of welfare differentials over time as market equalizes	Core–periphery structure reinforced leading to greater global inequality	New networks of inclusion/exclusion that are more complex than old patterns
Conceptualization of globalization	Borderless world and perfect markets	Regionalization, internationalization and imperfect markets	Time–space compression and distanciation that rescale interaction
Implications for the nation-state	Eroded or made irrelevant	Strengthened and made more relevant	Transformed governance patterns and new state imperatives
Historical path	Global civilization based on new transnational elite and cross class groups	Neo-imperialism and civilizational clashes through actions of regional blocs and neoliberal agenda	Indeterminate – depends on construction and action of nation-states and civil society
Core position	Triumph of capitalism and the market over nation-states	Powerful states create globalization agenda to perpetuate their dominant position	Transformation of governance at all scales and new networks of power

Source: Adapted from Held *et al.* (1999)

barrier than a bargaining resource for a politics characterised by complex transnational networks' (Keohane, 1995, cited in Held *et al.*, 1999, p. 9). Economic, cultural and political governance is diffusing away from nation-states to both higher and lower geographic scales simultaneously. According to Rosenau (1990) this has made states more 'activist', as they reach out to manage the transnational affairs that so intimately affect them. Much more will be said concerning the transformationalist theses in what remains of this book given that it is the perspective which sits best with a human geographical approach.

Geographical implications of the three discourses of globalization

From a geographical viewpoint the hyperglobalist position has important ramifications. The traditional 'core–periphery' model, together with variants on this theme such as 'North/South division' and 'developed/developing world', are seen as outdated given the decline in the importance of the nation-state as the unit of analysis. Rather, new hierarchies of power are evolving, particularly surrounding the *newer international division of labour* (see Chapter 4) and through networks. States still attempt to manage the new divisions that arise between social groups, localities and sub-national regions using redistributive and welfare measures. These, however, are becoming increasingly difficult to sustain and are being reformed. There are thus winners and losers in the new global economy. Echoing neoclassical trade theory, neoliberal hyperglobalists argue that the rise of the global free market, while creating losers, makes everybody better off in the long run, and will eventually create a homogeneous affluent and modern society. In contrast, radical hyperglobalists argue that the new patterns of inequality are likely to be permanent features. This harks back to neo-Marxist inspired structuralist and dependency theories developed in Latin America in the 1960s.

For the sceptics, as the vast majority of world trade and investment flows take place within and between the major blocs, the principal geographic outcome of this process is to further marginalize peripheral economies and to make it more difficult for semi-peripheral economies to compete. Significant barriers are often placed around the major blocs: the EU's Common Agricultural Policy, for example, discriminates markedly against agricultural exports from poorer countries. This means that the geographical embrace of the world economy is actually diminished from what it once was (Gordon, 1988; Weiss, 1998). It also implies an intensification of the structural inequalities that persist in the global economy (Gwynne et al., 2003). Sceptics question the rise of the new and newer international divisions of labour, and the existence of truly 'global' corporations (Krugman, 1996). Such companies, it is argued, remain deeply embedded in national territories (Thompson and Allen, 1997), which remain the principal 'containers' of economic activity (Daniels and Lever, 1996). Culture becomes increasingly differentiated along ethnic and racial lines and reflects the unequal distribution of modernization.

The most important spatial implication of the transformationalist perspective is that the global capitalist system forges increased unevenness. This does not replicate traditional core–periphery patterns and is not territorial in essence, but rather social in nature and articulated through networks. However, these divisions are not necessarily locked in for ever, as radical hyperglobalists argue. Within particular places, irrespective of where they sit in the world system as measured by reference to old divisions such as First and Third World, we see 'new hierarchies which cut across and penetrate all societies and regions of the world' (Held et al., 1999, p. 8). Overall, except in the case of neoliberal hyperglobalist accounts, each points towards the increased differentiation of global society, polity and economy as globalization unfolds. The concluding chapter of this book elaborates this discussion and draws up a framework for the spatial implications of globalization in the different spheres of concern (see Table 9.1). 353.

Four outstanding controversies

There exist significant sources of contention in the debate between the three globalization theses (Held et al., 1999, p. 10). Four of the most important ones are discussed below and responses contrasted.

- *Conceptualization*: What is the nature of globalization?
- *Causality*: What is driving it?
- *Periodization*: Is it new?
- *Impacts*: What are its outcomes?

Hyperglobalists and sceptics alike tend to *conceptualize* globalization as comprising a supposed end-state. This is generally argued to be a global market fulfilling the conditions of perfect competition, perfect integration and factor market equalization. Sceptics such as Hirst and Thompson (see Chapter 3) test the evidence for globalization by comparing empirical trends to this ideal type. Transformationalists, on the other hand, conceive of globalization as a historical process which is 'more contingent and open ended which does not fit with orthodox linear models of social change' (Held *et al.*, 1999, p. 11), and such changes cannot always be fully reflected or captured in terms of empirical data alone. Furthermore, transformationalists unlike hyperglobalists argue that the process of globalization is a highly differentiated one in which neither culture nor economics plays the sole determining role.

A second area of contention is *causation*. The central issue here is whether the process is mono- or multi-causal. Many accounts have reduced the phenomenon to the expansionist and, sometimes, imperial, tendencies of capitalism (see Harvey, 1989); others, for example, view technology as the driving force (see Ohmae, 1995). More recent literature has argued that a multiplicity of interlocking factors in the realms of culture, politics, economics and technology are relevant (Held and McGrew, 2002; Robertson, 2003). As the collection of readings assembled by Roberts and Hite (2000) attests, this debate relates to an older one concerning the diffusion of modernization. Should globalization be seen simply as the expansion of Western modernization, as world systems theory might argue, or is it something more complex that has different root causes and expression across the world? In brief, is globalization more than simply westernization?

A third controversy is whether globalization should be conceived of as something new, and this debate forms the focus of Chapter 3. Writers are divided on the *periodization* of the process. Some see it as a post-1970s phenomenon, some as a twentieth-century process, and others as beginning with the earliest human migrations. Held *et al.*, (1999) argue that growing evidence of global trade and cultural links in the pre-modern age force us to be more sensitive to history in our explanations and characterizations of the process. Defining the historical shape of globalization, of course, is as much an exercise in definitions as it is in teasing out complex historical geographies. Ultimately, as with many of the other controversies, it boils down to what we mean by globalization.

Inevitably, there is a great deal of contention over the *impacts* of globalization. At the broadest level there are two schools of thought.

The first sees globalization as eroding the power of the nation-state, leading to an end of the welfare state and the imposition of neoliberal policies across the world (S. Amin, 1997; Cox, 1997). The second is much more critical of the 'deterministic' argument and argues that national state power is adapting and transforming in response to the imperatives of globalization (Dicken, 2003; Ruigrok and van Tulder, 1995). This argument is consistent with a human geographical approach based on the co-determination of the global and the local, building on Giddens' ideas of *structuration* (see 'structuration' entry in *The Dictionary of Human Geography* (Johnston *et al.*, 2000)).

Analysing globalization – four important measures

How can we measure and interpret globalization and make coherent judgements with respect to the controversies outlined above? Four measures are critical.

- The extensity of global networks
- The intensity of global interconnectedness
- The velocity of global flows
- The impact propensity of global interconnectedness

The first is the *extensity* of social relations, which relates to the concept of time–space distanciation, and the way that change in one part of the world can have increasingly important ramifications for people at a distance. Together with the stretching implied above, globalization implies an increased *intensity* of trans-border and 'distanciated' flows so that the volume of interaction increases. Further, the increased intensity and extensity of such processes implies an increase in the *velocity* of social, economic and political flows. These three strands together imply a greater *enmeshment* of the global and local as the pace and intensity of the linkages rise and interpenetration increases.

What does the above imply for human geography? First, the agenda for human geography must be to contribute the base of knowledge which talks to these four categories above through detailed 'global/local' case studies. Second, it is important that human geographers are involved in *critiquing* the concepts we employ to describe and interpret globalization. Principally, we need to lead the exploration of what is meant by global and local, space and scale and the way these interact.

Geography and space

Much social theory has concerned itself with time (Gregory, 1994). There is now more attention to space – and this has arisen due partly to the difficult spatial questions raised by the globalization debate. Human geography is inherently concerned with space. Almost all definitions of the discipline incorporate the term – though in competing and variable ways (See Johnston *et al.*, 2000). Along with human–environment interaction, *spatial differentiation* is often the defining core. But the way human geography conceptualizes space has varied across time and place, and geography remains characterized by many different national and regional traditions. This has led to shifting concepts of scale also. This evolution in thinking has mirrored broader paradigmatic shifts in human geography especially in the Anglo-American world. We consider space below and scale in the subsequent section.

Three views of space

There are at least three ways of conceptualizing space – each of which has important implications for the way we understand the world.

- *Absolute space* – Units of territory which can be numerically measured and are ontologically given (i.e. they exist independently of the way they are perceived – they are real). Space is ready to be filled by the features that make up geographies. Sometimes this is referred to as cartographic space – measured in miles, kilometres and so on. *Absolute space is an external given which has neutral discursive meaning.*
- *Relative space* – Space is 'perceived' by humans and this perception may vary according to their culture, available technology and resources. Relative space does not correspond to a fixed unit – it is the 'sense of space' which becomes important. For example, a person in New Zealand may feel much 'closer' to his or her grandmother in the UK with whom he or she communicates via e-mail than with his or her neighbour to whom he or she has never spoken. A settlement that is on a railway line appears much closer than one that is not for those who use rail travel, even though the latter may be closer in an absolute sense. *Relative space is perceived. It constructs and is constructed by human activity and experience.*
- *Metaphorical space* – Space that does not refer to any territorial unit at all. The 'space of views' – thinkers in one paradigm might be

spread out but don't share an absolute space. The internet and the advent of cyberspace provide new metaphorical spaces for interaction. Virtual reality is a metaphorical space. *Metaphorical space does not exist 'on the ground' – but it can have real effects.*

Society and space – changing concepts in human geography

The way human geography conceptualizes space has played an important role in the evolution of the subject. In the 1960s and early 1970s a dominant approach in geography was that of *spatial science*. Growing out of spatial sociology and urban ecology, this was a largely quantitative endeavour based on the idea that the approach of the natural sciences (positivism) could be applied to the social sciences. By using sophisticated statistical measures of the distribution of phenomena it was believed that we could derive spatial laws that would explain and predict human activity. Space, then, could be read as a map of society. This approach gave precedence to the concept of absolute space. The paradigm was largely unsuccessful, partly because the techniques employed failed to capture the complexity of society and the importance of alternative concepts of space, although it is making something of a come-back now as new GIS technologies allow more sophisticated quantitative spatial analysis.

Following the critique of spatial science in some places (the UK in particular), ascendant humanist approaches began to inform a more sensitive reading of space. Relative space in particular was emphasized as the human perception of place was explored. Instead of considering how space could be measured and how society could be read off from this, it asked: How does space *construct* society? This is to say that the relationship between society and space was conceptualized as a two-way flow or co-determinist. Approaches in the 1980s included work on *gendered spaces, racialized spaces* and *spaces of poverty*, for example. In the 1990s postmodern approaches informed a more explicit engagement with metaphorical space, especially in the form of *Third Space* (Soja, 1996). In this construct, identities which are increasingly hybrid adopt marginal spaces as locations from which to challenge the dominant groups in society – cyberspaces being one such 'liberating' space (See Hillis, 1999).

All three concepts of space are important in shaping the way humans experience and interact with the world, and each plays a role in human geography. A major misconception that has caused some to doubt the continued relevance of geography is that it is concerned with absolute space only. Globalization is altering the way time–space operates and is perceived. Time–space geography has formed an important subdiscipline since the groundbreaking work of Hägerstrand (1968, 1975). Building on this foundation, there are three interlinked ways of thinking geographically about the relationship between space and time.

- *Time–space convergence*
- *Time–space distanciation*
- *Time–space compression*

Time–space convergence refers to the *decrease in the friction of distance between places.* Coined by Donald Janelle in the late 1960s (1968, 1969, 1973), it referred to the apparent convergence of settlements linked by transport technology. As transport evolved travel time would be reduced between them, giving the sensation that they had moved closer together. The velocity at which settlements are moving together may be called the time–space convergence rate. Using Edinburgh and London he noted that they had converged at a rate of thirty minutes per year over a 200-year period. This measure emphasizes the importance of relative distance and how this is shifted by technology (see Map 1.1 for an example of this approach).

Time–space distanciation refers to the *stretching of social systems across space and time.* The term was coined by sociologist Anthony Giddens (1990) and refers to the interpenetration of people and places over increasingly large distances. Echoing McLuhan's work (see global village section above) Giddens argues that people interact in two ways: face to face, and remotely through transport and communications technologies. The second has become increasingly important, 'distanciating' the relations between people. Thus people who are not actually physically present in absolute space can be important social actors. Giddens does not argue that this process leads to homogenization, indeed greater distanciation increases the potential for humans to restructure global scale systems. Massey (1991) builds on this concept when she discusses the 'global sense of the local' which pervades our everyday experiences.

Time–space compression – refers to the *annihilation of space through time that lies at the core of capitalism* (Harvey, 1989). Concepts of convergence and distanciation do not offer an explanation for why social relations are stretched across space. Geographer David Harvey, in *The Condition of Postmodernity* (1989), provided an argument which has been of central influence in the way geographers think about the relationship between time–space and globalization. He suggests:

> that we have been experiencing, these last two decades, an intense phase of time–space compression that has had a disorienting and disruptive impact upon political-economic practices, the balance of class power, as well as upon social and cultural life.

(p. 284)

He goes on to argue that time–space convergence and distanciation are results of the expansion of capitalist relations of production across the globe. Given that 'time is money', capitalists are constantly seeking ways to speed up the circuit of capital to reduce the 'turnover time of capital' – the amount of time it takes to convert investment into a profit. This search for reduced turnover time has led to the development of technologies and policies which facilitate time–space compression. Barriers have been removed through neoliberal discourse and technology has advanced to make capital, goods and people as rapidly transportable as possible. He notes the following outcomes for example:

- Between 1960 and 1990 operating costs per mile for world airlines fell by 60 per cent.
- Between 1970 and 1990 the cost of an international phone call fell by 90 per cent.
- In 1970 only 50,000 computers existed in the world. By 1998 the internet was being used by 150 million people worldwide, up from fifty million in 1995.

Harvey links this increased time–space compression to a crisis in accumulation of capitalism as we move from the modern to the postmodern era (see Chapter 4). He is careful to draw out the cultural roots and implications of this. Capitalism itself is a culture, and the impacts of time–space compression in the cultural sphere are important as identity 'melts into air' and people search for 'secure moorings and longer-lasting values in a shifting world' (p. 293). Ultimately the argument places an economic rationale at the core and this has been criticized by new cultural geographers especially. The links with his ideas and WST are apparent when he says: 'We have, in short, witnessed another fierce round in that process of annihilation of space through time that has always lain at the centre of capitalism's dynamic' (p. 293).

Geography and scale

Scale is a central concept in geography and yet it is something which is often used uncritically. What do we actually mean by 'global'? What do we actually mean by 'local'? If we are to attempt to understand globalization we must appreciate the roots and implications of these terms. These are not esoteric questions – how we interpret scale affects how we construct the world, and therefore what we do in terms of policy. In this sense there is a politics of scale – both personal and institutional. For example, what do members of the anti-globalization movement mean when they say they are fighting a global force? What does the environmental movement mean by 'think global act local'? What do transnational fruit companies mean when they advertise that they are bringing 'local' produce to the 'globe'? What does a 'global' war on terror mean when these actions are concentrated in very specific locations such as Afghanistan or Iraq?

The production of geographic scale

In a recent exposition on the role of scale in human geography, Herod (2003) notes that 'within human geography scale is usually seen in one of two ways: either as a real material thing which actually exists and is the result of political struggle and/or social processes, or as a way of framing our understanding of the world' (p. 229). Respectively these two points represent different ontological standpoints, the first being 'materialist' and the second being 'idealist'. In the case of the first, theorists such as Marx have argued that the real world exists and is created as a consequence of economic and political change. The idealists on the other hand, such as Kant, argue that the world is perceived in our minds and we make sense of it by constructing categories through which to view it. This corresponds to the broad division in human geography, which has seen the eclipse of positivist philosophy with a miasma of approaches often referred to as post-positivism. The idea that scale can be thought through critically is something that is really quite new in human geography. The result of this debate in human geography is that scale may be viewed:

> in terms of a process rather than in terms of a fixed entity. In other words the global and the local are not static arenas within which social life plays out but are constantly remade by social actions. This allows us to consider not only how a firm or a political organisation might 'go global' to engage with actors or opportunities not present within its own local

spaces of dependence, but also how a particular social actor such as
a TNC may attempt to 'go local' through tailoring . . . to appeal to
consumer tastes in different places or to reflect particular communities
cultural values.

(Herod, 2003, p. 233–234)

It is especially important to understand the relationship between the
global and the local. Free-market reform and neoliberalism (see Chapter
4) has either been strongly recommended, or imposed, by international
institutions such as the IMF and the World Bank, on the basis that it is
inevitable that 'global' forces will increasingly determine economic
change in any given place. This conception of the global as all-powerful
and inevitable has had incredible political force over the past twenty or so
years. One of the major pitfalls of many non-geographic interpretations
of globalization then has been a somewhat naïve concept of scale and
in particular a 'disabling dichotomy' of the global and the local. Dicken
argues (2004, p. 9) that '[F]ortunately geographers' conception of scale
is considerably more sophisticated than this'.

Herod (2003), drawing on the work of numerous geographers before
him such as Peter Taylor and Neil Smith among others, argues that it is
possible to view the concept of scale through five metaphors: *ladder,
concentric circles, Russian dolls, earthworm burrows, tree roots* (see
Figure 2.1). In the case of the first, scale it is seen like rungs on a ladder,
moving from the top (global) to the bottom (local). In this metaphor,
scales are seen as separate and arranged in a strictly hierarchical way.
In the case of the second metaphor the local is conceptualized as a
small circle encompassed by ever larger circles. Although scales are
still distinct in contrast to the ladder metaphor the global is not seen as
'above' the local. The Russian dolls metaphor is similar to the concentric
circles in that scales are seen as distinct and of different 'size'. However,
this concept gives the sense that the various scales fit together to make
a complex whole given that they nest inside each other and the doll is
not complete without fitting all of the pieces together. The metaphors
of earthworm burrows and tree roots provide two radically different
perspectives on scale compared to the other three concepts. They move
us away from 'layered' and 'territorial' notions of scale and suggest
the importance of networks. Thus all scales are interconnected through
the burrows or roots which penetrate to different strata of the soil. In the
latter two metaphors we move away from the idea of bounded scales.
When we conceptualize them at opposite ends of networks this alters
the way we think about the global and the local.

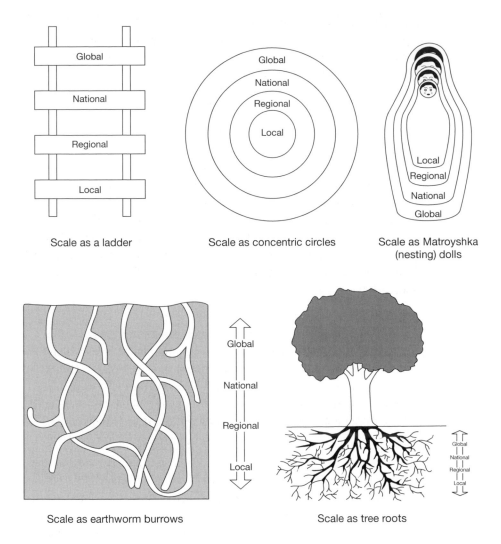

Figure 2.1 *Five concepts of scale*
Source: Adapted from Herod (2003)

Global–local interaction

In many ways, human geography is made distinctive by its simultaneous concern for both the global and the local. As geographers we want to know more about particular places 'in depth', at the same time as 'broadening our horizons' and learning about the wider world. Concern with the 'global' has peaked at certain times in the discipline. Thus we have seen the rise of geographies associated with imperialism and exploration in the early twentieth century, through to environmental and developmental concerns in the 1960s and 1970s, and more recently that seeks to tease out geographies of 'compression'. Concern with the local has permeated this transition and sometimes eclipsed globalism – especially in the 1980s with the rise of new cultural geography.

What has become clear, however, is that we cannot explain one without the other. The global and the local are two sides of the same coin. The call for focusing on big issues is welcome but it will not be successful without simultaneous concern for the local. Joe Painter points out that recent locality studies have been influenced by the 'upsurge of interest in globalization during the 1990s [which] has involved a recasting of the issue of local specificity in terms of global–local relations' (2000, p. 458). How, then, have geographers interpreted the local? Crang (1999) argues that this has come in three forms tied up with how we conceive of place:

- *Empirical significance of place* – Emphasizing actual existing differences. Every place is not the same – 'we are where we are'.
- *Normative significance of place* – The 'local' can stand for 'good' things. A celebration of difference and authenticity and resistance to homogeneity.
- *Epistemological significance of place* – How locality is important in knowledge formation. How ideas and perspectives are situated.

The importance of locality studies is emphasized when we consider what we mean by 'global'. The global is constructed from local action – and when we refer to the global process we are usually describing local-to-local flows that have become stretched across space to become global in extent. This is the process that is referred to in the definition of globalization provided in Chapter 1. In short, to be of use geographies need to be multi-scalar.

Crang (1999) provides a useful three-part framework in terms of how different thinkers have conceptualized global and local interaction (Figure 2.2; see also Gibson-Graham, 2002):

- World as mosaic
- World as system
- World as network

World as mosaic. This concept sees the world as resembling a jigsaw puzzle with different localities juxtaposed but independent. This might

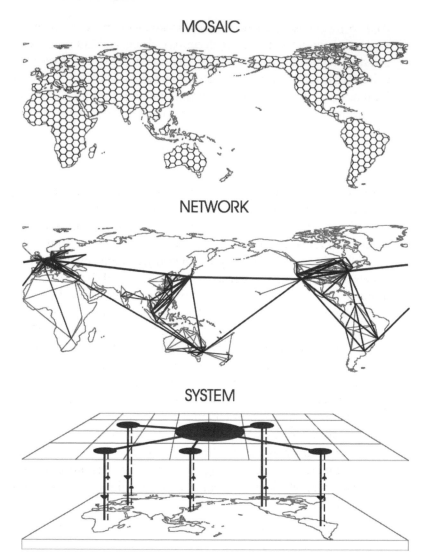

Figure 2.2 *Three concepts of global–local interaction*

refer to different neighbourhoods or countries, for example. This type of concept is promoted by tourism and other simplifying ways of experiencing and representing the world. There are three features

- It emphasizes borders.
- Each locality has a unique 'personality' and unified geographical identity.
- Intrusion is seen as a threat to authenticity and/or tradition.

Arguably this concept is no longer relevant as there are no totally isolated places left – it is a useful exercise to try and think of one. Crang argues that it tends to 'fossilize' difference and may be used as part of a defensive 'localism' which attempts to exclude 'outsiders'. Racism, for example, emanates from this worldview. Further, by fossilizing difference, places can actually be shaped in the image that the viewers desire. In the case of tourism it is often 'authenticity' and 'tradition' which tourists wish to see – and this is invariably 'constructed' and 'frozen' to meet this demand.

World as system. Many social scientists have looked at the interaction of global and local through a global system perspective (WST and dependency, for example). In this case, it is argued that local outcomes are produced through the particular location of the place within the broader global system at that point in time. This type of argument counters explanations which seek to explain local differences solely in terms of 'internal' characteristics (for example, modernization accounts of development theory). This kind of cultural determinism leads to problematic and sometimes racist explanations. For example, why is Europe rich and Africa poor? Because there is something inherent in European culture which makes it more productive and efficient? A systems-based explanation argues that it is that the history of the two regions in the context of the global system and the way this conditions their interactions, Europe as colonizer, Africa as colonized, that explains this inequality. Systems approaches do not always suggest that the 'global' conditions the 'local', however. The way the global unfolds is also influenced by pre-existing conditions. It is this process of mutual differentiation, which works both up and down the scale, which produces difference and diversity.

World as network. More recently geographers have focused on the concept of networks to explain links between the 'global' and the 'local', and in doing so have reflected critically on the nature of the former. This focuses on the connections between different people and institutions located on specific nodes across the world. In this sense the local becomes the global and the global the local – something which is sometimes called 'glocalization' (see discussion later in this chapter). For example, Massey (1991) talks of the 'global sense of place' she experiences walking down her home high street in Kilburn. She emphasizes the links between that particular place and other parts of the world, which have been transmitted down particular networks. Her conclusion is that locality is constructed from its interaction with the global and that the two are inextricably intertwined in character through such networks. As such, 'localities are always provisional, always in the process of being made, always contested' (Massey, 1991, p. 29). This argument supports the idea that what we are really referring to when we talk about the global is localized links over a large distance. Think of global money flows (see Box 2.1), international migration, world music, and even the worldwide web – they are all constituted of human action taking place in specific locations. Crang (1999, p. 33) sums this up by arguing that:

> [W]hat we have then is a global realm comprised of multidirectional, multi-fibred networks, the geographies of which are not mappable onto neat territories or systems. To grasp them geographers will have to get inside those networks.

In sum, although Crang's arguments concerning the importance of networks as opposed to systems and mosaic interpretations of global society are compelling, it could be argued that the reality is even more complicated than the startlingly complex 'architecture' he proposes. Networks are indeed important and becoming increasingly more so, of that there can be little doubt. However, is it not possible that networks might more appropriately be characterized as functioning within systems, principally the global capitalist system, but also the socialist system where this still exists? Mosaics are still existent; in world cities, for example (Knox, 2002; Sassen, 2001), one of the outcomes of globalization has been the creation of areas increasingly differentiated along socio-economic and sometimes ethnic lines. What we have is *a system of multi-fibred networks, characterized by multidirectional flows, nestled within a larger system (on the whole, capitalism), which creates at its nodes a mosaicing of space* – creating ever-differentiated places. As geographers then it is important that we continue to get inside systems and mosaics as well as networks that link them.

'Spaces of flows' versus 'spaces of places'?

The debate surrounding 'spaces of places' versus 'spaces of flows' illustrates some of the tensions between the different approaches outlined above. In the new 'informational age', which has replaced the 'industrial age', a new network society has evolved. This new society is seen as a 'space of flows' rather than as a 'space of places' which, according to Castells (1996), characterized society until the 1970s. Spaces of places

Box 2.1

Hyperglobalization in the financial sector – myth or reality?

The global financial sector is often held to be the epitome of globalization, and there can be little doubt that since the 1970s relative to other sectors transnational flows of finance have blossomed in unprecedented ways. As Pollard (2001) points out, a number of theorists talk of the role of money as a vehicle for the homogenization of space given its ability to transform different commodities and experiences into an abstract entity – a monetary value. A high level of integration between the different parts of the financial system has developed, embodied in apparently disembedded transnational banks that transfer money with extraordinary fluidity. These flows have woven local and national markets into what Martin (1994) refers to as a 'new supranational system'. In this new system, transactions – for example, in currency markets – in one part of the world can have almost instant material ramifications in another, making the regulation of financial flows by state institutions increasingly difficult and economic change in any one place less predictable.

The globalization of finance has come about through the facilitating role of neoliberal reform, the development of communications technology, and the evolution of tradable financial products that have made increasingly rapid movements possible. However, underlying the development of these regulations and instruments is the capitalist imperative to reduce the turnover time of capital through the shrinking of space (Harvey, 1989). However, is the financial sector as 'globalized' as many presume? Pollard (2001) debunks this myth through a number of observations. First, the global financial system is actually concentrated in a handful of Western economies, and the bulk of flows move between specialized districts in world cities, such as Wall Street, New York. Second, there are in fact distinct geographical circuits of money – or a 'patchwork of different financial spaces' (Leyshon, 1996, p. 62) – which are only loosely integrated. Finally, networks of money are actually constituted of 'materiality' and 'practices'. In this context, Pollard argues, 'although the global financial system may sometimes appear to be a rather intangible "thing", it is really an assemblage of consumers, workers, computers, telephones, office buildings, bits of paper, financial reports, and so forth' (p. 392). In this sense it occurs in specific places and is far from omnipresent. Overall, then, the global financial system is best conceived of as 'an intersecting web of networks – some fast, some slow, some long, some short . . . we are not talking about a machine operating "out there", but a socially, economically, culturally and politically constructed network of relations' (Pollard, 2001, p. 393).

are, to use Crang's terminology, those which give rise to 'world as mosaic' conceptions of the globe – delineated suburbs, villages, towns, regions and nation-states, for example. According to Castells this has been superseded by interrelationships of knowledge, power, information and other flows through space. Castells distinguishes between the 'wired world' and the 'social world' of networks (Taylor *et al.*, 2002, p. 8). The former refers to the system of telecommunications that makes e-transfers possible, and the latter to relationships between people that sustain these transfers. The latter is facilitated by the former and makes the network society operational. Geographers barely understand these spaces of flows – not least because they are evolving so rapidly. Central to this debate is the question of the relative importance of the two types of spaces. Places still exist, the nodes in the networks are important, and economic and social agglomerations are still significant responses and transmitters of flows, be they informational or otherwise. Furthermore, before the supposed advent of the network society, spaces of flows existed in the form of international trade, cultural exchange and political interaction, for example. However, these networks were less pervasive than they are today. The core geographical implication of this is that the simple divisions of the world along hierarchies such as core and periphery have become blurred and are likely to become more so. This produces *glocalized* outcomes, as we explore below.

Glocalization

There can be little doubt that localities are increasingly impacted by flows coming from 'higher' scales. A major error however is that this process has often been seen as deterministic. Geographer Swyngedouw (1997) coined the phrase *glocalization* to refer to a two-way relationship between the global and the local. As Dicken says, 'Such a term helps us to appreciate the interrelatedness of the geographical scales and, in particular, the idea that while the "local" exists within the "global" the "global" also exists within the "local"' (2000, p. 459). Through this concept geographers have been quite active in countering the 'pure' spaces of flows proposition. Dicken argues that many 'deterritorialization' arguments 'are based on a deep misconception of the nature of spatial processes which are deeply embedded in place' (2000, p. 458). Everything that takes places in the cultural, economic and political spheres is 'grounded' and localized, and requires what Dicken calls a 'spatial fixedness' to operate.

The concept of glocalization reminds us that the outcomes of globalization are path-dependent. Flows of information across the internet move from hub to hub in particular places, and, within networks of supposedly hyperglobalized sectors, such as finance, significant agglomerations of activity crystallize. The City of London, for example, by virtue of its history at the centre of the largest global empire yet known, functions as a major financial centre. Inertia of place is important and 'sunk capital', both in a socio-cultural and an economic sense, plays an important role in producing localities with specialized functions and character. At the supply end of the global economy, for example, the non-homogeneous distribution of natural resources such as oil, good land, forests, fishing grounds and so on continues to determine the comparative advantage of regions and localities and the way, therefore, that globalization influences 'development'. But historical patterns also play an important role as global flows move across them. Ask yourself why, for example, Latin America's economy largely comprises earnings from natural resource exports to the core economies of the world. Is it simply because Latin America has many natural resources, such as forestry, metals and minerals, or is to do with the pattern of trade that was established when the regional economy was brought into the orbit of the global capitalist system? (See Murray and Silva, 2004.)

Underlying geographies of agglomeration are thus physical, historical, socio-cultural and economic factors and processes. In the case of the latter, for example, economies of agglomeration flow from the economies of scale that are gained by certain functions clustering together – sometimes referred to as 'traded interdependencies'. There are also 'untraded interdependencies', that is to say, benefits (positive externalities to use economics jargon) which geographic proximity brings that cannot always be strictly measured. Amin and Thrift (1994) identify three sources of such interdependencies: (1) face-to-face contact; (2) social and cultural interaction which establish networks of trust; and (3) enhancement of knowledge and innovation. Underlying such ideas is the notion that human beings are essentially social animals who depend on the benefits that interaction with others yields. There is no direct evidence to suggest that this human social and organizational trait is diminishing in the age of the current communications revolution.

The above discussion leads to the conclusion that spaces of flows and spaces of places need not be mutually exclusive and both are simultaneously in action – spaces of flows operate within and between spaces of places. In this sense, Dicken (2004, p. 9) notes:

Bounded political spaces matter. Some, like the nation-state, matter more than others. In this sense, therefore, we have a very complex situation in which topologically defined networks (for example, of TNCs) both 'interrupt' – and are interrupted by – political-territorial boundaries.

Summarizing the distinct viewpoints on the interaction of the global and the local, Thrift (2000) suggests:

> There are three possible accounts. On one, global processes leave their footprints on places, allowing this little choice but either fall into line or be stamped out. On another, local places 'turn' global processes . . . global processes can only obtain a purchase by fitting in with local cultures. Finally, between these two views, is the one which argues for a process of glocalisation, a complex interaction between globalising and localising tendencies.
>
> (Thrift, 2000, p. 456)

In fact there are potentially more ways of conceptualizing the interaction of the global and the local; Gibson-Graham, for example, discusses six (see Box 2.2). There has been much talk of the prospect of globalization from above and globalization from below; if one accepts that globalization is a dialectical process (that it is local and global at the same time) the binary of globalization from above or below becomes redundant – it happens both ways at the same time.

Box 2.2

Global and the local: six ways of conceptualizing interaction

Gibson-Graham (2002) goes further than Crang (1999) or Thrift (2000), arguing that there are in fact six ways that the global and local, and the interaction between them, have been conceptualized in geographic writings:

1 The global and the local do not exist – they are just ways of 'framing' things.
2 The global and the local each get their meanings from what they are not; that is, in opposition to the other.
3 The global and the local offer different points of view concerning social networks.
4 The global is the local – all global things have local expression. Multinational firms are actually multi-local.
5 The local is the global – the local is where global process interact with the surface of the Earth.
6 All spaces are glocal – the global is constituted by the local and vice versa.

Conclusion – new geographies

According to a number of geographers there exists considerable confusion with respect to the conceptualization of the causal power of 'globalization'. In particular in many studies the lines between globalization as discourse, process and outcome are blurred. In this context Dicken notes (2004, p. 7): '[T]he problem is that . . . material processes are themselves enmeshed within a web of *discourses* from which material processes need to be disentangled (Kelly, 1999, p. 386) – although, of course, the discourses themselves influence material processes and outcomes.' In this context, according to Yeung (2002), globalization does not have determining power in and of itself, and this idea is similarly refuted by a number of commentators, including Urry (2003) and Hay and Marsh (2000). Rather its importance is arranged around two crucial themes. First, it reconstitutes scale. As we have seen in the above discussion globalization leads to a collapsing of scales, and increased interpenetration between them. This does not mean, as Thrift (2002) mischievously argues, that scale does not exist; but it has become more fluid. The dominant process in this respect then is that in the lived experience of humans the local grows (becomes more extensive) and the global shrinks (is compressed) leading to new 'glocal' outcomes. This requires that governments, businesses, people – agents in general – have to transform their activities and responses to cope with this. Second, there is a geography of globalization as discourse (i.e. how it is used – and abused – by different social groups to justify certain interventions and readings). In short, globalization is historically, spatially and politically contingent. To respond to it effectively in any given situation, it has to be understood within the context of that situation.

These arguments remind us that geography does indeed matter. In the preface to an excellent sociological survey of globalization, Waters (2001) argues: 'The constraints of geography are shrinking and world is becoming one place.' While it cannot be disputed that compression is occurring and that this has undermined the constraints of absolute space to both socio-cultural and economic flows, this statement misinterprets what is meant by 'geography', and 'place'. The constraints of geography are changing as space and scale are transformed; this creates new geographies of old and new places, which are undoubtedly harder to understand than ever before.

Further reading

Beck (2000): Provides a no-nonsense guide to globalization and some of the more pertinent questions surrounding it.

Castells (1996): This is the classic book on the network society which has been very influential in subsequent geographic thought.

Crang (1999): This chapter provides a useful overview of geographers' concepts of global local interaction.

Dicken *et al.* (2001): This article provides useful arguments on the concepts of space and scale as employed in economic geographies of globalization.

Gibson-Graham (2002): This chapter explores the links between the global and the local in a more detailed fashion than Crang and provides compelling arguments for glocalization.

Harvey (1989): It is in this book that the idea of time–space compression is most completely explored and used as a means of explaining capitalist expansion across the globe.

Held *et al.* (1999) Chapter 1: This is the classic chapter on the three theses of globalization and should be a starting point for anybody attempting to understand the competing perspectives that exist.

Herod (2003): This chapter provides an excellent and readable review of critical notions of scale as used in human geography.

Massey (1991): This article introduces the concept of a global sense of place and is one of the earliest which critically explores the relationship between the global and the local.

3 Globalization across time – contesting histories

Competing global histories

Interpreting the history of globalization is deeply contested, and different perspectives are greatly influenced by discourse, ideology and definition. Few works tackle the subject through the use of empirical evidence, although Held *et al.*, (1999) and Hirst and Thompson (1999) are notable exceptions. Below, four historical 'frameworks' are outlined. We begin by discussing a hyperglobalist and sceptical viewpoint respectively. This is followed by an analysis of two transformationalist histories – one sociological and one based in political economy. This section illustrates that the battleground over defining and conceptualizing globalization, let alone measuring it, is fraught. The chapter concludes with a simplified historical framework for interpreting the different waves of globalization that will be followed in the remainder of the book.

A hyperglobalist view – globalization as unprecedented

In 1990 Kenichi Ohmae, a business guru from Japan, wrote a book entitled *The Borderless World* (Ohmae, 1990) which, if nothing else, served to elevate the debates concerning the nature of globalization to the popular stage. Since that time the management analyst, who is head of a collective which seeks to reform the interventionalist Japanese government system, has written a number of books that have been influential beyond the management world. These works seek to support the neoliberal argument against government intervention. In *The End of the Nation State* (Ohmae, 1995), the author develops an argument against the state and offers a chronological interpretation of history. He begins the book by claiming that the end of the Cold War was a watershed in the history of the global economy and society. Contrary to Fukuyama's (1992) proclamation that the collapse of socialism led to the 'end of history', he argues:

> nothing could be further from the truth. In fact, now that the bitter ideological confrontation sparked by this century's collision of 'isms' has ended, larger numbers of people from more points on the globe than ever before have aggressively come forward to participate *in* history. . . . A generation ago, even a decade ago, most of them were as voiceless and invisible as they had always been. This is true no longer: they have entered history with a vengeance, and they have demands – economic demands – to make.
>
> (Ohmae, 1995, p. 1)

He goes on to argue that such demands cannot be met by nation-states that 'no longer possess the seemingly bottomless well of resources from which they used to draw with impunity to fund their ambitions' (p. 2). Should these new participants in history turn to international bodies, such as the UN, they will find that they are nothing more than collections of nation-states. Economic groupings such as OPEC, APEC or the EU are much the same and, according to Ohmae, incapable of managing and meeting the demands of the 'new global citizen'.

Underpinning Ohmae's argument against nation-states is the increased mobility of what he calls the four Is: Investment, Industry, Information technology and Individual consumers. He argues that each of these has become increasingly footloose and pervasive, and that only by tapping into the new flows that have emerged can welfare be maximized. He argues, then, for a world where the newfound mobility of the four Is is allowed to work in global space:

> Taken together, the mobility of these four I's makes it possible for viable economic units in any part of the world to pull in whatever is needed for development. . . . This makes the traditional 'middleman' function of nation-states – and of their governments – largely unnecessary. . . . If allowed global solutions will flow to where they are needed without the intervention of nation-states. On current evidence they flow better precisely because such intervention is absent.
>
> (Ohmae, 1995, p. 14)

Running through Ohmae's analysis is an argument for a new geography which transcends what he terms the 'cartographic illusion'. In the place of nation-states Ohmae argues for 'region states' – 'where real work gets done and real markets flourish' (p. 5). Such units are not defined in terms of their formal political borders, as this 'is the irrelevant result of historical accident' (p. 5). Rather, it is essential that they be 'the right size and scale to be the true, natural business units in today's global economy. Theirs are the borders and the connections that matter in the borderless world' (p. 5).

As noted above, neoliberal thought greatly influences Ohmae's argument, and the work has been seen as the epitome of hyperglobalist claims in that it imagines a world where flows of capital, people, information and goods are unrestricted. It is a normative argument, in that it suggests that such governance (or lack of) is preferable and will maximize welfare, for 'where information reaches, demand grows, where demand grows, the global economy has a natural home' (p. 25). This is the recipe to get countries on what Ohmae calls, harking back to Rostow, the 'ladder of development'. In terms of the policy implications that flow from his argument he offers the following:

> The evidence, then, is as exhaustive as it is uncomfortable: in a borderless economy, the nation-focused maps we typically use to make sense of economic activity are woefully misleading. We must, managers and policy makers alike, face up at least to the awkward and uncomfortable truth: the old cartography no longer works. It has become no more than an illusion.
>
> (Ohmae, 1995, p. 20)

Ohmae's chronology of the rise of globalization is straightforward. He divides history into two periods (see Figure 3.1) – the industrial age and the information age. The first, which corresponds to the nineteenth and twentieth centuries, was driven by nation-states, national sovereignty and centralization. This led to flows that were disrupted by borders and protected domestic capital above international capital. The centrality of the nation-state grew out of colonial models, and newly independent states adopted this model of governance – what Ohmae terms 'post-feudal relations' (p. 142). Ohmae views globalization as entirely new, beginning in the 1990s. To paraphrase the author, the 'economic' pendulum swung at the end of the twentieth century when the end of the Cold War and the triumph of capitalism paved the way for a new information age reducing the 'cholesterol' in the arteries of the global economy. This new age is driven by private capital and

Old game

Industrial age
Nineteenth to twentieth century

- Driven by nation-state governments
- National sovereignty
- Strong control by centralized forces
- Sensitive to borders
- Favours domestic capital and protects domestic companies
- Aims for on-state prosperity through development of export-led, manufacturing-driven economic growth
- Government initiatives
- Good government strengthens priority industries
- Changes occur gradually over decades
- Germany
- Japan/New Japans
- United Kingdom
- United States

Rise of the Region-State

New game

Information Age
Late twentieth to twenty-first century

- Driven by private capital and information
- Civil sovereignty
- Autonomous networks of interdependent private enterprises and regional entities
- Inherently borderless
- Welcomes foreign capital and world-class
- Companies/expertise, creating high-quality jobs
- Aims for harmonious regional prosperity based on interdependent, network-centric companies creating information-intensive services to capture value from customers
- Entreprenurial initiatives
- Good government nurtures regional development, not focused in specific industry
- Change occurs suddenly in months to years
- Hong Kong/Shenzhen
- Singapore/Johor/Batam
- Taiwan/Fujian
- Southern China (Pearl River Delta)
- Southern India (e.g. Bangalore)
- North Mexico/Southwestern U.S.A
- Silicon Valley
- New Zealand
- Lombardia
- Pacific Northwest of the United States

Figure 3.1 Ohmae's industrial and information ages

Source: Adapted from Ohmae 1995, p. 143

autonomous networks, and is characterized by citizen sovereignty. It welcomes foreign capital and is based on governments nurturing entrepreneurship. In terms of the winners and losers between the two periods, Ohmae argues that new region states such as Hong Kong/Shenzhen, Singapore/Johor/Batam, Silicon Valley and New Zealand benefit. The old winners – Japan, the UK, the USA and Germany – all lose out relatively speaking as their strait-jacketed governance systems become ineffective.

Criticisms of the hyperglobalist thesis

Ohmae's arguments have been criticized roundly from many quarters and much of what follows in this chapter, and indeed this book, takes issue with a number of the central points he makes. In particular, in Chapter 5 concerning political globalization, his attack on the nation-state is critiqued. A few of the more important criticisms are considered below:

- *Idealization or reality* – It is unclear as Ohmae writes whether he is talking about an ideal world (in his eyes) of borderless flows or if he is claiming that this is what is actually occurring.
- *Empirical data* – Linked to the above, the use of empirical data to back up his major points is scant and anecdotal. In *The End of the Nation state*, for example, he uses a journalistic article from the *Japan Times* (Peter Osbourne, 23 December 1994) as the main evidence for his argument that New Zealand is a winner in the new borderless world following its neoliberal reforms in the mid-1980s.
- *Unilinear model of development* – Ohmae explicitly argues for a ladder of development where economic growth leads to an ideal end-goal. In this sense his argument is a-historical and a-geographical.
- *Characterization of the nation-state* – In order to attempt to demolish the nation-state, Ohmae paints a characterization of it that is unfounded. He suggests that national governments are inherently bad, yet no evidence is offered to back this up.
- *Dualistic interpretation of history* – History is reduced to two periods with no particular rationale for the transition between them and with enormous generalization with respect to the conditions in each.
- *Favouring the networked elite* – Ohmae is way too optimistic with respect to the implications of the new information economy for those that are not on networks. There is no credible evidence to suggest that free markets close the wealth gap. The information gap, for example, is growing and a policy to reverse this is not suggested.

- *The region state* – The region state is poorly defined. What does it look like? How does it function with respect to other scales? This idea which is central to Ohmae's argument is barely developed and the concept of scale is not dealt with coherently.
- *Internal contradictions* – Ohmae talks of the role of governments in nurturing entrepreneurship, and of the 'borders that really matter'. The use of national governments and borders to illustrate his argument contradicts his attempted demolition of both.

A sceptical view – globalization in question

Hirst and Thompson (1999) question the very existence of globalization, some of the claims that have been made in its name, and argue that the world economy has become less integrated over time. Theirs is the sceptical position which offers a salient antidote to the 'hyperglobalist' approach. It is also one which, in the words of one reviewer, is 'systematically grounded in evidence; the literature on globalization contains all too few examples of this' (Perraton, 2001, p. 670). In *Globalization in Question*, the authors make five central points that highlight the exaggeration of claims made by the hyperglobalists (Hirst and Thompson, 1999, pp. 2–3):

1 Today's levels of integration fall short of the period between 1870 and 1914 (what is referred to as the classical Gold Standard era). In this respect they argue: 'The present highly internationalised economy is not unprecedented: it is one of a number of distinct conjunctures or states of the international economy that have existed since an economy based on modern industrial technology began' (p. 2).

2 Contrary to the literature from some areas of business studies, truly transnational corporations are scarce. Such companies are embedded nationally, and do their trade internationally.

3 The vast majority of FDI is concentrated in advanced capitalist economies and this trend is becoming exacerbated. Integration is not leading to the diffusion of investment from such nations to peripheral ones – apart from a few leading 'emerging nations'.

4 The majority of the world's economic activity remains concentrated in a 'triad' of regional blocs, namely the NAFTA, the EU and Japan. Much global trade flows within and between these areas and 'this dominance is set to continue' (p. 2).

5 The core economies maintain the ability and wherewithal to regulate the global economy, its firms, financial markers and other less

powerful nation-states. To do so they often have to act in cohorts,
but this has been fruitful. The authors refer to the triad powers as
the G3 and argue: 'Global markets are by no means beyond
regulation and control even though the current scope and objectives
of economic governance are limited by the divergent interests of the
great powers and the economic doctrines prevalent among their elites'
(pp. 2–3).

Hirst and Thomspon use two major measures to illustrate that the global
economy is no more open today than it was in the early nineteenth
century. The first is the ratio of trade levels to GDP: in 1914 trade/GDP
ratios were higher in the major economic powerhouses than they were in
1973. By 1995, the major economies of Japan, Netherlands and the UK
were still less open using this measure (while France and Germany were
slightly more open) (see Table 3.1). They do concede, however, that the
USA was substantially more open in 1995 than it was in 1914. They
make the important point that if analysis concentrates on the period after
the Second World War only, then growth in trade openness is steady and
consistent. This, they claim, is underpinned particularly by the rise of the
East Asian countries as trading economies.

Hirst and Thompson argue that 'the evidence also suggests greater
openness to capital flows in the pre-First World War period compared
to more recent years' (1999, p. 27), a finding which is supported by
Grassman (1980) and Lewis (1981). There is however, evidence to
suggest that, again, since the Second World War, levels of capital
integration have indeed risen (see Figure 3.2). In conclusion, building
in evidence for decreased per capita migratory moves since the start of
the twentieth century, the authors argue:

**Table 3.1 Ratio of Trade to GDP, advanced economies,
1914 and 1995 (exports and imports combined)**

	1914	1995
France	35.4	36.6
Germany	35.1	38.7
Japan	31.4	14.1
Netherlands	103.6	83.4
UK	44.7	42.6
US	11.2	19

Source: Adapted from Hirst and Thompson, (1999, p. 27)

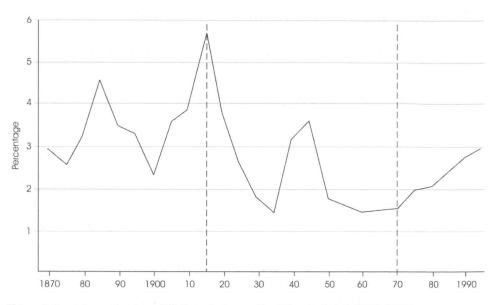

Figure 3.2 *International capital flows between the G7 economies 1870–1995 (as a percentage of GDP)*

Source: Adapted from Hirst and Thompson, 1999 (p. 28)

> [W]e can say that the international economy was in many ways more open in the pre-1914 period than at any time since, including from the late 1970s onwards. International trade and capital flows, both between the rapidly industrialising economies themselves and between these and their various colonial territories, were more important relative to GDP levels before the first world war than they probably are today. . . . Thus the present position is by no means unprecedented.
>
> (Hirst and Thompson, 1999, p. 32)

As Perraton argues (2001), there is a clear normative agenda in what Hirst and Thompson seek to establish through their work. In arguing that current levels of integration are a result of political choice and change rather than technological 'progress', they seek to establish that claims of 'inexorable globalization' are merely a smoke-screen to further the neoliberal agenda. This implies that 'globalization' can be reversed and governed. As Perraton (2001) argues, 'They repeatedly emphasise links between globalization and advocacy of neoliberalism as the only viable policy option' (p. 671).

Criticisms of the sceptical thesis

There are numerous critiques of Hirst and Thompson's thesis, but the most complete is that undertaken by Perraton (2001). While broadly supportive of their work, and in particular their recourse to empirical fact rather than speculation, his central critique is that the methodology is faulty. Hirst and Thompson draw their major conclusions by comparing existing trends with an imagined 'hyperglobalized' economy. Perraton, as well as Held *et al.*, (1999), argues that this is teleological, and that the evidence should be redirected to provide 'a more nuanced assessment of the impact of contemporary integration' (Perraton, 2001, p. 678), rather than to dismiss the concept of globalization altogether. He argues that by comparing current levels of integration with the hyperglobalist thesis, the work is:

> if not quite a straw man conception of globalization, GiQ [*Globalization in Question*] still focuse[d] on a particular version of globalization analysis that commands very limited academic support. . . .
> Fundamentally it is unhelpful to study economic processes in terms of single implied end-point.
>
> (Perraton, 2001, p. 672)

The most important criticism is that of the proposition that global levels of integration are not unprecedented. Perraton argues that the measures employed by Hirst and Thompson are too crude to capture the changing nature of the global economy and, in particular, its *qualitative* aspects. One of the most notable trends has been the rise of cross-border transactions in bonds and equities, which have risen exponentially since the 1970s (see Table 3.2). Overall, Perraton argues, 'there is substantial

Table 3.2 *Cross-border transactions in bonds and equities as percentage of GDP, 1980–1998*

	1980	1990	1998
USA	9	89	230
Japan	8	119	91
Germany	7	57	334
France	5	54	415
Italy	1	27	640
Canada	9	65	331

Source: Adapted from Perraton (2001, p. 674)

evidence of unprecedented levels of trade, FDI and international financial activity relative to national economic activity. . . . Simply comparing crude indicators across time periods misses the changing character of contemporary integration, notably the growth of MNCs and the rise of short-term international financial flows' (p. 675).

Perraton offers five counterpoints to Hirst and Thompson's central claims:

1 Financial and trade integration moves have advanced further than ever before; although integration is not unprecedented global markets have emerged.
2 True TNCs are rare, and although they retain a national base (they have to), their mobility has increased making the sharing of the profits, through taxation, for example, within nation-states harder.
3 There are uneven patterns of economic activity, but no bloc is insulated from the rest of the world.
4 Many poorer countries are indeed peripheral and networks bypass them. However, this does not imply that global markets are non-existent. Global markets are not likely to be perfect and therefore incomes are not likely to converge globally.
5 Nation-states still regulate markets but they are increasingly constrained by market shifts and international agreements.

Perraton argues in conclusion that 'global markets have evolved for goods, services and financial assets. As such, this goes beyond 'internationalisation' and can reasonably be characterised as globalization' (2001, p. 682). This conclusion is echoed most explicitly in human geography in the work of Dicken (2004), who argues that Hirst and Thompson fail to distinguish between the shallow integration of the pre-1914 period and the deep integration of the contemporary era, arguing, 'it is in a qualitative sense that the material processes of globalization must be analysed' (p. 8). To Ash Amin, the argument fails because 'it does not provide any sense of the trends and changes in the world economic system which might genuinely be challenging the balance between national and global influences' (1997, p. 124).

A transformationalist view – dematerializing globalization

There are numerous 'transformationalist' views on the history of globalization, although only two are discussed in detail here – Waters'

(2001) 'dematerialization' thesis and Held *et al.*'s (1999) fourfold 'historical forms' chronology. Particularly useful alternatives are offered in Rosenau (1990), Harvey (1989) and Robertson (1992, 2003). Generally, transformationalist perspectives are more holistic than hyperglobalist or sceptical accounts in that they emphasize the importance of culture and politics as well as economics. One of the most accessible and succinct expositions of globalization comes from Australian sociologist Malcolm Waters. In his book *Globalization* (Waters, 2001), he proposes an historical theorem to explain increased interconnectedness. This sociological perspective, which explicitly builds in notions of space and place, is particularly useful for human geographers.

Waters argues that globalization should be traced through transformations in three 'regions' of social life – the economy, the polity and culture. The author adopts the views of Weber *et al.*, (1978) and Bell (1974) that each of these arenas is 'structurally autonomous' (Waters, 2001, p. 17), although 'relative efficacy of the arenas can vary across history and geography' (ibid.). This contrasts to the Marxian view which gives ascendancy to the economic sphere of life (i.e. It determines culture and politics) and the Parsonian view which gives determining power to culture. It is argued that each of these arenas comprises three kinds of exchange:

- *Material exchange* – which includes trade, tenancy, wage-labour, services and capital accumulation.
- *Power exchange* – which includes party membership, elections, command and leadership, coercion and social control, legislation, redistribution and international relations.
- *Symbolic exchange* – what Waters refers to as 'exchanges of signs' through, for example, oral communication, publication, performance, teaching, ritual, display, propaganda, advertisement, and research.

His key argument is that 'each of these types of exchange organises social relations in space in a particular fashion' (p. 19). First, *Material exchanges localize,* in that the production of items requires concentrations of the factors of production (labour, skills, capital, raw materials). Trade often takes place over long distances, but it is mediated by specialist intermediaries, such as merchants, sailors and financiers, which link each node of a commodity chain. Second, *Power exchanges internationalize*, in that they are geared towards controlling a population which occupies a given territory, culminate in national boundaries, and international relations are practised generally in order to exert and/or confirm sovereignty. Finally, *Symbolic exchanges globalize*, in that they release 'social arrangements from spatial referents' (p. 19). He argues that it is much more difficult to monopolize such exchange in local space as they can be 'proliferated rapidly in any locality' (p. 19). Thus they are generally more easily transportable and can claim global significance more easily.

Claiming that these arenas, or spheres as they are termed in this book, are structurally autonomous does not imply that they operate in isolation from one another. Waters qualifies this point (p. 20), arguing:

> The apparent correspondence between the three arenas of social life – economy, politics and culture – and the three types of exchange – material, power and symbolic – should not mislead us into thinking that each type of exchange is restricted to a single area. For example, firms, resolutely located in the economy obviously include many material exchanges but they also involve power exchanges between managers and workers and symbolic exchanges about such matters as work norms, dress, sets of rules and so on. However, there is a general tendency for material exchanges to originate in the economy, for power exchanges to originate in the polity and for symbolic exchanges to originate in culture.

Waters' claim is that globalization will proceed where the emphasis in society moves from exchanges based on the material, through power exchanges, to symbolic exchanges. That is to say, exchanges become increasingly 'dematerialized'. In this sense he argues that 'we can also expect that if globalization is highly advanced it will be more highly developed in the cultural than in the other two arenas' (p. 21). These ideas are then applied to explain the historical evolution of global society. According, three processes in each of the main arenas of life 'support' globalization:

- *Economies tend towards marketization* – that is, freedom from command, constraint and class monopolization.
- *Polities tend towards liberalization and democratization* – this leads to the deconcentration of power.
- *Culture tends towards universalization* – the abstraction of values and standards to a very high level of generality that will permit extreme levels of differentiation.

Waters' three-part periodization

Waters conceptualizes three periods – the early modern, the modern and the global – punctuated by major crises leading to the eclipse of one set of processes and exchanges by another. In the early modern period, the major event was the evolution of capitalism and economic marketization between the 1600s and 1900s. The system of material exchanges that evolved broke the 'traditional ties' of medieval society and led to the establishment of a capitalist class which undermined monarchies and other traditional governance forms. Trade, together with exploration and military conquest, were the main globalized outcomes during this period but 'although they constituted a beginning they were relatively ineffective in establishing global integration' (p. 23). As capitalism hit a crisis and workers were less willing to accept exploitation and misery, the collective action of the working class led to new political realities. In the modern phase, 'inter-nationalization' became a dominant transition given political liberalization and democratization. This led to the rise of the welfare state in many cases and, in some, to the rise of fascism and communism. In general then the emphasis shifted to the state to mediate these new relationships. The state performed the function of governing the power relations between corporations and workers, and also actively developed national traditions – often at the expense of local ethnic differences. During this period the main globalized outcome was the internationalization of state action 'under the development of such phenomena as alliances, diplomacy, world wars, hegemons and superpowers' (p. 23). Capitalism as both economic practice and culture was diffused under 'hegemonic sponsorship' colliding with socialist and fascist ideologies.

In the current global period the state itself has hit a crisis, unable to ensure sustainable prospects for societies or provide economic growth. Thus, individuals are less willing to surrender their autonomy to national scale organizations and have instead made increased reference to 'universalizing' cultural symbols. Waters argues that 'These accelerated and increasingly effective cultural flows indicate an oncoming culmination of the globalization process' (2001, p. 25). The new symbols include: human rights, the environment, liberal democracy, consumption rights, religious traditionalism, ethnic diversity and cosmopolitanism. Accordingly, '[C]ultural action is disrupting states, especially where they are most highly organised, and party politics is being disrupted by universalising and diffuse social movements' (Waters,

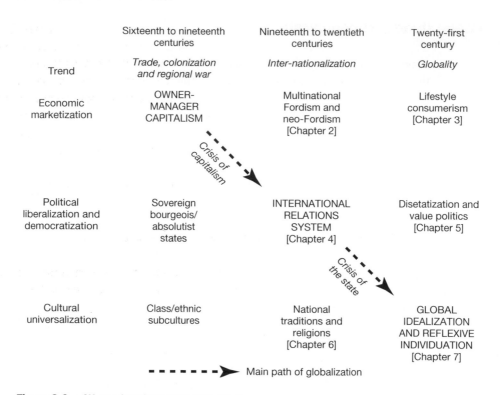

	Sixteenth to nineteenth centuries	Nineteenth to twentieth centuries	Twenty-first century
Trend	*Trade, colonization and regional war*	*Inter-nationalization*	*Globality*
Economic marketization	OWNER-MANAGER CAPITALISM	Multinational Fordism and neo-Fordism [Chapter 2]	Lifestyle consumerism [Chapter 3]
Political liberalization and democratization	Sovereign bourgeois/ absolutist states	INTERNATIONAL RELATIONS SYSTEM [Chapter 4]	Disetatization and value politics [Chapter 5]
Cultural universalization	Class/ethnic subcultures	National traditions and religions [Chapter 6]	GLOBAL IDEALIZATION AND REFLEXIVE INDIVIDUATION [Chapter 7]

Crisis of capitalism

Crisis of the state

- - - - - - ▶ Main path of globalization

Figure 3.3 *Waters' globalization through time*
Source: Waters (2001, Fig. 1.1)

2001, p. 25). In this sense the new symbolic exchange renders traditional boundaries less meaningful, the economy becomes detached from localities as it is increasingly 'symbolically mediated', and consumption replaces production as the main economic activity. Particularly important in this respect has been the rise of postmodernism, which has led to the collapsing of the distinction between high and popular culture. This is combined with the opening of boundaries to symbolic flows that cultures have become more permeable and globalized. Waters (2001, pp. 24–25) characterizes the current period, arguing that;

> [c]ultural products become more fluid and can be perceived as flows
> of preference, taste, and information that can sweep the globe in
> unpredictable and uncontrolled ways. Even the most casual inspection
> of such preference issues as environmental concerns, Pokemon games,
> investment in high tech shares, skirt lengths, roller blading and the AIDS
> panic can confirm this development.

Criticisms of the dematerialization thesis

Waters himself leads the criticism of his own ideas (p. 21) arguing:

> This summary . . . will inevitably make sweeping and occasionally offensive claims, brushing aside the particularities of individual corners of the planet and the raggedness of social transformations in an effort to make generalised, perhaps over generalised, sense out of daunting complexity.

Two further criticisms can be made. First, in arguing that each type of exchange operates at a particular scale, he overlooks the complexity of the interaction of processes at different scales. As was discussed in Chapter 2, geographers have recognized that the distinction between scales is problematic; every exchange – be it material, symbolic or power-related – has its source in a locality, and only becomes global when local-to-local interactions are stretched across space. In this sense, all exchanges localize and potentially globalize at the same time. It is meaningless to talk about the 'universalization' of culture, for example, since cultures are embedded in localities. Similarly, the evolution of new social movements leads to the practising of politics in a way that does not internationalize but transcends borders. Second, the argument that one set of processes has been eclipsed by another from economic, to political, to cultural – gives the impression of a neat unilinear transition which is not backed up empirically, although feasibly it could be. This transformation to a universalized end-state plays down geographical contingencies. Overall, however, Waters' thesis is groundbreaking in that it at least provides a *rationale* for globalization. Furthermore, it stretches the analysis beyond the economic, something which is to be welcomed in the social sciences in general.

A chronology of globalization – global transformations

Held *et al.*, (1999) take a broadly transformationalist view of the evolution of globalization overtime and make two central points. First, that 'globalization is neither wholly novel, nor a primarily modern social phenomena. Its form has changed over time and across the key domains of human interaction, from the political to the ecological' (p. 415). Second, that the processes of globalization do not unfold according to some unilinear or inherent logic. They say; 'globalization as an historical process cannot be characterised by an evolutionary logic or an emergent teleos. Historical patterns of globalization have been punctuated by great shifts and reversals, while the temporal rhythms of globalization differ between domains' (p. 415). With this in mind, the authors develop a fourfold periodization of globalization:

- Pre-modern globalization – pre-1500
- Early modern globalization – 1500–1850
- Modern globalization – 1850–1945
- Contemporary globalization – 1945–

Pre-modern globalization

This period began around 9,000 to 11,000 years ago with the rise of distinct centres of settled agrarian civilization in Eurasia, Africa and the Americas. These civilizations developed relatively long-distance trade and, crucially, the projection of power across space – although these two things were limited by transport and communications technology. During this period it was the Eurasian realm that was characterized most by interregional and/or intercivilizational flows, while Oceania and the Americas remained largely autarkic.

Globalization during this period was evident in four areas: military/political empires, world religions, migratory movements, and – to a lesser extent – trade. In the case of empires, the earliest known example was that of the Sumerian empire (3000 BC), and subsequent empires differed greatly in their extent and endurance. In contrast to the later periods of 'global' empires, most were relatively small, although some went on to form important regional level organization which set the stage for encounters in the subsequent centuries – including the Indic civilizations, Han China and the Roman Empire. EarlyΔ world religions formed during this period too, crossing imperial boundaries with universalized messages and texts – thus unifying many disparate cultures. The most important religious development at this time was the rise of Islam in the sixth century CE which some see as the first globalized world religion.

Migratory movements of those in nomadic empires including the Germanic people (who eventually settled in Western Europe having conquered the Roman Empire in the 5th century CE) and later the Mongols caused major upheavals through conquest during this period. Held *et al.*, make the point that such movements, creating devastation and plague in their wake, did not leave as great an imprint as the slow diffusion of settled agriculture into hunter-gatherer societies which took place throughout the period.

Some forms of economic interaction such as trans-Eurasian and Indian Ocean trade assumed importance in this period. For example, Muslim traders from the Indian subcontinent established an enduring presence on the east coast of Africa. It is of course highly difficult to quantify such flows. What is important though is that such flows represented conduits for the diffusion of cultural and social ideas, as well as technology, as interaction among elites intensified. In general, the velocity of interaction was greatly impeded by the low level of technology and the poor infrastructures that existed. The Romans and the Incas, for example, had roads that only extended to the edge of their respective empires, and canals in China were in their infancy. In general then the vast majority of the world's population remained isolated. As such, one-off interactions were often spectacular, but sustained political and military interactions and extension of political control were difficult to sustain. In this sense, the degree to which global interactions were institutionalized and regulated remained minimal. According to the authors this period could be termed one of *thin globalization* which was limited in its global extensity.

Early modern globalization – 1500–1850

The sixteenth century is the period that many see as the starting point of the 'rise of the West' (Held *et al.*, 1999, p. 418). It was during this period that Europe developed the technologies and institutions that facilitated modernity and led, eventually, to the establishment of the European global empires. The period is sometimes referred to as the mercantilist phase of colonialism and witnessed the first stage of the diffusion of merchant capitalism across the globe. In the early part of this period the most notable expansion was that of the Spanish and Portuguese into the 'New World' in search of precious metals and minerals and other natural resources to trade (see section on Latin America in Chapter 7). By the mid-1500s virtually all of South and Central America was colonized. Map 3.1 shows the extent of the Iberian empires in 1800.

Held *et al.*, (1999) are sceptical of 'conventional accounts', arguing that European expansion was opportunist and fragile and that much of the innovation which facilitated this was in fact imported and adapted from

ATLANTIC
OCEAN

VICEROYALTY
OF
NEW
SPAIN

UNITED
STATES

CUBA

SANTO DOMINGO

Mexico City

JAMAICA
BELIZE   St. DOMINGUE

VICEROYALTY
OF
NEW
GRANADA

Bogota

PACIFIC
OCEAN

VICEROYALTY
OF
BRAZIL

VICROYALTY
OF
PERU

Lima

Rio de
Janeiro

VICEROYALTY
OF
LA PLATA

CAPTAINCY-GENERAL
of CHILE

Buenos Aires

LEGEND

	Spain
	Portuguese
	French
	Dutch
	British

Indian Frontier

0   1,250   2,500   5,000   Kilometers

Map 3.1   *Colonial Latin America, 1800*

other regions. From a geographical point of view, they also emphasize that the European 'global' empires were in fact spatially partial; Africa and Asia remained largely untouched. However, the new global reach of Europe, together with changes in technology and politics in the heartlands, led to major transformations across the world. As Europe's political and military empires flourished, new forms of economic globalization were forged. Examples included the trading organizations of the Dutch and English East India companies, which were early precursors to TNCs. In general, European expansion into other lands took the form of nodes between which 'ran the thickening sinews of global economic interaction' (p. 419). Particularly important in terms of global economic flows during this period was the Atlantic slave trade (see Map 3.2), which only began to dwindle starting at the end of the seventeenth century as the European heartlands were industrialized and colonial corporations were replaced by formal government. This trade was made illegal by many European nations and the USA by the 1820s.

As in the pre-modern period the largest interactions were sporadic, and concentrated at the levels of elites. This is evidenced by the fact that by the end of the era the Spanish and Portuguese empires in Latin America (see Map 3.2) and the British and French empires in North America had gained independence. In other words, 'spectacular European thrusts' could be made but it was difficult for the Europeans to retain full control over their colonies. Notwithstanding this the cultural impacts of colonialism, especially in the form of the diffusion of the Catholic Church, were profound (Gwynne and Kay, 2004) (and see Plate 3.1).

The infrastructure which facilitated this period of globalization remained much the same as it had been in the pre-modern period. By the end of the eighteenth century railways, iron ships, canal and road building were more extensively developed – although their existence did not often stretch beyond the borders of the heartlands of the Western empires themselves. The velocity of these flows, however, was ultimately limited by the speed of shipping and other less advanced technological transport forms such as the horse. The institutionalization and regulation of global relationships, as in the pre-modern period, remained limited. There was a nascent system of diplomacy in Europe between states and the Roman Catholic Church increased its influence – but the ability of such institutions to regulate the periphery was relatively minimal. As in the previous period one-off encounters often had devastating effects but long-lasting impacts were often yielded by less spectacular interactions.

Map 3.2 *The Atlantic slave-trade, c.1500–c.1820*

0.2–0.3 imported
into the Old World

3.5–5 million
imported into
Brazil

4–5 million
imported into
the Caribbean

0.5 million
imported into the
Guianas

0.4–0.5 million
imported into the US

0.2 imported
into middle
America

0.5 imported
into Spanish
South America

LEGEND

→ Shipping routes

CONGO Slave-exporting
regions

Plate 3.1 *A colonial church in Santiago, Chile, nestled within modern architecture*

Arguably however, the greatest impact of the new global relationships was felt in Europe itself where the wealth from the New World, transatlantic migration, and the stimulation of inter-imperial rivalry led to new institutional power rivalries, and a military technology race. Indeed, it is during this period that the seeds of the great European conflicts of the twentieth century were sown. During this period, too, the development of formal states, with demarcated territories and fixed borders, began in Europe (see Chapter 5 for details). The creation of the nation-state was to become one of the major defining processes which would underpin the subsequent period of globalization.

Modern globalization 1850–1945

This period 'witnessed an enormous acceleration in [the] speed and entrenchment' (Held *et al.*, 1999, p. 421) of global networks and flows principally under the control of the European powers. The era has sometimes been referred to as industrial colonialism – based on the expansion of the industrial core party in search of raw materials for their new production techniques (Daniels and Lever, 1996). Four processes in Europe, which to a large extent were underpinned by events in the previous period, were instrumental in this respect: the development of capitalist economies; the rise of advanced weapons technology; naval technology; and, the rise of powerful state institutions. Of these the rise of capitalism based around the industrial revolution was of greatest importance. These changes were put into motion to the effect that, by the mid-twentieth century, European powers had 'exploded' across the globe taking in Africa, much of Southeast Asia directly and making important controlling inroads into East Asia – an area from which Europe had been formerly excluded. This was the age of imperialism. As global trade and investment increased massively the extensity, intensity, velocity and impact of globalized flows increased. There were large migrations across the Atlantic and the end of the slave-trade led to the export of indentured labour from India to many parts of the world. This was underpinned and facilitated by the rise in transport and communications technology – especially more advanced railway systems and the telegraph. This period also saw the beginnings of the rise of globalized environmental effects as transboundary pollution in Europe rose and tropical deforestation began in the colonized parts of the world (See Chapter 8 for more on this subject).

It was the global reach of the European and American empires and their political and military operations that characterized this period (see Map 3.3). Given communications advances, political influence could be exerted at great distance. This led to economic globalization at unprecedented levels as trade and financial investment, especially between the European core and its colonial periphery, increased. There was, relative to the current period, limited trade between richer nations. Furthermore, the massive migrations from Europe were paralleled by large migrations across the world from other societies, including the Japanese to Latin America and Indians to the Pacific Islands under the indenture system. Migrations within and to the core stimulated the expanding industrial complex.

In the cultural realm during this period the emphasis shifted away from the diffusion of Western religious discourses to the diffusion of secular discourses including science, Marxism and nationalism which eventually filtered down to the masses and transformed lives everywhere (see Chapter 5). Great advances in transport and communications technology made possible partly through the industrial revolution underpinned all of this. The unit costs of transport fell due to the extensive development of railroads and the steamship, and transoceanic telegraphy reduced communications times enormously. Thus elites at a distance could be kept well informed of events far away. Increased literacy and the advancement of the mass media meant that some sections of the population developed a global consciousness (Robertson, 2003). What set this period apart from the earlier period in particular was the roles of institutions in underpinning globalization processes. For example, migratory flows were heavily regulated by states for the first time. In the economic arena multinational banks' rise in importance and the Gold Standard (see Chapter 4) regulated trade and financial sectors. There was a range of other common global standards that evolved in this period, emanating from the West, in order to facilitate increased global interactions.

It is during this period that the 'core–periphery' structure of the world economy theorized by dependency and World Systems thinkers crystallized. There existed a relatively affluent industrialized core and a relatively impoverished periphery from where raw materials were sent (see Chapter 4). There was a range of impacts across the periphery of the spread of colonialism, according to the particular contingency of place. There were major politico-military impacts (the decline and destruction of many non-European states and the modernization of those not directly

Map 3.3 *European imperial coverage of the globe*

Legend

Europe

Territory controlled by
European states at some time
between 1500 and 1950

colonized), trade and investment impacts (especially the locking in of many regions as primary product suppliers), migratory impacts (especially to North America but elsewhere too) as well as cultural impacts (principally the spread of Christianity in some areas and the circulation of Western secular discourses in others).

The period of modern globalization was shattered by the First World War, which came about partly due to the rising rivalry between the European powers which imperialism both reflected and perpetuated. The end of the war led to a collapse in global trade and investment, and the Great Depression which followed saw an end to the Gold Standard and other systems of global regulation. The period of openness was then replaced by imperial preference in Western empires and autarky in the USSR and Germany. Ironically then it was this imperial war, which inspired unprecedented imperial war efforts, which laid the seeds of the destruction of the great empires.

It was during the inter-war period that the first attempt at a truly global system of governance was attempted, in the form of the League of Nations established in 1919. However, imperial rivalry meant that it was not as effective as had been hoped. It was unable, for example, to prevent the Second World War, which arose due to continued rivalry in Europe and the mistakes made during the post First World War settlement. It was this second 'global' war that finally led to the collapse of the European empires. Weakened and exposed by the war the colonial powers were unable to prevent a tide of decolonization from taking place. Ironically, a generation of independence leaders and their movements employed concepts of nationalism diffused by the colonizers themselves. Decolonization led to a new order – which arguably pervades until today:

> In their place [the European empires] arose a hegemonic USA which would establish the formal and informal structures of global governance alongside which a renewed wave of globalization, supported by new technologies and infrastructures of interaction, would cross a world dominated, not by amorphous empires, but by territorially demarcated nation-states.
>
> (Held *et al.*, 1999, p. 424)

Contemporary globalization

The remainder of this book deals largely with this period, which is outlined only briefly in what follows. Held *et al.*, (1999, pp. 424–425)

argue that contemporary globalization is a distinctive phase in a range of
ways. Counter to sceptical theses, they argue:

> in terms of the extensity of global linkages, the appearance is one of
> catching up. . . . But we argue that in nearly all domains contemporary
> patterns of globalization have not only quantitatively surpassed those
> of earlier epochs, but have also displayed unparalleled qualitative
> differences.

Innovations in transport and communications technology have been
extraordinary, the emergence of institutions of global governance
is unprecedented, and there has been a 'clustering' of patterns of
globalization across many domains. New migratory patterns, such as
refugees and asylum-seeker movements, have been established. There
has been a major globalization of environmental impacts of industrial
production and consumption as well as concern about it (see Chapter 8).
The velocity of social-cultural flows has increased enormously as
communications technology has advanced. The rise of global tourism
to become the major economic sector, as well as unprecedented levels
of migration, are testimony to the changes in facilitating technologies.
In the previous periods of globalization cultural interactions were often
elite-to-elite, but there has now been a proliferation of popular culture.
The dominance of English as the major lingua franca across the globe is
an element of this cultural shift which also sets it apart from the other
periods, together with the increased role of private capital in cultural
diffusion to the masses.

As noted above, the major historical moment defining the start of
contemporary globalization was the Second World War, leading to
the end of empire and the emergence of the Cold War. Military power
relations between the two superpowers, the USA and the USSR, took
on global proportions (see Chapter 5), pulling in many of the world's
new nation-states on to one side or the other. At the same time, the
League of Nations' successor, the United Nations, was established,
and set a new precedent in terms of global governance. Working in
parallel with these trends, the Bretton Woods institutions (see Chapter 4)
– the World Bank, the IMF and the GATT (succeeded by the WTO) –
have fostered economic globalization. Following the 1973 oil crisis
in particular, these institutions played a crucial role in the spread of
free-market neoliberalism across the world. The deregulated governance
model has diffused globally, even to the last remnant of communist
societies in China and Vietnam, for example, propagated by GATT's
successor the WTO. TNCs benefited and flourished in this new

economic environment and have been key agents in the diffusion of neoliberalism.

From a geographical point of view one of the most important arguments of Held *et al.*,'s transformationalist argument is that simple global dichotomies such as 'North–South' and 'core–periphery' have been eclipsed by more complex patterns of economic distribution and political power relations. These flows are uneven, however:

> This pattern of interconnectedness is unique to this era. That said, the enmeshment of different states and of different social groups within those nation-sates, and their relative levels of control over those flows, remains highly uneven.
>
> (p. 427)

The end of the Cold War seemed to affirm the dominance of the democratic, capitalist and generally economically liberal state, which has been propagated as the desirable form of governance by a host of global institutions. The age of empires had come to an end as we saw a worldwide system of nation-states overlain by regional, multilateral and global systems of governance and regulation. Even the world's only superpower and hegemon the USA did not practise overt colonization – through formal political means at least. This factor, that patterns of globalization are not driven by 'expansionary logic or the coercive institutions of empire', sets the current period apart from earlier periods.

Held *et al.*, describe the current era as *thick globalization*, which is distinct from previous phases in a range of ways (See Box 3.1). Notable points are that the current period is characterized by decreased Westernization compared to modern globalization, territories are more fixed relative to the past, and governments play a bigger role in national affairs than ever before. These points contradict hyperglobalist arguments and exemplify the benefits of a long-run historical perspective. Ultimately, the authors do not see the phenomenon unfolding along a predetermined path – underlying these transformations is no unifying logic, but rather a collection of forces that may or may not converge at any given time.

Reflections on the global transformations chronology

Held *et al.*, provide a very detailed and historically rich interpretation of the evolution of globalization. In doing so they seek consciously to avoid repeating the well-trodden histories of the development of Europe and

---

Box 3.1

Distinguishing features of the current era of globalization

- *Spatio-temporal* – Unprecedented extensity, intensity, velocity and impact of global interactions and flows in all domains.
- *Organizational* – Unprecedented worldwide social, political and economic regulation of power relations through new institutions of multi-layered and multilateral governance.
- *Reflexivity* – A worldwide elite and popular consciousness of global interconnectedness as never before. National elites versus transnational social forces are the main groups of contestation as opposed to the coercive empires of the past and their subjugated colonial territories.
- *Contestation* – A new global politics of agenda settling, coalition building and multilateral regulation often designed to contest globalization. This was an internal imperial affair up until this era.
- *Regionalization* – Regional blocs have replaced imperial blocs. These blocs are not autonomous but link into and mutually reinforce each other.
- *Westernization* – Until the current period, globalization was largely synonymous with the spread of Westernization.

Arguably, today's forms of globalization are less Eurocentric.

- *Territoriality* – Contemporary globalization is associated with a different kind of politics of territoriality. Territorial control and governance has always been at the core of globalization. The threats to territory are not just military or external; territories are now *more* fixed.
- *State form* – Compared to earlier phases of globalization states are characterized by 'big government' (despite recent reforms which have sought to reduce this size) where states spend a large proportion of GDP and have responsibility for management and welfare of populations. States are thus more visibly impacted by globalization today.
- *Democratic governance* – Individual states are democratic (or most claim to be) while global governance structures have limited elements of democracy. Combining democratic ideals of national governance with the global organization of social and economic life raises unique dilemmas.

Source: Adapted from Held *et al.*
(1999, pp. 430–431)

---

how this has underpinned what we have come to know as globalization. In this sense to a certain extent, they de-centre the debate, shifting the focus away from Europe and to the other parts of the world which have simultaneously, and in very different ways, been exposed to the forces of globalization. Further to this, their approach is multi-disciplinary, combining commentary, analysis, measurement and interpretation of economic, social, technological and other forces of changes and how these have combined to create the grand sweep of global history. This is a very ambitious undertaking – one that is bound to be fraught with difficulties regarding overgeneralization. However, they do escape much of the determinism of other interpretations from single disciplines.

A number of criticisms can be made of the thesis, however. There is no force underlying globalization and no theoretical rationale for the increased velocity, intensity and extensity of global interactions. Their definition of it is so broad, one might argue, that it effectively captures everything. Every migratory movement, every trade flow, becomes evidence of 'globalization' under this schema. Geographers should be critical too of their dichotomous use of the terms 'global' and 'local'. Notwithstanding this, there can be little doubt however that the periodization set out in *Global Transformations* is the most complete and coherent interpretation of globalization we yet have.

Waves of globalization – a framework

There can be no doubt that we are in an accelerated period of globalization. However, that globalization is not fully complete is also obvious, and whether it ever will be is doubtful. In this sense it is perhaps more accurate to talk of a globalizing world as opposed to a globalized world. The processes and agendas that are leading to this are contradictory, dialectical, complex and heterogeneous. This is why mapping the impacts is both a necessary and highly challenging task for geographers, and one which has only just begun.

A framework of the waves and phases of globalization is offered below which draws together many of the ideas discussed so far in this book. Particular emphasis is placed on transformationalist accounts influenced, partly, by world systems theory, and by the periodization work of Robertson (2003) and McMichael (2004). For this framework the definition of globalization developed in Chapter 1 is used. The framework begins in the early 1500s with the rise of the Hispanic empires. This is not to suggest that processes like globalization did not exist before that point or that some civilizations, given limited horizons, believed they were acting in a 'globalist' fashion (Robertson, 2003). These were, arguably, forms of nascent globalization that did not attain the stretching of systems and networks that took place with colonization of the New World. The framework sees the process divided into two waves (dates approximate): colonial globalization (the first wave), *c*.1500–*c*.1945, and postcolonial globalization (the second wave), *c*.1945–. Each of these waves is broken down into two further phases. The first wave proceeds through the mercantilist phase (*c*.1500–*c*.1800), and the industrialist phase (*c*.1800–*c*.1945). The second wave moves through the modernization phase (*c*.1945–*c*.1980), and the current neoliberal phase (*c*.1980–) (see Table 3.3). Between each of these waves are major restructuring events that disrupt the dominant world order, an approach which builds on ideas from the regulation school (see Chapter 5). As the waves proceed, time–space compression intensifies though not necessarily evenly across time and space.

In the first wave the transition between mercantilism and industrialism was caused by major shifts in Europe, based partly on processes unleashed by mercantilist colonialism that led ultimately to the industrial revolution and the consolidation of capitalism. This cultural shift had profound economic and political impacts. It led to intensified competition between newly industrialized powers and the search for resources and locations for the investment of surplus capital. The two World Wars, punctuated by the Great Depression, led to a crisis in the colonial world order and eventual decolonization. This marked the shift to a postcolonial period where control was effected through new cultural and economic means rather than through overt political ones, fostered in part by the development of globalized political institutions which were designed to support this hierarchy. In the first period of the second wave – the

Table 3.3 *Waves of globalization – a framework*

Wave	Period (approximate dates)	Restructuring crisis
Wave 1	**Colonial Globalization (*c*.1500–*c*.1945)**	
	Mercantilist phase (*c*.1500–*c*.1800)	
		Industrial revolution
	Industrialist phase (*c*.1800–*c*.1945)	
		Great Depression and the Second World War
Wave 2	**Postcolonial globalization (*c*.1945–)**	
	Modernization phase (*c*.1945–*c*.1980)	
		Oil Crises
	Neoliberal phase (*c*.1980–)	

modernization phase – state developmentalism promoting TNCs was the primary motor for the diffusion of the culture of capitalism (McMichael, 2004). A restructuring crisis, brought about by the oil hikes of the 1970s, culminating in the debt crisis, led ultimately to the rise of global neoliberalism. In this neoliberal phase the power of the state is challenged by global institutions that seek to further the penetration of capitalism, and also by civil society that grows increasingly frustrated by the state's ineffectiveness in a multi-scalar world. At the core, then, is the diffusion of the culture of capitalism that seeks to reduce turnover time of capital through the development of technologies of time–space compression.

It should be stressed that this is not economic determinism – capitalism is a *culture* which produces economic outcomes and political change. The ideologies of capitalism and the traits which comprised it evolved from the Enlightenment period in Europe. It is also highly variable in its spatial expression, as the recent rise of alternative capitalisms in East Asia, Latin America and Eastern Europe attest (see Gwynne *et al.*, 2003). The outcome of each period is to create a system of imperialistic power relations (Harvey, 2003) – the first wave involving explicit territorial imperialism and the second more insipid neo-imperialism (these ideas are discussed in subsequent chapters). In the first wave asymmetries in power were articulated largely between nation-states. The second wave is much more complicated spatially, involving partially disembedded networks of power which are less bounded by territorial space. Globalization then comprises both agendas and processes and is fuelled by asymmetries in power and well-being. Cores, semi-peripheries and peripheries still exist; they are just not as neat as they once were. We now see more core *within* the periphery in the form of networked elites and peripheries within the core in the form of those who are marginalized from networks.

There is no reason why the process of globalization cannot be reversed – although there are powerful vested interests that are likely to resist any attempts to do so. But globalization does not have its own inexorable trajectory – it is the result of collective human actions, agendas, desires and perceptions. Although globalization to date has been driven by capitalism, to move to a post-capitalist globalization would require new 'globalized imaginations' that are only now beginning to form. Reducing the almost infinite complexity of the world to such periodization is a risky business, but such frameworks allow us to untangle the inherently messy business of history. The terminology outlined above will be referred to at various points in the book – although the model is not *tested* in a formal sense and provides a framework only. Much of what follows deals with the second wave of globalization – and especially the neoliberal phase. This is because the intensity, extensivity and velocity of globalization processes are unprecedented in this period and also because this is the period that we live in and need to know more about if we are to rebuild globalization in a way that promotes equality and sustainability rather than threatening it.

Conclusion – two myths dispelled

Dicken (2004) argues that 'we need to . . . break out of the increasingly sterile argument over whether what we are experiencing is either totally new and unprecedented or simply more or less life as we have known it' (p. 7). The idea that globalization is something new has been effectively dispelled above and in the literature in general. It has been clearly shown as a collection of processes and outcomes which are 'manifested very unevenly through both time and space. It is neither an inevitable, all pervasive, homogenizing end-state nor is it unidirectional and irreversible' (Dicken, 2004, p. 8). Rather, it is 'multiple non-linear interconnected *trajectories*' (Hart, 2001, p. 655). This chapter adopts the position that the depth and breadth of the interconnections are indeed unprecedented, although the roots of that interconnection can be traced back to the first global European empires. The logic underpinning the expansion of the World System has changed little, being that of capitalist accumulation through time–space compression, although the conditions and facilitating technologies have evolved beyond recognition. The globalization of today is not as new as the popular imagination might believe, but it is certainly qualitatively very different.

In the last two chapters we have touched upon a range of ways of looking at the concept and history of globalization. It is hoped that at the very least two common myths have been dispelled. The first is that globalization has been cast over the world like a blanket. This is clearly not the case; it is a net which does not affect everybody equally or make everywhere the same – there are holes in the net and some parts are woven more finely than others. However, everything *is* increasingly defined in relation to it. The second myth is that globalization is new. Clearly, it is not unprecedented – but there are important distinctions between the waves that we have identified above (see Box 3.1). These historical and geographic points are returned to throughout the remaining chapters of the book which now moves on to deal with shifts in the three spheres of human experience – economy, polity and culture – and to analyse some of the challenges presented by such transformations.

Further reading

Held *et al.*, (1999), ch. 12: The chronology explored in this chapter builds from Held *et al.*,'s classic work which is comprehensive and coherent.

Hirst and Thompson (1999): This study is one of the few that sets out to empirically test the concept of globalization. The sceptical viewpoint it takes has been influential.

McMichael (2004): This book is one of the only texts that explicitly combines globalization and development. The chronology used is especially helpful and the case studies are illuminating.

Ohmae (1995): This is the classic hyperglobalist account and reading it will reveal the roots of many popular arguments concerning the positive impacts of globalization.

Robertson (2003): Robertson's three-waves argument deals with the evolution of the global consciousness from a multi-disciplinary perspective.

Waters (2001): This is an excellent intermediate-level book, which presents a sociological argument for the evolution of globalization.

Part II

Shifting spheres

Globalizing economic geographies

- New economic geography and spatial divisions of labour
- The evolution of the global economy
- Contemporary patterns in the global economy
- Flexible accumulation – the newer international division of labour?
- Chains and networks in the new global economy
- Transnational corporations
- Neoliberalism and free-trade reform
- Regional economic integration
- The globalization of agriculture
- Conclusion – a transformed economy

New economic geography and spatial divisions of labour

Geographers have defined economic geography in many different ways, but it essentially refers to the *spatiality of the economy.* The most straightforward, and yet surprisingly rich, definition is offered by Roger Lee in *The Dictionary of Human Geography* (2000a, p. 195) as 'geographies of peoples' struggle to make a living.' Stutz and De Souza (1998, p. 41) offer another useful definition, seeing economic geography as 'concerned with the spatial organisation and distribution of economic activity, the use of the world's resources, and the distribution and expansion of the world economy'. In the 1990s there was something of a revolution in the subdiscipline. Critics charged that the old economic geography was economically deterministic; a-historical; over-descriptive and quantitative; and incapable of dealing with moral and ethical questions (see Bryson and Henry, 2001; Daniels, 2001; Peck and Yeung, 2003). Reflecting broader changes in human geography in general, and in particular the *cultural turn*, the subdiscipline now attempts to offer holistic explanations for change; is more sensitive to history and the uniqueness of place; and is more willing to engage with the cultural roots driving economics; and, incorporates qualitative analysis to make sense and interpret activity. Most importantly, economic geographers are now incorporating ethical analysis into their work (i.e. the normative 'rights and wrongs' of change). This is an important step in the evolution of human geography, which allows it to contribute to globalization debates more effectively.

The globalizing economy and spatial divisions of labour

We need a solid historical backdrop for the study of the contemporary world of economic globalization. In this chapter we use the concept of *spatial divisions of labour*. A division of labour in general refers to the separation of tasks in the production process. There are two generally recognized divisions of this nature: (1) *social divisions of labour* – where different social groups concentrate on particular activities, and (2) *technical divisions of labour* – where distinct stages in the production of one good/service are separated. Social divisions of labour have been evident in society from prehistorical times, for example, in precolonial Fijian society different tribes were charged with different socio-economic functions such as fishing, hunting and speech making. Technical divisions of labour were greatly advanced from the industrial revolution onwards when technology allowed the establishment of the first 'production lines'. Geographers and others have identified four other division-of-labour types, and these often overlap with each other: *gender divisions of labour, cultural/racial divisions of labour, international divisions of labour and spatial divisions of labour* (Painter, 2000). Suggested first by Massey (1984), spatial division of labour refers to the concentration of specific stages, nodes or sectors of any given production chain or complex in different geographic locations. As capitalism has expanded outwards so divisions of labour have become increasingly stretched across space and in many sectors new globalized divisions of labour link together localities at different ends of the world. These globalized divisions of labour are both a product of globalization and play a role in reproducing it. A subsequent section on the evolution of the global economy traces the evolution of spatial divisions of labour through the various waves of globalization as defined in Chapter 3.

Circuits of capital

The concept of circuits of capital is a Marxist-influenced idea which helps us understand globalization. Under this schema the capitalist process is seen as an infinite circuit (See Figure 4.1). The capitalist imperative is to ensure that the money at the end of each circuit is greater than the money at the beginning (i.e. that surplus capital is produced). There are three types of capital: Money (M), Commodities (C) (these refer here to raw materials and labour) and the means of production (P). Commodities acquire increased value (C') through the production process and are exchanged for money (M'), this creates profit and this value can be used to purchase another round of inputs. The central idea with respect to globalization is that each of these circuits has become expanded spatially in the following ways:

- *Commodity capital circuit* – This was the first to become globalized through trade.
- *Money capital circuit* – This globalized subsequently through the rise of investment capital.
- *Productive capital* – This globalized most recently with the rise of TNCs and global networks and chains.

Other circuits could be added to this model. For example, it is feasible to argue that cultural capital is now being globalized through the technological revolution. From a geographical point of view, the global capitalist system and the circuits that constitute it have explicit spatiality. This is to say that the creation of value or surplus takes place in specific economic spaces. These may be broken down into spaces of production, investment and consumption, all of which are highly complex and have particular geographies. As noted in Chapter 2, Harvey (1989) provides the most compelling explanation for globalization related to the circuit of capital. Inherent to capitalism is the desire to make the circuit operate at the highest possible velocity, maximizing the accumulation of capital and surplus in any given period. This requires the development of space-shrinking and time-saving technologies which reduce the turnover time of capital. It is the development of such technologies, based on the capitalist imperative of maximizing profit, that, according to Harvey, has led to time–space compression and globalization as we know it.

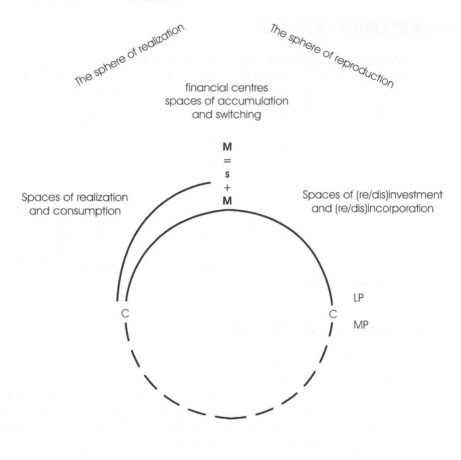

Figure 4.1   *Circuits of capital*
Source: Adapted from Cloke *et al.* (1999)

The evolution of the global economy

By tracing spatial divisions of labour and how they have influenced the circuit of capital across time and space through history we gain a better understanding of the patterns and issues that define the contemporary world economy. The periods discussed below draw on the chronology of globalization developed in Chapter 3.

Mercantilist roots 1550–1880

The 'classic' period of growth in the world economy was in the nineteenth century, but a visible world economy evolved earlier than that. Before the colonial wave of globalization (i.e. before the late 1400s; see Chapter 3) trade was largely localized, although some regional networks existed. With the emergence of Hispanic colonialism the first world economic system was forged – based initially on mercantilism. This led some localities to specialize in the production and distribution of particular items based on comparative advantage – i.e. the items they produce relatively cheaply (see argument for free trade later in this chapter). The main actors in this were merchants who became the richest class and who acted as intermediaries between nodes in the core and periphery. The expansion across the Atlantic to fuel trade underpins the evolution of colonialism there and lies at the core of the evolution of the global market.

Industrialist expansion 1780–1900

Based on the increased rural to urban migration that took place in Europe in the 1600s a cheap labour force was made available. This, combined with efficiency gains and the resulting surplus in agriculture, stimulated the industrial revolution (See Plate 4.1). This development further fuelled rural to urban migration and thus set a virtuous circle of capitalist growth in motion. The industrial revolution involved increased levels of specialization, and divisions of labour were set up in factories. These employed new forms of technology and pursued *economies of scale*. Capitalism needs these economies of scale, which are realised when increasing the volume of production lowers unit costs, because *growth* is absolutely central to the system's survival. Industrial complexes and chains were extended across space forging a worldwide division of labour which largely favoured accumulation in the ruling core. Space-shrinking technology was used to search out cheap labour and resources and to foster flows of trade and investment that reduced capital's turnover time. By the early nineteenth century Great Britain was far and away the world's greatest economic power – due partly to the legacy of colonial history and the fact that the industrial revolution began there. An important point to reiterate is that capitalism is a cultural system – other societies have ways to manage and regulate people's need and desire for accumulation. It is not inherent that people wish to make ever-increasing profits; the *growth ethic* emerged out of capitalism during this period.

Plate 4.1 *'Fresh air for the potteries' – Stoke-on-Trent, England c.1900*
Source: Blake Collection, City of Stoke-on-Trent museum and art gallery

The twentieth century – colonial divisions of labour?

The principles of capitalist expansion altered little between the eighteenth
and nineteenth centuries, although the pace did. In the late nineteenth
century the economic power of the USA expanded enormously based
on cheap resources, immigrant labour, innovation and – in particular
– the emergence of *global corporations*. These organizations were
unprecedented as they were single firms producing both within and
outside their home country. As such, the USA became the leader in terms
of *foreign investment* which was funding an ever-increasing proportion of
economic output. A virtuous circle of outward foreign investment and
profit repatriation by core economies led to the further skewing of an
already highly concentrated global economic system, and this led to a
concentration in political power also. In 1914, for example, 71 per cent
of total manufacturing output was concentrated in just four countries
(and 90 per cent in eleven) – the most powerful being Germany, France,
the UK and the USA. Dependency theorists, and other neo-Marxists,
have argued that the early evolution of the global economy produced
a core–periphery system underpinned by a *colonial division of labour*.

This saw industrial production located in the core and raw material production based in the periphery (Kay, 1989). Although this simplifies the world system greatly, it does capture the essence of global economic structure until at least the mid-twentieth century. The Great Depression and the Second World War were the major events which led to economic restructuring. The former led to a collapse of the liberalized economy and a retreat into protectionism, and the latter was followed by an attempt to return to free trade and reconstruct the world economy along modernist lines.

Post-Second World War – new divisions of labour?

The geography of the world economy shifted profoundly after the Second World War (1939–1945). A number of processes stand out, including: the rebuilding of industrial capacity following the war; the evolution of new technologies and products through the war effort which underpinned subsequent growth; the wave of decolonization across the world from 1945 onwards; and, the new international division of labour. However, the major economic imperative to emerge in the West following the war was the pursuit of *free trade* regulated through the General Agreement of Tariffs and Trade (GATT) (now the WTO). This was attempted largely as a means of avoiding the worst implications of the Great Depression, which had led to a massive collapse in world trade and investment flows. The rise of Japan as a reformed economic power was also an important process in the post-war economy, growing rapidly to become one of the largest in the world with overwhelming influence in its own region. In the socialist bloc there was an attempt to remain aloof of capitalist development and there were moves to internationalize socialist systems between like-minded countries. During the modernization phase, which lasted until the late 1970s (Chapter 3), a strong role for the state was envisaged. In the context of the poorer world, aid and loans for industrialization were seen as the major component in stimulating growth. The oil crises of 1973 and 1979 altered the nature of the global economy profoundly. Neoliberal policies first adopted by Chile, and then by the UK and the USA, were diffused across the poor world in the wake of the debt crisis provoked by the oil price hikes. This recession led to the eventual triumph of neoliberalism, which has become the hegemonic global economic paradigm ushering in a new phase of globalization, as argued in Chapter 3.

The new international division of labour (NIDL)

One of the most important trends in the post-Second World War economy was the rise of the new international division of labour (NIDL). Lee (2000b, p. 552) defines this as '[a]n emergent form of worldwide division of labour associated with the internationalisation of production and the spread of industrialisation'. As argued above, until the end of the Second World War the global economy was characterized by a colonial division of labour, revolving around a basic core/periphery dichotomy in trade and investment. The NIDL evolved particularly in the 1960/1970s as profits in the core declined and companies increasingly invested and located abroad pursuing cheap factors of production (land, labour, capital) and new markets. Frèobel *et al.* (1980) argue that three things facilitated the NIDL:

- Developments in transport and communications technology;
- Developments in processing technology that can standardize production and use unskilled labour;
- The emergence of a world reservoir of labour.

To the above list should be added the increased mobility of capital from the neoliberal reforms of the 1980s. Some have argued that the powerful control which TNCs have gained over poor world nation-states is a form of neocolonialism – that is to say, economic dependence combined with political sovereignty (Murray, 2004a). In this context, Frèobel *et al.*, bemoaned the impacts of the NIDL in 1980 (quoted in *The Dictionary of Human Geography*, 2000, p. 553):

> So far export-oriented industrialisation has failed to achieve any improvement in the social conditions of the population of the developing nations, not even as far as their most fundamental needs such as food, clothing, health, habitation, and education are concerned.

Arguably, a new semi-periphery has been added to the world economic structure through the operation of the NIDL (Taylor and Flint, 1989). The major recipients of investment during the forging of the NIDL were those that became known as the NICs (newly industrializing countries), particularly Taiwan, Hong Kong, Singapore and South Korea. These countries experienced rapid economic growth in the 1980s and early 1990s based on the spread of capital from the USA and Japan. Figure 4.2 shows the economic growth record of the Asia 'tigers' until the crash of 1997 as compared to three non-core regions. Theorists are divided over what lies behind the so-called 'Asian Miracle'. Some

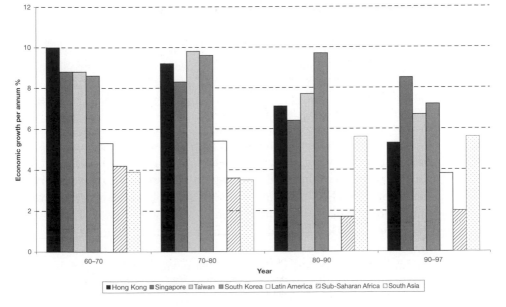

Figure 4.2 *Economic Growth in East Asia, 1960–1997*

Source: Calculated from Gwynne *et al.* (2003)

have argued that open economic policies based on neoliberal and neoclassical ideas underpin development success there. Others have argued that strong state interventionalism regulating capital and trade flows and designing 'competitive advantage' are the explanatory factors. As Rigg (2002) notes, it is probably the combination of the two in different ways in different places that really lies behind the heady economic ascent of the region in the late 1990s when it was interrupted by the Asian Crash.

Based on the high rates of economic growth and reductions in absolute poverty in East Asia, and some other semi-peripheral countries (such as Chile), it could be argued that Frèobel *et al.*,'s (1980) critique of the NIDL is now inapplicable. However, a range of persistent negative impacts has been posited which include:

- Rural to urban migration and explosive urbanization and industrialization.
- Environmental degradation and pollution resulting from industrialization.
- Cultural shifts – particularly towards potentially unsustainable Western-style consumerism.

- Increased economic vulnerability as economies open themselves up to investment flows.

The latter criticism became especially relevant following the Asian Crash beginning in 1997 that some blamed on open-economy and neoliberal policies (Rigg, 2002). The intricate links constituting the global economy allowed negative impacts of the crash to travel rapidly across the world as the contagion became globalized.

There is some evidence pointing to a second wave of NICs emerging in Southeast Asia and, to a certain extent, Latin America based on successor waves of the NIDL. In East Asia, for example, outward foreign investment by TNCs from first-wave NICs has played an important role in the economic development of countries such as Malaysia, Thailand and Indonesia. Despite the diffusion of industrialization to the NICs made possible through the NIDL, it should be noted that the majority of investment, trade and growth takes place within and between the core economies of Japan, the USA and Europe. The semi-periphery is left largely outside of this triad – although the potential for increased East Asian participation as it recovers from the 1997 crash is considerable and growing. As we consider later in this chapter, the *newer international division* has the potential to further marginalize economies in the semi-periphery and periphery.

Contemporary patterns in the global economy

Structure of the global economy

The global economy transformed rapidly over the past two centuries as capitalism spread across the planet in its search for profit. During the neoliberal phase its structure in terms of the proportional role of different sectors has shifted fundamentally. Across the world, primary (exploitation of natural resources) and secondary sectors (the manufacture of value-added products from primary goods) are shrinking in terms of their relative contributions to both total employment and GDP. The tertiary (retail and wholesale trade, transport and personal services) and quaternary sectors (banking, finance, business services, media, insurance and administration) now form the most important component in virtually all of the world's economies. When the sources of world output are divided into agriculture, industry and services the overwhelming importance of the latter becomes apparent. In 2003, services accounted for 68 per cent of the value of total world output. In low-income countries (based on World Bank classifications) services are playing an increasing role, rising from 30 per cent of total income in 1965 to 50 per cent in 2003. In high-income countries the contribution of services to output has risen from 54 per cent to 71 per cent in the same time period. Agriculture is now of relatively little economic importance, representing 4 per cent of income for the world as a whole. In lower income countries, agriculture remains important however, contributing 25 per cent to total income in 2003 (see Table 4.1). Daniels (2001) outlines six further central trends in the contemporary global economy:

1 *A marked increase in the power of finance over production* – Capital moves almost uninterrupted across the globe and has the power to transform companies, economies, and even entire regions almost instantly.
2 *Knowledge is a vital factor of production* – Skills and the production of an educated workforce have become major policy priorities.
3 *Technology and its diffusion have become transnationalized* – Managing technology has become essential for economic success and the technology gap which exists further compounds the global welfare gap.
4 *Global oligopolies have grown* – Successful TNCs have come to dominate production and marketing in any given industry (e.g. Microsoft or McDonald's).
5 *Transnational institutions have arisen to regulate the global economy* – Including the UN, IMF and WTO.
6 *The 'wiring' of the world* – Has led to more rapid cultural flows, what Amin and Thrift (1994) call 'deterritorialised signs, meanings and identities'.

Table 4.1 *Structure of world output by sector, 2003*

	Agriculture	Industry	Services
Low income	25	25	50
Lower-mid income	12	40	48
Upper-mid income	7	32	61
High income	2	27	71
World	**4**	**28**	**68**

Source: World Bank Data (2005)

Globalization, and the increased role of networks in particular, has made the world economy more uneven than ever. The discussion of global investment and trade patterns that follows illustrates this asymmetry clearly.

Geography of foreign investment

As previously discussed, the NIDL changed the geography of foreign investment, leading to significant outflows from the core. In general terms the range of recipient countries has broadened, and while North America, Europe and East Asia are the main destination regions for FDI, investment in Latin America and Southeast Asia is growing. Notwithstanding the broadening in the range of recipient countries, at the aggregate level, inflows into 'developed' countries (based on UNCTAD classification) accounted for a greater proportion of the total stock in 2002 (64.5 per cent) than in 1980 (56 per cent) (see Table 4.2). In this sense, there has been a reconcentration in global investment patterns. The USA, UK, Germany, France and Japan are currently the

Table 4.2 *Foreign direct investment by region, 1980 and 2002*

| | Inward FDI stock | | | | Outward FDI stock | | | |
| | 1980 | | 2002 | | 1980 | | 2002 | |
	US$B	%	US$B	%	US$B	%	US$B	%
Africa	32.2	4.6	170.9	2.4	6.8	1.2	43.6	0.6
Latin America	50.4	7.2	762.2	10.7	51.5	9.1	173.2	2.5
Asia	223.7	32	1402.5	19.7	6.2	1.1	632.1	9.2
Developing countries	**307.5**	**43.9**	**2339.6**	**32.8**	**64.6**	**11.5**	**849.5**	**12.4**
North America	137.2	19.5	1527.6	21.4	239.2	42.4	1775.1	25.9
Europe	232.7	33.3	2779.9	39	237.7	42.2	3771.5	54.9
Developed countries	**391.9**	**56**	**4594.9**	**64.5**	**499.3**	**88.5**	**5987.7**	**87.2**
World	*699.4*		*7122.4*		*563.9*		*6866.4*	

Source: Calculated from UNCTAD (2004)

largest source countries, accounting for nearly 60 per cent of FDI in 2002, and TNCs from the EU are now responsible for approximately 55 per cent of all FDI activity. Approximately 70 per cent of outflows from the 'developed' world stay within the 'triad' grouping of the EU, USA and Japan. Overall, the 'developed' countries accounted for 87.2 per cent of all FDI stock outflows in 2002, which represented a marginal fall from the figure of 1980 (88.5 per cent). This fall was attributable to the fact that outward investment from a handful of 'emerging' economies has increased, including Hong Kong, Singapore, Taiwan, China, South Korea, Malaysia and Brazil. In general, however, outflows from the developing world remain very low, standing at 12.4 per cent in 2002. Africa's position in this respect has worsened significantly over the past two decades; the continent now accounts for only 0.6 per cent of global FDI outflows and 2.4 per cent of inflows.

Geography of trade

Trade has become an increasingly important component in national economies as borders have opened across the world. From 1950 to the present, merchandise trade increased approximately fifteenfold, while output increased only fivefold. Trade flows are increasingly concentrated however. In 2002, for example, a small number of large countries accounted for an overwhelming proportion of world exports, the most important being the USA (10.7 per cent), Germany (9.5 per cent) and Japan (6.5 per cent). The world's twenty largest importers and exporters and their shares of world trade are shown in Table 4.3, and the geographical distribution of exporting countries is illustrated in Map 4.1. Taken together, the top twenty countries account for approximately three-quarters of world merchandise imports and exports. Despite globalization trends, countries still tend to trade with territories that are located closest, so much of global trade is intra-regional. Western Europe is the largest trader, although two-thirds of that trade takes place within the region. Asia is the second largest with around 50 per cent of its trade being intra-regional. North America is the third largest region conducting about 40 per cent of its trade internally. In terms of the relative importance of exports and imports the largest surpluses are found in East Asia, the Middle East and some European nation-states. The USA has by far the largest trade deficit on the planet and this is growing. In 2002, the USA accounted for approximately 18 per cent of total global imports.

Map 4.1 *Merchandise exports by country, 2003*

Source: UNCTAD (2004)

Table 4.3 *The world's twenty largest merchandise exporters and importers by value (US$ millions), 2002*

		Exporters			Importers	
Rank	Country	US$ billions	%	Country	US$ billions	%
1	USA	693.9	10.7	USA	1202.4	18
2	Germany	613.1	9.5	Germany	493.7	7.4
3	Japan	416.7	6.5	United Kingdom	345.3	5.2
4	France	331.8	5.1	Japan	337.2	5
5	China	325.6	5	France	329.3	4.9
6	United Kingdom	279.6	4.3	China	295.2	4.4
7	Canada	252.4	3.9	Italy	243	3.6
8	Italy	251	3.9	Canada	227.5	3.4
9	Netherlands	244.3	3.8	Netherlands	219.8	3.3
10	Belgium	214	3.3	Hong Kong	207.2	3.1
11	Hong Kong	201.2	3.1	Belgium	197.4	2.9
12	S Korea	162.5	2.5	Mexico	173.1	2.6
13	Mexico	160.7	2.5	Spain	154.7	2.3
14	Taiwan	135.1	2.1	S Korea	152.1	2.3
15	Singapore	125.2	1.9	Singapore	116.4	1.7
16	Spain	119.1	1.8	Taiwan	112.6	1.7
17	Russia	106.9	1.7	Switzerland	83.7	1.3
18	Malaysia	93.3	1.4	Malaysia	79.9	1.2
19	Ireland	88.2	1.4	Austria	78	1.2
20	Switzerland	87.9	1.4	Australia	72.7	1.3
Top 20		4902.5	75.8		5121.2	76.8
World		6455			6693	

Source: Based on UNCTAD (2004)

As noted above, services have grown in proportional economic importance over recent decades. Services exports grew at around 11 per cent per annum between 1989 and 2000 while merchandise exports grew at 10 per cent. Trade in services is also concentrated, although not to the extent of merchandise exports. Approximately 65 per cent of service exports come from fifteen countries. The balance of service imports and exports is different geographically to that of merchandise exports, with the USA being a major surplus exporter and Japan being a major services importer. When trade in goods and services is taken together, a clear

Table 4.4 *International trade in goods and services by region, 2002*

Region	Exports		Imports	
	US$ billions	%	US$ billions	%
North America	7900	15.9	1640	21.0
Latin America	414	5.2	408	5.2
Less Mexico	241	3	222	2.8
Western Europe	3336	42.2	3147	40.3
Central and Eastern Europe/				
Baltic States/CIS	379	4.8	358	4.6
Africa	173	2.2	165	2.1
Asia-Pacific	2097	26.5	1913	24.5
Less Japan	1637	20.7	1505	19.3
Less Japan and China	1272	16.1	1177	15.1

Source: UNCTAD, UNCTAD *Handbook of Statistics* online,
http://stats.unctad.org/restricted/eng/ReportFolders/Rfview/Explorerp.asp?&IF_Embedded=1, 30 September
2004

regional concentration emerges. In 2002, Europe accounted for 42.2
per cent of total goods and services exports, and 40.3 per cent of total
imports. The Asia Pacific is the second most dynamic exporting region,
accounting for 26.5 per cent of global exports and 24.5 per cent of
imports – although this is reduced significantly if Japan and China are
taken out of the calculation. The surplus in service exports from the
USA is not great enough to offset the deficit in goods trade, with North
America accounting for 15.9 per cent and 21 per cent of total global
exports and imports respectively. Table 4.4 illustrates the minute
diminutive proportional role played by Latin America, Africa and Eastern
Europe in global trade.

Summary – the global economic triad and shifting economic patterns

In terms of economic flows there is considerable evidence for the
existence of a 'global triad'. Figure 4.3 illustrates this for trade and FDI
respectively. However, it is important to remember that such macro trends
hide important variations across regions and within countries. When the
geographical scale of analysis is reduced we see that most investment and
trade actually takes place between specialized industrial spaces. These
clusters of economic activity increasingly define the global economy and

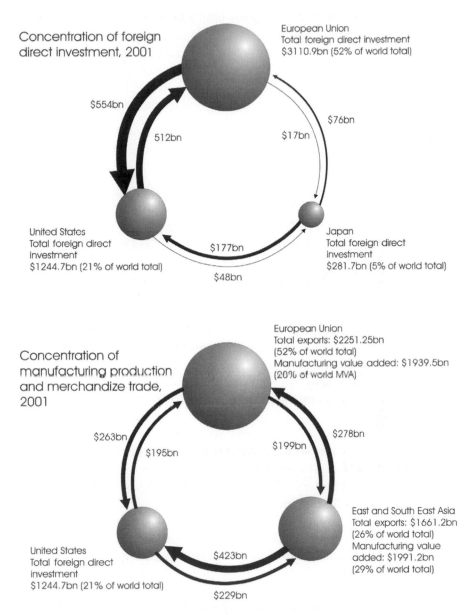

Concentration of foreign direct investment, 2001

European Union
Total foreign direct investment
$3110.9bn (52% of world total)

$554bn

512bn

$76bn

$17bn

United States
Total foreign direct
investment
$1244.7bn (21% of world total)

$177bn

$48bn

Japan
Total foreign direct
investment
$281.7bn (5% of world total)

Concentration of
manufacturing production
and merchandize trade,
2001

European Union
Total exports: $2251.25bn
(52% of world total)
Manufacturing value added: $1939.5bn
(20% of world MVA)

$263bn

$195bn

$278bn

$199bn

United States
Total foreign direct
investment
$1244.7bn (21% of world total)

$423bn

$229bn

East and South East Asia
Total exports: $1661.2bn
(26% of world total)
Manufacturing value
added: $1991.2bn
(29% of world total)

Figure 4.3 *The global economic triad*

Source: Adapted from Dicken (2003)

imply even greater spatial concentration of activity than national macro-economic figures suggest. Examples include Europe's 'vital axis' from the English Midlands to Northern Italy, the urban growth corridors of East Asia (e.g. the South China Economic Zone) and the US–Mexican border. In analysing and describing the evolving global economy it is important that we are able to see beyond national borders.

Summarizing the most important geographical trends in the world economy map over the past thirty years, Dicken (2003, p. 45) outlines seven points of note:

1 Japan's rise to the second largest economy and third largest single exporter;
2 The dominance of the USA as the world's largest economy despite intense competition from Japan over the recent past;
3 The uneven performance of the Western European economies;
4 The rise of a number of East Asian NICs to become global players in merchandise exports;
5 The heady rise of China over the past few years to become a major economic power;
6 The weak performance of most 'developing countries' except for a few exceptions, such as Chile;
7 The appearance of a number of transitional economies in Eastern Europe following the collapse of the USSR.

Flexible accumulation – the newer international division of labour?

The rise of 'flexible accumulation' is complicating the global economic map further, giving rise to a *newer international division of labour*. This restructuring has potentially profound spatial and socio-economic implications for the unfolding of globalization.

From Fordism to flexible accumulation

'Fordism' is the name used to describe the 'regime of accumulation' (see discussion in Chapter 5) that dominated from the Second World War until the 1970s and underpinned the NIDL. Fordism involves the mass production of standardized goods, pursuit of economies of scale, and a rigid division of labour that sees different parts of the task clearly demarcated. This conceptualization is based on the production techniques first introduced by Henry Ford in car manufacture. A linked concept is 'Taylorism'. The 'going abroad' of many TNCs was based on the search for low-cost factors of production, especially labour, for the assembly part of their chains. In the late 1960s there was a crisis of Fordism, and firms experienced a fall in profits due in part to rising wages and competition from low-cost economies. Further to this, tastes were shifting away from mass-produced items towards niche products. Lay-offs of labour to cut wage bills together with outflows of FDI led to a process of deindustrialization in the core. In the place of Fordism – some argue – a more flexible regime of accumulation has been forged, sometimes termed 'post-Fordism' (see Amin, 1994; Malecki, 1991). Among other things, this regime has the following characteristics, developed largely in Japanese corporations and later diffused to Europe and North America:

- More versatile computer programmable machines
- Labour that is more flexibly deployed
- Vertical disintegration of large firms
- Greater use of contracting and strategic alliances

Plate 4.2 *Flexible accumulation: automated car assembly in a German factory*
Source: Still Pictures

- Just-in-time production
- Closer integration of product development, marketing and production
- More flexible labour training.

Combining the above characteristics, flexible accumulation caters for niche markets, is often consumer driven, and places product innovation at the core. Correspondingly, R&D and proximity to market are now more important elements in locational decisions of many TNCs than traditional factors such as resources and labour. This has led to the evolution of a number of industrial districts in advanced capitalist countries, such as Third Italy in Northern Italy and Motor Sport Valley in the UK (see later example).

Box 4.1

Nike – flexible shoes?

There are numerous firms we could use to explore the concept of flexible accumulation – Toyota, Hyundai, Mitsibushi for example. The Nike case illustrates a number of important points made by sceptics of the flexible accumulation thesis however. Nike is both Fordist and flexible simultaneously. For example, in the case of the former Nike's *Air Max Penny* basketball shoes are made up of fifty-two component parts from five different countries, touched by 120 hands during production, on a clearly demarcated global production chain where timing is extremely important. Nike is notorious for searching out cheap labour pools. Nike has 150 Asian factories and over 350,000 Asian workers. Wages in China are reported in some cases to be as low as 11c per hour (Nike's chairman earned a salary of US$2.7 million in 1997). These are truly Fordist characteristics indicating the operation of the NIDL. But

elements of flexibility are also exhibited. For example, product research is extremely important and Nike aims to produce a new shoe every week to cater to niche markets. To do this it employs a just-in-time innovation structure where it can buy in expertise at short notice. This entails the use of short-term subcontracts that are often given to firms located close to the R&D headquarters in Oregon, USA. These are flexible characteristics, which suggest a post-Fordist approach. In sum, Nike, like many other global corporations, displays elements of both 'regimes of accumulation' as is evidenced in Figure 4.4, which shows Nike's 'commodity circuit'. Examples of this nature suggest that 'flexiblity' arguments may in fact be exaggerated.

Box 4.1
continued

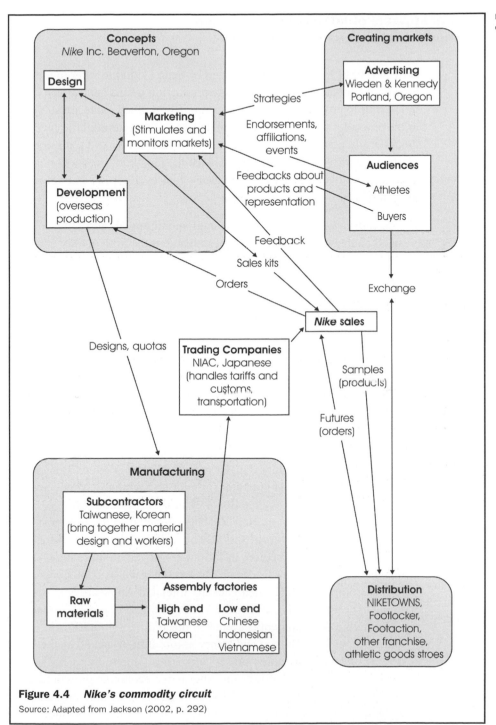

Figure 4.4 *Nike's commodity circuit*

Source: Adapted from Jackson (2002, p. 292)

Spatial impacts of flexible accumulation – new industrial spaces

New industrial spaces is a term coined by Scott (1988) to refer to the growth of economic nodes of flexible production which have arisen due to the shift to post-Fordist/flexible accumulation. Such areas are subject to the processes of re-agglomeration and, as such, are thought to reverse the NIDL. In new industrial spaces economies of scope and scale act to cluster activities as the relative priority of the factors of production is altered. Bryson and Henry (2001) discuss four overlapping examples of such spaces:

1   *New industrial districts* – 'Dense agglomerations of, normally, small and medium-sized firms specialising in the high quality production of a particular good or service' (p. 368). Examples include Jutland in Denmark (furniture, shipbuilding), Småland in Sweden (metalworking) and 'Third Italy' in Northern Italy (textiles, ceramics, shoes) (Malecki, 1991). These are also sometimes referred to as Marshallian districts and have been the subject of work by geographical economists such as Krugman (1998).

2   *High technology areas* – These are also sometimes referred to as 'sunbelt' areas. They are described as 'A diverse range of technopoles, corridors and innovative complexes, often in previously un-industrialised areas, which have grown through their capture of the "sunrise" high technology industries of information technology, electronics, pharmaceuticals, and R&D' (Bryson and Henry, 2001, p. 369). Examples include Japan's technopolis programme, Boston's route 128, the UK's M4 Corridor (London to Swindon), Motor Sport Valley, Oxfordshire, UK (see Box 4.2) and 'Wellywood' film production in Wellington, New Zealand (See Plate 4.3). The most well-known example however is Silicon Valley, Palo Alto, California. A former agricultural region, in 1950 it employed no more than 800 manufacturing workers. Today nearly 400,000 high-technology workers in such companies as Hewlett Packard, Intel and Apple are employed.

3   *Flexible production enclaves within old industrial regions* – A range of economic developments have taken place in more traditional industrial regions. Bryson and Henry argue: 'Often rocked by the decline of previously core "sunset" industries (shipbuilding, coal, steel, heavy manufacturing) some alternative economic growth has been achieved through firms and sectors applying flexible production techniques' (Bryson and Henry, 2001, p. 369). Examples include

South Wales, which restructured away from coal and steel towards car production based on Japanese and Korean investment in the 1980s. The US Midwest has been largely restructured through what is known as 'transplant' investment from Japan.

4 *World cities* – As the advanced economies have moved largely to service-based activity global cities have become the 'pinnacle of the global urban hierarchy once again' (Bryson and Henry, 2001, p. 369). Major world cities such as Tokyo, New York and London are the control centres of the global economy in a world city network where the links between such sites far outweigh the links to other cities. They house the headquarters of major global corporations as well as providing financial and business services that manage the global corporate world. The district of Canary Wharf in London, Wall Street in New York and downtown Los Angeles are the physical manifestations of this process. Such centres also include the 'servicing' industries – cleaners, couriers, guards, nannies, maids and so on – that service the new global elite as well as clustered cultural and retail outlets that service the same population. The world city hypothesis and its structure is outlined in Box 4.3.

The rise of flexible accumulation has implications for theses of globalization. Supporters of the idea of a new regime argue that it is

---

Box 4.2

Sunbelt: Motor Sport Valley, UK

Henry has undertaken research on the rise of an important high-technology sunbelt area in the UK known as Motor Sport Valley. This industrial cluster is located in a 70-kilometre radius around Oxford in Southeast England. It employs around 50,000 people in hundreds of small and medium-sized firms. As Bryson and Henry (2000) argue, 'Motor Sport Valley has become the place in the world to produce your motor sport product – no matter how large the company and the global choice of locations they offer'

(p. 371). Three-quarters of the single-seat racing cars used in over eighty countries are UK-built. This area is home to the vast majority of world rally teams including Ford, Mitsubishi, Hyundai and Subaru. Over 50 per cent of the supply chains that won the major rally championships of 1995 were sourced in the Valley. These functions cluster, taking advantage of localized R&D, production technology, expertise, flexible labour and parts to produce highly advanced and differentiated products.

Box 4.2
continued

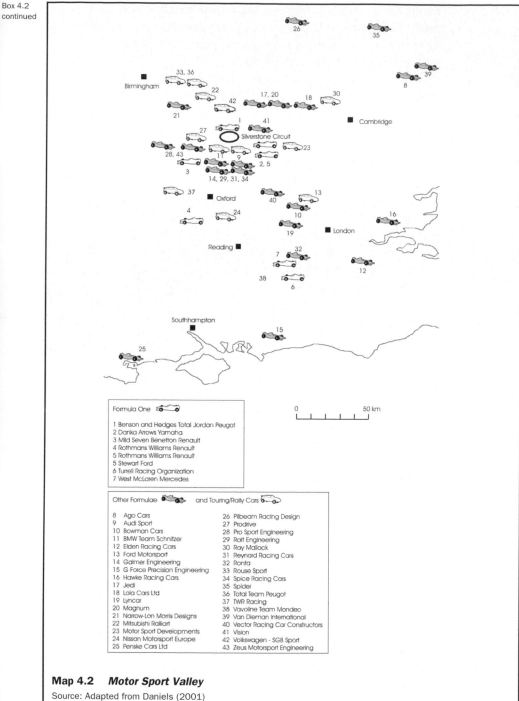

Map 4.2 *Motor Sport Valley*

Source: Adapted from Daniels (2001)

Plate 4.3 *Wellywood, Wellington, New Zealand, where a new high-technology film-making district is evolving which sees complementary functions – such as special effects, computer graphics and sound studios – clustering in suburban Miramar and Central Wellington close to Peter Jackson's New Line studio pictured here.* **The Lord of the Rings** *trilogy and* **King Kong** *were largely shot and produced in Wellywood.*

leading to a reconcentration of investment away from the periphery to the core, resulting in a reversal of the NIDL (Gwynne *et al.*, 2003). For some, this is a bad thing, as the NIDL diffused industrialization and created growth in the Third World (Chandra, 1992). Critics of the notion of flexible accumulation argue that much production is still carried out using Fordist techniques or something that closely resembles them. In many sectors, and in relevant parts of commodity chains, 'going abroad' remains a central strategy. This is particularly the case in labour or natural resource-intensive industries, such as garments in the case of the former and fruit production in the case of the latter. Communications possibilities over the worldwide web add another dimension to the global division of labour, as skilled labour in low-cost economies such as India can be employed in sectors such as computer programming to produce for markets anywhere. In general, work remains to be done on the empirical outcomes of the newer international division of labour. This research would do well to focus on social impacts. The casualization of the workforce and the rise of flexible labour markets in the core, and the impact in terms of socio-economic change in the Third World require greater consideration for example.

Chains and networks in the new global economy

Two useful ways of thinking geographically about the organization of the globalizing economy are commodity chains and networks. The first to elaborate the chain concept was Porter (1990) in his exposition of the 'value chain'. Gereffi (1994, 1996), a sociologist, has been the leader in seeking to further our understanding of commodity chains. He argues that the concept of the 'global commodity chain' is the best organizational unit through which to map the complexity of the global economy. The approach provides a new paradigm for the investigation of the global economy – one which transcends nation-states and other formerly assumed 'containers' of economic activity. This new framework;

> permits us to more adequately forge the macro–micro links between processes that are generally assumed to be discretely contained within global, national and local units of analysis. The paradigm that GCCs embody is a network-centred and historical approach that probes above and below the level of the nation-state to better analyse structure and change in the world economy.
>
> (Gereffi, 1994, p. 2)

Any commodity chain is characterized by four features: (1) the flows of commodities from each point in the node (an input–output structure); (2) the utilization of labour at each point on the node – what is sometimes referred to as the relations of production (for example, feminization, use of unskilled labour and so on); (3) the mode of production at each node, in technological and scale terms (governance); (4) the geographic location of the nodes (a territoriality). From a regulation theory point of view, some geographers (Goodman and Watts, 1997; LeHeron, 1993; Whatmore, 2002) have argued that we must also consider the forms of regulation that exist between the nodes.

Gereffi distinguishes between two types of commodity chain that link together various parts of the production system – the *buyer-driven chain* and the *producer-driven chain*. These have been globalized to the extent that they have stretched so that different nodes are located in various national territories. There are a number of problems with the concept of Gereffi's chains, but they do provide a useful starting point for analysing the global economy. Below, we distinguish between producer- and consumer-driven commodity chains (see Figure 4.5).

Producer-driven chains

These are found in industries where TNCs generally control the mode of production and the overall direction of the chain. This tends to be the case in technology-intensive industries such as computers and motor vehicles where a high proportion of costs are associated with advanced capital. Gwynne et al., (2003) use the example of the global motor industry, which 'provides a good example of a producer driven chain, with multilayered production systems that involve thousands of firms' (p. 167). In such systems, assemblers are linked with component suppliers at different tiers of organization. The Japanese production system in the late 1980s, for example, involved the linking of one assembler to an average of 170 first-tier suppliers, 47,000 second-tier suppliers and 31,600 third-tier suppliers – all linked through contracts (Hill, 1989). The rise of flexible accumulation methods as noted above means that these suppliers are often located within the core. However, there are still examples of component suppliers in lower wage cost economies – as was more traditionally the case under Fordism.

Buyer-driven chains

These are different from producer-driven chains in that the agents downstream control the development of the chains, leading to the development of 'decentralized' networks where subcontracting at different stages in the chain is more common. Commodities such as toys, garments and footwear fit into this model. As Gwynne et al., (2003, p. 168) say, 'essentially they are manufacturers without factories, at the centre of a highly flexible and global network of production distribution and marketing'. According to some authors these kinds of systems offer greater scope for emerging market economies, with Taiwan and Singapore in the 1980s being good examples. Such systems are generally geared towards labour-intensive industries, and thus production is often located in lower wage economies. As products are altered and go through cycles, the required factors of production (land, labour, capital, knowledge) change (see Gwynne (1990) on the Product Life Cycle model) and this can shift the optimal location of production. The TNCs in control of such chains are able to strictly direct through the subcontracting systems the production and supply of components, are able to capture much of the surplus value that is added in the chains, and to pass on the risk to the subcontractors. According to Gereffi, this type of commodity chain is becoming dominant in the world economy – although this has not been shown empirically in any satisfying way.

Producer-driven Commodity Chains

Buyer-driven commodity chains

Figure 4.5 *Producer- and buyer-driven commodity chains*

Source: Adapted from Gwynne *et al.* (2003)

Beyond chains?

Dicken *et al.*, (2001) are critical of Gereffi's concept of global commodity chains – despite 'the very substantial contribution of his work to our understanding of the structures of the global economy' (p. 98). They argue that the concept offers only a partial framework of analysis dealing with a limited number of commodities. The input–output nature of the chains remains relatively unexplored, and their linear nature is over-emphasized. Further, the regulatory aspects in between the nodes of the chain – the institutional context – is under-analysed (Whatmore, 2002). Gereffi accords too much power to TNCs in governing the chains, *vis-à-vis* the nation-state. Some geographers then suggest that a more useful concept is that of the 'network'. This is based on the following view of the global economy (adapted from Dicken *et al.*, 2001, p. 97):

1 *The economy is constituted of spaces of network relations.* Networks allow us to build relational views of the global economy – that have to do with the relations between agents within them.
2 *Agents act intentionally and this has effects on power structures.* The way social actors – which include individuals, households, firms, industries, states, unions – behave has what may be referred to as 'intentionality' – there are motives behind actions.
3 *The relationships between actors are 'embedded' or take place within particular spaces.* This does not imply that it takes place within territorially delimited space but network relationships do have a spatiality.

Summing up their argument, Dicken *et al.*, (2001, p. 97) claim:

> [T]he global economy is thus made up of social actors engaged in relational networks within a variety of 'spaces'. The analytical lens we adopt can thus vary widely. It may be geographical, it may be sectoral, and it may be organisational. It may be some combination of these.

They go on to argue the case for actor-network theory which better addresses the three characteristics above. This theory is useful for assessing the globalizing economy in three main ways:

1 *Networks ground globalization* – They provide a real time-and-space face to the process which is sometimes seen as universalizing and inevitable. It is played out in networks by the intentional action of individual actors with certain motives. It has both discursive and material outcomes but it is an agenda as well as a process. This also means that globalization can be resisted, as Whatmore and Thorne

(1997) point out in their work on 'fair trade' coffee production networks.

2 *Networks allow space for ethical analysis* – The rights and responsibilities of all network participants can be considered – be they human or otherwise. In addition, because network theory explicitly increases the range, or reach, of our individual actions it means that we can take responsibility for outcomes that are distant. It also provides justification for representing distant others by virtue of our links to them (Corbridge, 1993).

3 *Actor-network theory transcends the problems of grand theory* – Unlike Gereffi's commodity chains, actor-network theory does not

---

Box 4.3

Networks in action – world cities

Geographers argue that cities are the principal points through which the processes of globalization have been articulated and resisted. Armstrong and McGee (1985, p. 41) propose that '[C]ities are the crucial elements in accumulation at all levels and the locus operandi for transnationals, local oligopoly capital and the modernising state'. The concept of *world cities* refers to a tier of cities that dominates global economic and political affairs, acting as controlling nodes on a network of flows of capital, goods, people and knowledge (Friedmann, 1986; Sassen, 2000). The idea originates in the work of Geddes (1915), was later pursued by Hall (1984), and is now associated with Knox and Taylor (1995) in particular. As 'basing points' for global capitalism world cities function as major manufacturing centres, provide advanced service systems, operate transport and communication hubs, and house TNC headquarters. They also play a major role in the diffusion of cultural symbols, especially those associated with capitalism and Westernization. In sum, they are 'both the cause and effect of economic, political and cultural globalization' (Knox, 2002, p. 329).

In summing up the debate, Friedmann (1996) noted that by the 1990s there were a number of points of convergence with respect to the research on world cities:

1 World cities articulate economies into the global economy and serve as nodes in the global complex.

2 There is a space of global capital accumulation represented by world cities; most areas are off this network however.

3 World cities are large, urbanized areas of intense social and economic interaction.

4 World cities can be arranged hierarchically based largely on their economic power; this is determined by their ability to attract investment.

5 The controlling strata of people in world cities have a shared vision and form a transnational capitalist class

Box 4.3
continued

Plate 4.4   *World city, London's evolving financial district, Canary Wharf*

(Sklair, 2001). They have a shared vision of the effective functioning of the global economy, consumerism cosmopolitanism. They often clash with more localized groups around them.

World cities have been classified into a hierarchy based on the political/economic power they command and the extent of their articulation into global networks. The most complete classification is that undertaken by the Loughborough Globalization and World Cities Research Group (see Knox, 2002). This separates world cities into three tiers: Alpha, Beta and Gamma. Alpha world cities include Los Angeles, Chicago, New York, London, Paris, Frankfurt, Milan, Hong Kong, Tokyo and Singapore. Beta cities include San Francisco, Toronto, Mexico City, Sao Paulo, Brussels, Madrid, Moscow, Seoul and Sydney. Gamma world cities include Montreal, Boston, Dallas, Santiago, Buenos Aires, Johannesburg, Jakarta, Shanghai and Melbourne. Note that under this classification there are no world cities in South Asia or the Middle East, and only one in Africa, illustrating that population size is not the principal determinant of world city status. Although the world city is an important concept in globalization studies at least two things need to be built into future studies. First, differentiation within world cities themselves; second, change in large non-world cities – or what Rennie-Short (2004) refers to as 'Black Holes'.

make grand claims about the structure of the global economy and does not suggest any overriding logic. This takes economic geography beyond the impassé of thinking about the economy in a structuralist way.

In reality the functioning of the global economy is somewhere in between the structural relationships that Gereffi suggests and the 'intentionality' or subjectivism of actor-network theory. As is often the case with the major problems in human geography, geographers are grappling with the relationship between structure and agency. Dicken *et al.*, (2001) talk of 'attempts to draw together structural understandings of power relations with relational approaches to human agency' (p. 107), while Larner and LeHeron (2002) discuss the possibility of 'post-structural political economies'. This debate will be an important one in economic geography over the next ten years.

Transnational corporations

One of the most controversial debates in globalization studies surrounds the rise and role of transnational corporations. TNCs are loathed by the anti-globalists and seen as saviours by pro-globalists. They are defined as:

> A firm which has the power to co-ordinate and control operations in more than one country, even if it does not own them . . . typically involved in a spider's web of collaborative relationships with other legally independent firms across the globe.
>
> (Dicken, 2003, p. 7)

In the above sense Lee (2000c) argues that TNCs cannot be merely reduced to foreign direct investment, and what is crucial is the ability to command and control production complexes and chains. This may be achieved through direct ownership of the different parts of the production complex, but may also involve subcontracting as well as other less formalized means of influence. Such enterprises operate at a worldwide scale and account for an increasing proportion of the world's flows of trade, investment, knowledge, skills and technology transfer. Between one-fifth and one quarter of all production in the advanced capitalist economies is undertaken by TNCs, and a large proportion of world trade is intra-firm exchange (over 50 per cent in the case of USA–Japan trade, for example). In sum TNCs have the following characteristics (though there are many different types):

- *Functional integration* – in terms of R&D, production, investment, marketing and sales at a global scale. Essentially, within a TNC you find the location of various parts of the production network or complex in different global locations. In the past, this spatial division of labour may have happened at the national scale only.
- *Global sourcing* – an ability to take advantage of geographical differences in the distribution of factors of production (labour, capital, resources, knowledge) and state policies (regulations, tariffs and taxes).
- *Geographical flexibility* – an ability to switch its resources and key operational facilities to different locations.

Overall, TNCs evolve through either vertical or horizontal integration – and some may evolve through both:

- *Horizontal integration* occurs when a given company attempts to gain greater control of the market for its product by taking over or expanding production at the same stage of production that comprises its core business. This may come about through purchasing companies at that same stage or linking to them through less formal means. Firms will usually prefer to trade abroad than invest directly, and will do the latter only when the benefits of having a base location overseas outstrip the benefits of exporting there. Such advantages may include cheaper labour costs or reduced tariffs. Often, companies expand in this way in order to gain larger markets (Gwynne *et al.*, 2003). Horizontal integration and expansion of this nature is responsible for the growth of 'global consumerism' through the expansion of service TNCs – such as fast food/beverage corporations like KFC, Starbucks, McDonald's and Pepsi (see Plate 4.5). However, horizontal integration has also driven the global expansion of the following sectors: motor vehicle industry (Nissan, General Motors, Ford); international accounting services (PricewaterhouseCoopers); newsagents and booksellers (Waterstones and WH Smith); retail outlets (Walmart); and, banking (Lloyds, HSBC).
- *Vertical integration* occurs when companies attempt to gain control over different stages in the production process either upstream (supply) or downstream (distribution or outlet). Vertical integration is driven by TNCs' attempts to reduce risk. Some vertical integration takes the form of direct ownership of the different stages of production, as this can reduce the 'transaction cost' of contracts. In the fruit sector, for example, Del Monte owns orchards in Chile, packing and shipping facilities in Chile and Holland, refrigeration

plants in the UK, and its R&D and headquarters are in the USA. Other examples of vertical integration include the oil industry as well as the aluminium sector. Many TNCs, in the face of the increased flexibilization of the global economy, attempt to reap the benefits of vertical integration while dissipating risk by developing contract-based networks. This greatly reduces fixed costs and makes reaction to markets more flexible and rapid. The Nike case in Box 4.1 illustrates this strategy.

TNCs have a relatively long history and are rooted in the expansion of the colonial trading companies. The USA was the leader in TNC development, and invested in Latin America and Asia early in the twentieth century. As noted elsewhere, TNC growth blossomed in the 1960s driven by the NIDL. The rise of TNCs has been facilitated by two major factors during the neoliberal phase of globalization. First, the increased mobility of capital in deregulated markets clearly favours these institutions. Second, progress in communications and transport echnologies has made corporate decision-making possible at

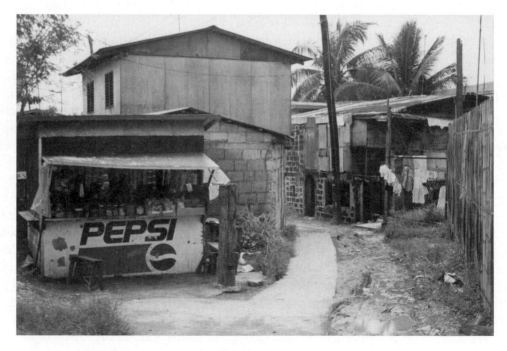

Plate 4.5   *Advertising for Pepsi Cola in a Manila slum*
Source: Donovan Storey

increasingly long distance. Data and control have become centralized, while operations have been increasingly decentralized. In this way, corporations have been able to reduce their risks while increasing flexibility.

The 'home' locations of the world's largest TNCs are concentrated in the global triad – the USA, EU and Japan (see Map 4.3). Both the USA and EU are each home to approximately 30 per cent of the largest hundred TNCs, while Japan is home to 25 per cent. Over the past three decades there has been some diversification in terms of the home location of TNCs as non-Japanese East Asian TNCs have grown in particular. The range of destinations for TNC investment has broadened considerably over the past fifty years however. There has also been a diversification in the nature of TNCs in terms of their social and cultural organization. In South Korea the *chaeabol* based on the *zaibatsu* of post-war Japan represents a particularly state-linked form which differs from the Western models (see Gwynne, 1985; Gwynne *et al.*, 2003). Japanese *keiritsu* are characterized by 'transactions conducted through highly symbolic alliances of affiliated companies based around long-term relationships within "families" of related firms and founded on highly complex links' (Lee, 2000c, p. 853). Such organizations may be centred around a bank or a central trading company (e.g. Mitsubishi) or around a parent company (e.g. Sony). It could be argued that the vertical and horizontal linkages of such corporations make restructuring difficult, and that this partly explains the inability of Japanese firms to react quickly to the 1997 Asian crisis.

Controversies in the geography of TNCs

As argued previously, TNCs have been central players in the evolution of globalization since the Second World War. In terms of sheer size, and thus economic command, a number of facts warrant mention. Many of the largest fifty TNCs have a higher turnover than the majority of the world's countries, while the larger ones are greater in absolute size than some medium-sized countries. In 2003, Walmart recorded turnover that was greater than the national income of Austria. In the same year there were twenty-six TNCs larger in economic size than New Zealand (see Table 4.5). The ten largest TNCs have a total income greater than the world's 100 poorest countries. In general, the largest TNCs have the ability, by virtue of their vast size, to stabilize and/or destabilize national economies.

Map 4.3 *Home location of TNCs*

Source: Adapted from Potter *et al.* (2004, p. 152)

Table 4.5 *The eighty largest transnational corporations and nation-states by Gross Domestic Product (GDP) or sales, 2003*

Country/company	Annual sales/ GDP ($US billions)	Name	Annual sales/ GDP ($US billions)
1. USA	10881.61	41. Toyota Motor	135.82
2. Japan	4326.44	42. General Electric	134.19
3. Germany	2400.66	43. Royal Dutch/Shell Group	133.50
4. United Kingdom	1794.87	44. Total	131.64
5. France	1747.97	45. Argentina	129.74
6. Italy	1465.90	46. ChevronTexaco	112.94
7. China	1409.85	47. Mitsubishi	112.76
8. Spain	836.10	48. Mitsui & Co.	111.98
9. Canada	834.39	49. Israel	103.69
10. Mexico	626.08	50. Malaysia	103.16
11. Korea, Republic of	605.33	51. Carrefour Group	96.94
12. India	598.97	52. Allianz Worldwide	96.88
13. Australia	518.38	53. ING Group	94.72
14. Netherlands	511.56	54. Citigroup	94.71
15. Brazil	492.34	55. Nippon Tel & Tel	92.41
16. Russian Federation	433.50	56. Singapore	91.34
17. Switzerland	309.47	57. Volkswagon Group	91.33
18. Belgium	302.22	58. ConocoPhillips	90.49
19. Sweden	300.80	59. AXA Group	90.10
20. Wal-Mart Stores	256.33	60. IBM	89.13
21. Austria	251.46	61. Itochu	88.51
22. Turkey	237.97	62. Siemens Group	86.62
23. BP	232.57	63. Czech Republic	85.44
24. ExxonMobil	222.88	64. Venezuela	84.79
25. Norway	221.58	65. Hungary	82.81
26. Denmark	212.40	66. Egypt, Arab Republic of	82.43
27. Poland	209.56	67. Philippines	80.57
28. Indonesia	208.31	68. Sumitomo	78.08
29. Saudi Arabia	188.48	69. Colombia	77.56
30. General Motors	185.52	70. American Intl Group	76.66
31. Greece	173.05	71. New Zealand	76.26
32. Ford Motor	164.20	72. Marubeni	74.39
33. Finland	161.55	73. Hewlett-Packard	73.06
34. South Africa	159.89	74. Chile	72.42
35. Hong Kong	158.60	75. United Arab Emirates	70.96

Table 4.5 *continued*

Country/company	Annual sales/GDP ($US billions)	Name	Annual sales/GDP ($US billions)
36. DaimlerChrysler	157.13	76. Ahold	70.57
37. Portugal	**149.454**	77. Hitachi	69.30
38. Ireland	**148.55**	**78. Pakistan**	**68.86**
39. Thailand	**143.16**	79. Peugeot Group	68.23
40. Iran	**136.83**	**80. Puerto Rico**	**67.90**

Source: Sales figures are from the Forbes (2000) list
(http://www.forbes.com/lists/2004/03/24/04f2000land.html); GDP figures are from the World Bank Group quick reference tables, (http://www.worldbank.org/data/quickreference/quickref.html)

There are many controversies surrounding the role and impacts of TNCs both in the West and the Third World, though particularly in the case of the former. Two issues are dealt with below: the race to the bottom, and the 'embeddedness' or otherwise of TNCs.

The race to the bottom? Effectively, states bid for the investment of TNCs, often offering economic incentives to attract them. To remain competitive states must maintain, among other things, low tax regimes, flexible labour markets (and low levels of unionization), and flexible environmental laws. TNCs can take advantage of these conditions in two ways: first, in their production location decisions; second, through 'transfer-pricing'. The latter sees TNCs allocating sales and profit returns in ways that exploit local conditions, such as low taxation, irrespective of the actual sales and profits at the real point of production. States have invested billions of dollars in attracting TNCs through, for example, the provision of infrastructure as well as the establishment of tax-free export processing zones (EPZ). It could be argued that this erodes the sovereignty of nation-states *vis-à-vis* TNCs.

The embeddedness of TNCs? Linked to the above is the issue of the supposed 'placelessness' of TNCs. To some, TNCs are so powerful and pervasive that they are increasingly bypassing nation-states. The argument of hyperglobalists is that such firms operate on a 'global surface' (Dicken, 2004). There has been some groundbreaking work in geography, informed by a number of the discussions concerning scale and space (see Chapter 2), which seeks to go beyond this simplistic notion. The first counter to the 'placelessness' argument is built on quantitative analysis. If firms were truly 'global' it would be expected that the majority of their assets and employment be located outside of their country of origin. A useful measure in this case is the transnationality index (TNI), which calculates a combined ratio of foreign assets to total assets, although these measures cannot always capture the true extent of the control that large firms have over global production complexes. In 2003, only 7 per cent of the largest fifteen TNCs had TNIs greater than 50 per cent, and only one of those exceeded 75 per cent (British Petroleum) (see Table 4.6). Some of the most transnational companies (such as Nestlé, Electrolux and Lever) come from relatively small economies, while TNCs from larger countries tend have lower TNIs. In this sense, TNCs remain remarkably tied to their home base. Quantitative measures provide only a partial picture of 'embeddedness' or otherwise. The particular configuration of economic, political and cultural traits of both home and destination states can have an important influence on the transnationality of TNCs. Although TNCs are global in scope they often take on some of the characteristics of the host economy or society in order to further their business – Japanese firms in the UK, for example, have adapted Japanese production practice to suit local labour forces and markets. Marketing is also increasingly differentiated to reflect local conditions – as in the case of the Kiwi Burger in New Zealand, which comes with beetroot! There is certainly a move among large globalized corporations to insert the 'local' into their operations. Slogans such as 'your local global bank' used by HSBC exemplify this. Overall, many TNCs are classic examples of glocal organizations.

Costs and benefits of TNC diffusion

There are highly divergent opinions with respect to the costs and benefits of TNC activity for both home and, especially, host countries. Gwynne *et al.*, (2003) separate theoretical opinion on the net impacts on the latter into two camps as explored below.

Neoclassical approaches – Lewis was the first to write of the benefits of TNC investment in peripheral countries in the context of his influential modernization paradigm in the 1950s (Lewis, 1955). He argued that TNC investment would promote technology transfer, better labour practices, managerial know-how, and increase domestic income and foreign exchange. Although Lewis favoured domestic capital, he argued that in its absence it is better to have foreign capital which would lead to improvements in consumption and education through raised incomes.

Table 4.6 The world's fifteen largest non-financial transnational corporations ranked by total assets (millions of US dollars), 2003

Corporation	Country	Industry	Assets Total	Assets Foreign	Transnationality index[a] %	Rank
General Electric	United States	Electrical and electronic equipment	495,210	180,031	39	83
General Motors	United States	Motor vehicles	323,969	75,379	29.8	87
Ford Motor Company	United States	Motor vehicles	276,543	81,169	38.4	85
Vodafone	United Kingdom	Telecommunications	207,458	187,792	83.2	13
DaimlerChrysler AG	Germany/US	Motor vehicles	183,765	25,795	22.1	97
Verizon Communications	United States	Telecommunications	170,795	10,159	4.6	100
Deutsche Telekom AG	Germany	Telecommunications	145,802	90,657	40	82
Toyota Motor Corporation	Japan	Motor vehicles	144,793	68,400	59.3	47
Exxonmobil Corporation	United States	Petroleum expl./ref./distr.	143,174	89,426	64.8	39
British Petroleum	United Kingdom	Petroleum expl./ref./distr.	141,158	111,207	80.5	15
Deutsche Post AG	Germany	Transport and storage	138,837	20,840	22.3	96
Vivendi Universal	France	Diversified	123,156	91,120	66.3	36
Electricité De France	France	Electricity, gas and water	120,124	28,141	27	91
Royal Dutch/Shell Group	Netherlands/UK	Petroleum expl./ref./distr.	111,543	73,492	59.3	48
Volkswagon Group	Germany	Motor vehicles	92,520	47,480	57.4	51

Note: [a] The transnationality index is an average of the following three ratios: foreign assets to total assets, foreign sales to total sales, and foreign employment to total employment.
Source: UNCTAD (2003) *World Investment Report*, UNCTAD, New York.

Lewis' neoclassical assumptions have been criticized by authors such as Jenkins (1987). The idea that in the absence of foreign investment there may be no investment at all is seen as true for some sectors only. This may be the case, for example, in capital-intensive industries with high entry and running costs – such as mining. In contrast, in sectors where smaller scale operations are more prevalent such as agriculture and services (e.g. hotels, supermarkets) foreign investment could have the effect of crowding out domestic capital. Furthermore, in protected markets, such as those operating import substitution industrialization policies, the entry of large foreign firms can lead to what economists term an 'oligopsonistic reaction' leading to higher prices and lower wages.

The neoclassical approach has evolved from a macroeconomic perspective to one which has become more microeconomic in its focus, particularly at the scale of the corporation itself. Led by authors such as Dunning (1988) and Vernon (1977), a diffusionist model has evolved. Diffusion theory argues that the diffusion of technology has been of particular benefit for host countries (see Malecki, 1991). In general then, these approaches 'see TNCs as efficient allocators of resources internationally and as providing benefits for the host and home nations' (Gwynne *et al.*, 2004, p. 153). Benefits in terms of trade, foreign capital inflows, employment and linkage effects are emphasized. It is this perspective that underpins neoliberal ideas which are supportive of TNC FDI in institutions such as the WTO.

Radical approaches – These approaches have their roots in Marxist and neo-Marxist theory. Broadly speaking, such thinking sees TNCs as agents of imperial and neo-imperial power attempting to control resources, capital, labour and markets to the advantage of elites in core nation-states. These processes are accentuated where host countries are economically small. The *neo-imperialistic* or *dependency* perspective has been influential in academic, activist and socialist policy-making circles. With respect to dependency thinking on the role of TNCs (see Kay, 1989), there are at least three ways in which TNC investment is linked to the underdevelopment of the periphery:

1 *TNCs are the main vessels through which surplus capital is appropriated back to the core.* There is an unequal exchange that characterizes this process. Some have extended this analogy to culture where TNCs act as the main lines of the diffusion of Western cultural models, designed to promote consumerism and the growth of capitalism.

2 *Monopolies and oligopolies tend to spread from the core countries to the periphery*. Large companies that have gained significant market share face little competition in relatively small economies and are thus able to capture significant proportions of economic activity, reducing the space for local capital.

3 *The emergence of what Frank (1969) called a 'comprador' class*. This 'class' is dependent on TNC patronage and acts as a conduit for its demands. Such elites will influence national decisions on the basis of what is required for the TNCs rather than the host nations.

Overall, Lee (2000) argues that assessing the balance sheet of TNC activity is a difficult analytical task. This is partly because it is a counterfactual problem (i.e. we do not know what would have happened in any particular case had the TNC not invested). For the circuit of capital as a whole it is uncertain whether TNCs make it more or less efficient. From the perspective of peripheral countries, however, Lee is quite unequivocal, claiming:

> Their [TNCs'] influence is malign insofar as at the same time that they connect economies into the global circuit of capital and so increase supply lines and markets and introduce new technologies (which have both positive and negative developmental effects), they displace control over the making of history and geography from people struggling to make their living and define their identity in such economic geographies by imposing a particular understanding and measure of value and progress.
>
> (Lee, 2000c, p. 853)

Notwithstanding all of the above criticisms, and the current popularity of a populist dependency-type approach evident in the works of journalists such as Michael Moore and John Pilger, it is the neoclassical approach that is hegemonic in policy-making. Left-leaning critics might argue that this is the case because TNCs have purchased enormous influence over the policy agendas of Western governments. Those on the Right argue that the role of TNCs is largely beneficial and that poorer countries are not liberalised enough to make the most of what TNCs can bring. The arguments of the Right are based upon theories that assume perfectly competitive markets. In reality, the imperfect competition of monopoly capitalism is becoming increasingly pervasive through the activities of TNCs, and even neoclassical analysts would argue that this does not bode well for society.

Neoliberalism and free-trade reform

It was during the second half of the twentieth century that the global free-trade agenda reached its zenith. During the latter half of the nineteenth century trade was relatively spatially limited, involving a handful of colonies and their imperial superiors. By the second half of the twentieth century, however, virtually all economies were enmeshed in the global system. This was heightened after the end of the Cold War as many former socialist economies were launched into the orbit of the global capitalist economy. The growth in world trade between 1950 and 2000 outstripped the growth in world production significantly (Gywnne *et al.*, 2003). The Bretton Woods institutions were critical in the pursuit of free-trade goals in the post-war era (see Chapter 7), GATT being the major initiative in this area. This was set up in 1947 with the goal of reducing tariffs and removing other barriers to free trade. It had four major principles (Gwynne *et al.*, 2003):

1 *Non-discrimination* – no one country should be given preference over another.
2 *Reciprocity* – tariff reductions should be reciprocal between countries.
3 *Transparency* – trade measures should be made clear.
4 *Fairness* – practices such as anti-dumping and unfair protectionism were regulated against.

GATT increased its influence steadily throughout the latter half of the twentieth century with membership rising from thirty in 1950 to 120 in 1995, with seven trade negotiation rounds having taken place. Often the sticking point in these negotiations was agriculture. Richer countries maintained, and continue to maintain, high direct and indirect subsidies for their agricultural sectors, citing political, social and cultural reasons. In 1995, GATT was replaced by the World Trade Organization (WTO), which was more powerful than GATT (see subsequent section). Despite the fraught debate over the socio-economic impacts of these institutions, they have been relatively successful in achieving their stated objectives. By the end of the twentieth century, tariffs were at an all-time low and barriers to trade in services had been reduced.

The pursuit of free trade is the cornerstone of neoliberalism, the now dominant economic paradigm. This approach, rooted in the neoclassical economics of the eighteenth century, argues that markets allocate resources in the most efficient way and that state intervention distorts allocative processes and, ultimately, reduces total global welfare. Neoliberalism also implies political and cultural shifts, arguing for smaller government, unfettered capital flows and TNC penetration, and the benefits of individualistic welfare maximization. Adopted first in dictatorial Chile in the mid-1970s and then in the form of Reaganomics in the USA (1980) and Thatcherism in the UK (1979), it has spread rapidly subsequently. There is a growing literature within geography which analyses and describes contemporary spaces of neoliberalism and the impacts of this from economic, cultural and political perspectives (see Brenner and Theodore, 2003). In this sense, then, the discourse of neoliberalism has become synonymous with globalization as currently practised, and it is against neoliberalism that the anti-globalization protests are aimed.

The World Trade Organization

Together with the IMF and the WB, the major international institution driving the neoliberal agenda is the WTO. The WTO was formed in 1995 after the completion of the Uruguay round of GATT talks. The WTO covers trade in manufactured goods, raw materials, agricultural services and intellectual property rights, and has over 150 member countries. The most important recent addition was China, whose accession in December 2001 was considered a major moment in the history of global capitalism given the government's continued adherence to state communism. The organization is ostensibly democratic and each country receives one vote in major matters – although it is well known that the larger economies do have a greater say in practice.

Essentially, the WTO monitors whether member countries are following free-trade rules. If they do not, the organization has the power to impose fines and sanctions. The WTO has the right to make legally binding agreements, which sets it apart from the GATT regime. As noted above, agricultural discussions have been the most fraught of all. In 2004, the WTO announced significant reductions in barriers to agriculture in core economies, which if they are implemented will come into force by 2012. We are yet to see if the new WTO will move to see countries in the periphery and semi-periphery to gain non-discriminatory access to agricultural markets in the West however.

The World Trade Organization is extremely powerful and pro-neoliberal. It has become the source of major controversy and there is increasing criticism, in academia and civil society more generally, that it is too deterministic and too skewed towards the interests of TNCs. As such, it has been the focus of many protests by anti-globalization groups, and has even been criticized by insiders such as Joseph Stiglitz (a former World Bank chief economist) who claim that it has become undemocratic, bureaucratic and is undermining the sovereign rights of national governments (Stiglitz, 2002). A number of areas of policy have caused special concern; under WTO rules there have been a number of cases where environmental protections and welfare subsidies within any given state have been ruled as obstructive of free trade and have been removed (see e.g. Box 4.4).

Box 4.4

The WTO in action: banana wars

Bananas caused one of the largest trade disputes in history in 1997, and tested the WTO in action. In terms of trade, that of bananas is the largest fruit sector and the fifth largest agricultural sector. Three agribusiness TNCs dominate the industry – Chiquita, Dole and Del Monte. They are vertically integrated; owning farms, transport and distribution networks. The commodity chains linking the plantations to the consumers are top-heavy in terms of the distribution of profits along the chain; approximately 90 per cent of the final price stays in the destination countries. Wages in plantations are very low – in Guatemala, for example, workers receive around US$0.60 an hour.

There are two major production zones in the Americas: the Caribbean (e.g. St Vincent, Dominican Republic) and Central America (e.g. Guatemala, Honduras) (see Map 4.4). Many of these countries, especially the Caribbean ones, are extremely reliant

on these exports. Higher production costs in the Caribbean – due to smaller production sizes and less intensive input use – mean that they are not able to compete in the global free market. Aware of these competitive pressures the EU granted the Caribbean special access to their market under the Lomé Convention of 1975. This convention was established ostensibly to assist former colonies and represented an extension of 'imperial preference'.

Under the new WTO rules US firm Chiquita, the major banana TNC in Latin America, appealed for this development assistance to be removed. The firm asked the USA to represent it in the WTO tribunal and President Clinton himself headed up the protest. The EU was ruled against, and the USA threatened $500 million dollars-worth of sanctions if preferential access was not withdrawn. The WTO ruled that $191 million-worth of sanctions could be applied.

Box 4.4
continued

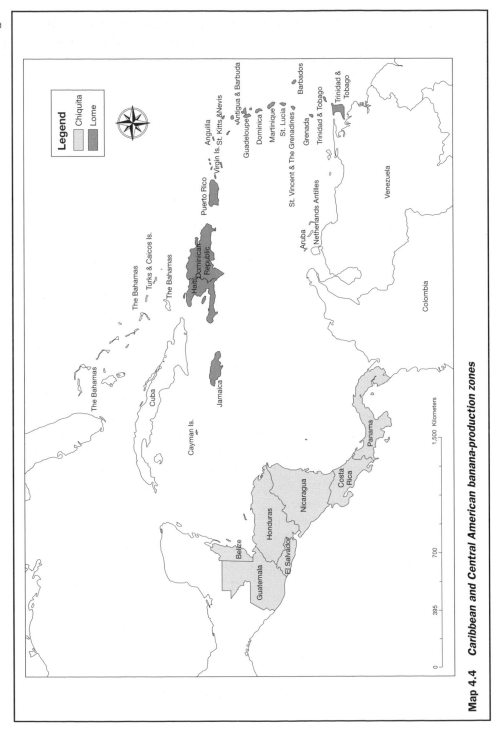

Map 4.4 Caribbean and Central American banana-production zones

Box 4.4
continued

The Lomé Convention has been replaced by the Cotonou Agreement – which substantially reduces EU assistance to former colonies. As a result of this case a number of Caribbean national banana sectors and hundreds of thousands of individual livelihoods have been undermined.

Arguments for and against free trade

The argument for free trade is based on ideas developed by two leading neoclassical economists David Ricardo and Adam Smith. The former developed international trade theory based around the concept of comparative advantage. Although the nature of the world economy has shifted dramatically, the arguments of the proponents of free trade, such as the WTO, are remarkably similar to early neoclassical ideas. It is argued that where the opportunity cost of producing goods varies between two parties (countries) trade will result in welfare gains for both. A country has comparative advantage in producing a good or services relative to others if the relative cost of producing that good (i.e. its opportunity cost in terms of other goods forgone) is lower than in the other country. Note that this refers to the relative cost of producing the item not the absolute cost, therefore, even if one country can produce all goods at a cost which is absolutely lower than another there will still be gains from trade. If all countries specialize in producing the commodities in which they have a comparative advantage, global living standards will be raised. Essentially then, in theory, free trade allows a maximization of world production, making it possible for every household to consume more goods than it could without free trade. Proponents argue that the long-run welfare maximizing potential of free trade offsets short- and medium-term restructuring costs.

Free trade has certainly raised consumption possibilities in some cases, but whether it has raised global living standards for all is clearly not the case (see also Chapter 7). This is the case for a range of reasons:

1 Many poorer countries have often been unable to persuade richer ones to open their markets to the extent that they open theirs.
2 Free trade can lock natural resource-exporting nations into a primary product export trap from which it is difficult to evade the long-run decline in the terms of trade (relative cost of exports to imports).

3   There are high restructuring costs of moving to a free-trade regime which often result in structural, and often localized, unemployment and other social impacts.

4   Many of the theoretical assumptions underlying the argument that free trade maximizes global welfare are not met in the real world. Factors of production are not mobile and markets do not operate perfectly. Monopoly capitalism practised by corporations in already wealthy nations can erode any benefits of free trade.

5   Free-trade theory does not take into account environmental factors and the long-run implications of any given nation specializing in resource-intensive activities (as is the case in much of the periphery). Furthermore, the concept of maximizing resource consumption may not be compatible with the limits of global ecosystems (see Chapter 8).

6   Specialization brings its own dangers and makes economies, especially small ones, highly vulnerable to changes in global conditions. Linking world economies has the potential to spread growth, but it also has the potential to spread downturns (such as during the Great Depression and the Asian Crisis of 1997).

7   The current policy dogma of free trade does not take into account the historical evolution of global capitalism. Richer industrialized countries built up their economies over centuries, often behind protectionist walls. Poorer economies are not afforded the same luxury and thus the wealth gap is structural/historic in nature.

Proponents of free trade argue that the problem with the current global economic complex is that there is not *enough* free trade and that the long-run benefits of reform will take time to trickle through. Sceptics argue that the model is flawed and that free trade in itself brings socio-economic and environmental costs which outweigh the benefits. There are of course many shades of opinion in between these two perspectives (see Cho (1995) for a review of some of them). In reality, global trade is anything but free, and the system is highly skewed to benefit core economies which maintain high levels of protectionism while encouraging, and sometimes coercing, poorer nations to open up their economies.

Summary – reforming the globalized system

Does free trade have the potential for universal benefits as promised? Or is it likely to go on concentrating wealth? Free-trade proponents argue that globalization will increase the participation of poorer nations in the world economy. However, the share of world trade accounted for by poorer nations remains very low (see above section on contemporary trade patterns). The terms of trade for poorer countries exporting primary products have worsened since 1995. There is no doubt that the gap between rich and poor has increased (as is elaborated in Chapter 7). The open question is: Is this a result of World Bank and WTO strategies, or is it because they have not been pursued with enough vigour?

In 1993, an OECD report argued that 70 per cent of the tariff-reduction benefits of liberalization would go to OECD nations. A *New Internationalist* article (May 2001) argues that the developed world gained $141.8 billion through the 30 per cent tariff cuts imposed by the 1994 GATT meeting. By the same calculation, Africa lost US$2.6 billion. According to such analyses, free-trade reform is absolutely failing to reduce inequalities (see McMichael, 2004). There are calls from across the political spectrum for the reform of free-trade rules and the broader regulation of the globalizing economy. In Latin America, for example, there is growing popular unrest and grassroots action against the IMF and WTO, and neoliberalism more generally (Kay, 2001).

There is growing consensus that the inequality of the world economy is not inevitable and that the system can be reformed in order to distribute the benefits in a more equitable and, possibly, efficient, manner (Amin, 2004; Gray, 1998). The World Bank and WTO argue counter to this that it is isolation from global capital circuits that causes poverty and that peripheral countries that have become enmeshed have improved their economic performance. The question is not whether globalization should be allowed to occur; it is not possible to roll back the networks, ideologies and trends that have unfurled. However, '[T]he debate is instead over the forms that globalization should take. It is in other words over the priorities, rules and regulations governing globalization. It is also over which bodies should have authority over the process' (Gwynne *et al.*, 2003, p. 226). Although a considerable amount of literature has been drawn up with respect to such reform (see Amin, 2004; S. Amin, 1997; Bello, 2002; Stiglitz, 2002), alternative 'bottom-up' regulation remains poorly articulated. This debate will be a defining feature of global politics over the coming decades. To those who doubt the possibility of building an alternative to the current hegemony Amin (2004) retorts, 'just look at the spell cast by neoliberalism' (2004, p. 231).

Regional economic integration

The thesis of economic globalization is made controversial, to sceptics at least, by the recent increase in regional economic integration. The WTO estimates that by 2001 there were 170 Regional Trade Agreements (RTAs) in force covering 40 per cent of world trade. The nature and scope of such agreements vary but essentially they represent attempts among often (though not always) neighbouring countries to reform the rules that govern trade and investment flows such that preferential access be granted to signatories. This usually involves lowering tariff barriers and non-tariff barriers such as subsidies and environmental controls. In this sense they are liberating and discriminatory at the same time. Examples abound including the European Union, the North American Free Trade Agreement, the Mercosur, the ASEAN and APEC (see Table 4.7 and Map 4.5). The agreements vary in size and coverage,

Table 4.7 *Important regional trading blocs and value of exports, 2003*

Regional bloc	Members	Formation	Type	Exports 2002 (US$ billion)
Andean Community	Bolivia, Colombia, Ecuador, Peru, Venezuela	1969 (restarted 1990)	Customs union	5.7
APEC (Asia Pacific Economic Cooperation Forum)	Australia, New Zealand, PNG, Indonesia, Malaysia, Singapore, Brunei, Philippines, Vietnam, Thailand, China, Taiwan, South Korea, Japan, Russia, Canada, USA, Peru, Chile	1989	Pre-free-trade area	2169
ASEAN	Brunei, Cambodia, Indonesia, Laos, Malaysia, Myanmar, Philippines, Singapore, Thailand, Vietnam	1967 (ASEAN) 1992 (Asean Free trade agreement)	Free-trade area	95
Caribbean Community and Common Market (CARICOM)	Antigua & Barbuda, Bahamas, Barbados, Belize, Dominica, Grenada, Guyana, Haiti, Jamaica, Monserrat, Trinidad	1973	Common market	1.2

	& Tobago, St Kitts & Nevis, St Lucia, St Vincent & the Grenadines, Surinam			
Common Market for Eastern and Southern Africa (COMESA)	Angola, Burundi, Comoros, Democratic Republic of Congo, Djibouti, Egypt, Eritrea, Ethiopia, Kenya, Madagascar, Malawi, Mauritius, Namibia, Rwanda, Seychelles, Sudan, Swaziland, Uganda, Zambia, Zimbabwe	1994	Common market	1.4
European Union (EU)	Austria, Belgium, Cyprus, Czech Republic, Denmark, Estonia, Finland, France, Germany, Greece, Hungary, Ireland, Italy, Latvia, Lithuania, Luxembourg, Malta, Netherlands, Poland, Portugal, Slovak, Republic, Slovenia, Spain, Sweden, United Kingdom	1957 (European Common Market), 1992 (EU)	Economic union with limited political union	1700
MERCOSUR (Common Market of the Southern Cone)	Argentina, Brazil, Paraguay, Uruguay (Chile – associate member)	1991	Common market	16.5
NAFTA (North American Free Trade Agreement)	USA, Canada, Mexico, Chile	1994	Free-trade area	612
Southern African Development Community	Angola, Botswana, République Démocratic du Congo (DRC), Lesotho, Malawi, Mauritius, Moçambique, Namibia, Seychelles, South Africa, Swaziland, United Republic of Tanzania, Zambia and Zimbabwe	1980	Pre-free-trade area	3.9

Source: Export values from UNCTAD (2004)

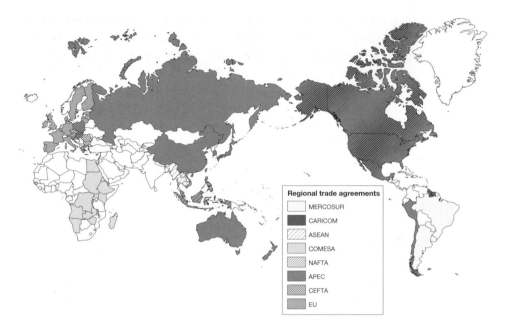

Map 4.5 *Major regional trade blocs, 2004*

Map 4.6 *Association of Southeast Asian Nations (ASEAN)*

and in some cases – especially in Latin America – membership overlaps. Dicken (2003, p. 146) identifies four 'waves' of regionalization: during the second half of the nineteenth century; after the First World War; from the 1950s until the 1970s; and the most recent wave following the end of the Cold War. We must be careful not to see regionalization as a homogeneous process. The nature of integration in Europe is very different to that which has taken place in ASEAN, for example (see Map 4.6); the latter is often seen as a purely economic body which does not intervene in the politics of member states. The free-trade agreement of 1992 was not prevented, for example, by the poor human rights records of numerous governments within the grouping. There are thus 'alternative regionalisms'.

Classical trade analysis allows for two broad outcomes in terms of the impacts of regional integration:

- *Trade diversion* – where, due to the formation of the new agreement, trade with a former partner outside the bloc is reduced and replaced by trade with a partner inside it.
- *Trade creation* – where, as a result of the agreement, growth creates new trade or replaces domestic production.

An RTA may be considered a 'success', in purely economic terms, where creation outstrips diversion. In general, RTAs shift flows of trade and investment in ways that have significant geographical impacts on the ground as networks connecting places are created or destroyed. There are at least five different types of RTA according to different levels of integration:

- *Free-trade area* – where the barriers to trade between members are reduced but each country retains sovereignty with respect to external parties in terms of trade policy.
- *Customs union* – a free-trade agreement plus a common external policy towards non-members.
- *Common market* – a customs union plus the free movement of factors of production – labour, capital, resources, knowledge – between members.
- *Economic union* – a common market plus the harmonization of economic policies, including monetary and fiscal policy. Such things become governed by extra-national bodies.
- *Political union* – an economic union plus harmonization of social policy and the surrendering of national sovereignty to the regional body.

Most agreements are of the *free-trade area* type, and thus much of the regionalism that is occurring has been 'open' regionalism. Examples of free-trade areas include NAFTA, EFTA and AFTA-ASEAN. Customs unions include the examples of ANCOM and CARICOM. In contrast, MERCOSUR is a common market. The EU represents a different example, and is a customs union with elements of economic and political union also (see case study in Chapter 5). Of the various regional 'blocs' three stand out as 'megaregions' (McMichael, 2004, p. 188). NAFTA, the EU and the APEC account for two-thirds of all global manufacturing and three-quarters of world exports. Of the three megaregions, the EU is the least reliant on trade with the outside world. Some critics have labelled it *Fortress Europe*, around which protectionist walls, especially in agriculture, are built. In the case of Europe, and to a lesser extent in the case of other examples, the issue of regional integration throws up tensions with regard to the sovereignty of nation-states in conducting their economic affairs. Such groups are often subject to sets of WTO-type rules, sometimes implemented by unelected officials.

Regional integration has inspired an enormous literature that deals with the political, economic and cultural impacts of such change. Thinkers are divided as to whether integration is a defensive reaction to globalization and other regional integration moves, or a precursor to further global interconnectedness. The discriminatory aspects of the formation of RTAs contravene GATT and WTO rule in principle but article XXIV allows for the formation of such groups under certain conditions. As already pointed out, most of the world's trade takes place between three major groupings: EU, NAFTA and East Asia. APEC, however, arose out of a particular concern among potential members that regional blocs were eroding the potential trade benefits of globalization. Does regionalism confirm or deny the thesis of economic globalization? To sceptics such as Hirst and Thompson (1999), the regionalization of the world economy is seen as further evidence that globalization does not actually exist. Furthermore, they argue, it is a process that compounds the core/periphery nature of the world economy. For others, it represents a *creeping globalization*, which will eventually encompass all regions. Individual countries are under great pressure to take part in such ventures in order to face the enormous economies of scale and scope that existing regional groupings possess. In South America there is an active process of integration taking place (see Box 4.5). Chile is currently the most aggressive country in the world in terms of forging new bilateral and multilateral trade and integration agreements.

Box 4.5

Regional integration in action: South American MERCOSUR

Regional integration is a theme that has a long heritage in Latin America. Following the conclusion of independence in the 1820s there were some efforts to create a unified Spanish America. Following the Great Depression, and the theories proposed by the structuralists to explain its devastating consequences, the idea of economic integration was once again raised. Structuralists argued that it was the nature of the insertion of the Latin American economies into the wider economy – as resource peripheries established during colonialism – that explained relative impoverishment. Crucially, the structuralists wanted to stimulate industrialization through import substitution (ISI) in order to break out of the primary product export trap. This would require economies of scale that regional economic integration would create – fostering autonomous capitalist, as opposed to dependent, development. Furthermore, such agreements would be geared towards restructuring the insertion of the economies into the global economy as manufacturers. Efforts towards this end included the Latin American Free Trade Area, The Andean Pact, and the Central American Common Market (see Gwynne and Kay, 2004). Many such models were not especially successful, which, in part, led to the eclipsing of structuralist ideas by more radical 'semi-autarkic' dependency theories (see Chapter 7) (Kay, 1989). More recently, however, a more 'successful' example of regional economic integration has been constructed on the Continent.

The *Mercado Comun del Sur* (Common Market of the Southern Cone) was established in 1991 to facilitate trade between member countries Argentina, Brazil, Paraguay and Uruguay. During the 1990s intra-MERCOSUR trade grew by a multiple of five and trade barriers were removed consistently. In 1995 a common market of the four countries came into effect and a cooperation agreement with the EU was signed. In the same year, Chile and Bolivia expressed their interest in joining and have since been admitted as associate members. MERCOSUR has perpetuated the rise of neoliberalism in the region, where it is now hegemonic. When the agreement was set up in 1991, the era of ISI had ended in the continent and governments were committed to outward orientation. Governments appealed to interest groups who might otherwise oppose the negative implications of structural adjustment policies by offering the supposed benefits of a greater regional market. In 2006 the members of the agreement are due to enforce a Common external tariff, which will expand the existing one between Argentina and Brazil. Despite the heralding of the agreement as a success story, there are many imbalances between and within countries in the context of free trade reform. Marginalized groups (such as peasant farmers, and agriculturalists in general) have voiced considerable concern and protested the impacts of the MERCOSUR widely (Murray, 2004a).

Impacts of regionalism case study – NAFTA and the maquiladoras

The North American Free Trade Agreement was signed in 1994 by the USA, Canada and Mexico. Under the agreement, flows of goods, capital and services are governed by WTO rules. Trade volumes between the three countries in the decade since the agreement was signed have tripled. Mexico has grown rapidly to become the ninth largest economy in the world. Canada and the USA have benefited, it is argued, from cheaper manufactured imports in particular. Spectacular regional economic growth along the US/Mexican border has been promoted, especially on the Mexican side, where hundreds of assembly branch plants (maquiladoras) of US TNCs have sprung up in cities such as Ciudad Juarez and Tijuana. The economic transformation of the border zone actually builds on Mexican policy beginning in 1965 when the Border Industrialization Programme was established. Under this plan, capital and parts were imported duty free from the USA to Mexico, assembled into final products, and exported back to the former. This represented one of the earliest examples of an export-processing zone in the Third World. In the 1970s, the factories were largely labour-intensive operations populated by unskilled workers, many of whom were women. With the arrival of high-technology TNCs from the mid-1980s, skill requirements have shifted somewhat (Gwynne et al., 2003).

There are many critics of the apparent macroeconomic success which NAFTA has brought to its participants. In the USA, unions from manufacturing sectors, such as car production, have protested the shipping out of jobs to cheaper labour locations in Mexico resulting in large-scale structural unemployment in many 'rust belt' cities of the US North West (such as Detroit and Chicago). Liberal US critics have argued that the rules governing NAFTA have stopped the prevention of the import of products where environmental standards have not been met and where labour conditions, such as the use of children, are allowed. In Mexico, in order to gain reciprocal reductions in US protectionism, wage rates were forced down in the 1990s. In 1995, hourly rates were 9 per cent of equivalent rates in the USA. Environmental protection laws were also eased in order to push the deal through. Despite high economic growth in much of the 1990s (aside from the peso crash in 1994), levels of poverty remain stubbornly high. In an absolute sense, it is estimated that some fifteen million individuals joined the ranks of poverty through the decade, and income inequality rose considerably (McMichael, 2004). Social conditions on the Mexican side of the border are marginal for

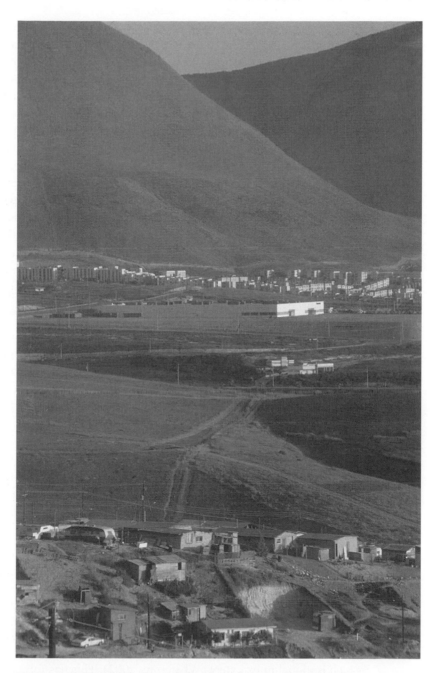

Plate 4.6 *Maquiladoras: a shanty town in the foreground is a source of labour for the assembly factories on the Mexican/US border pictured in the middle distance*

Source: Robert N. Gwynne

many, as crime and deprivation in local economies based on cheap labour exploitation has risen dramatically. Protest in Mexico has been widespread, the most famous being the Zapatista uprising against NAFTA in the traditional agricultural Chiapas region of the South. Indeed, some see this event as the genesis of the anti-globalization movement (see Chapter 5). Currently, in Southern Mexico, a new 'region state' is being constructed, the *Puebla de Panama*, to take advantage of relatively low wages in the periphery of the country. This will function as an export-processing zone linking Mexico with Central America. As such, NATFA is leading to the widespread maquilization of the Mexican economy and the diffusion of unskilled and non-unionized employment across the country. According to Hayden (2003, cited in McMichael, 2004, p. 192) there is evidence of an increasingly intense race to retain low-wage competitive advantage:

> Relocating the crisis-ridden maquiladora industry to Southern Mexico, where wages are half those of the Mexican maquilas on the US border, is a desperate effort to prevent the haemorrhage of jobs to China where [the wage] is only one-sixth of the Mexican wage.

Despite these criticisms, free-trade proponents see NAFTA as a model of regional integration that lays the basis for the wider Free Trade Area of the Americas (FTAA), which will include thirty-four countries and 800 million people. Many neoliberal Latin American governments (with the current exceptions of Cuba and Venezuela only) are keen for this scheme to progress, although they face being locked into the hemispheric economy as suppliers of cheap labour and cut-price resources. Chile signed a free-trade agreement with the USA, one which was largely favourable to the USA, in 2003 and is leading the way in the diffusion of the FTAA.

The globalization of agriculture

Agriculture is transforming as rapidly as any other sector in the global economy. Over the past century, the definition and nature of farming and food production has moved from the traditional model of family-based production for local and national markets to a very complex activity which is global in its extent (Whatmore, 2002). This has forged what some have called a 'new political economy of agriculture' both driving and driven by processes of globalization where systems, chains and networks link the four corners of the planet in unprecedented ways

(LeHeron, 1993). The new global agriculture, however, is – as is the case with all sectors touched by globalization – not homogeneous or on a predetermined path to a particular end-state. If anything the processes that have been unleashed by globalization have made global agriculture a more uneven and contested terrain than ever. For, while in a number of cases we are seeing the evolution of industrialized and truly globalized agricultural production complexes (for example, in the case of fresh fruit), paradoxically we are seeing the evolution of new niche and localized forms (sometimes referred to as post-productive agriculture). At the same time, subsistence agriculture remains of critical importance to the livelihoods of many in poorer countries. This section looks at one of the most understudied of the world's sectors in terms of the nature and impacts of globalization.

Whatmore (2002, pp. 57–58) conceptualizes the contemporary agricultural system as an *agro-food complex*. Drawing on an OECD definition, this is defined as:

> The set of activities and relationships that interact to determine what, how much, by what method and for whom food is produced and distributed.

This complex has been transformed by two important, and overlapping, trends: the growth of 'agribusiness' and its globalization.

The growth of agribusiness

In an attempt to increase profits the agricultural sector has industrialized and commercialized, with the effect of subsuming non-capitalist agriculture. This expansion has taken place through processes of horizontal and, especially, vertical integration (see section above on TNC growth). The latter has involved the linking up of the different nodes of the agro-food complex – from research laboratory, to the field, in transport and to the supermarket (see Figure 4.6). This process has allowed companies to internalize risk and reduce transaction costs, thereby increasing profits. In general, integration has led to oligopsonistic competition as TNCs take over significant proportions of the industry to control where production, marketing and distribution function. There are numerous examples of large-scale agribusiness TNCs that play an overwhelming role in their respective agricultural sectors, including Monsanto (technological innovation and GM), Nestlé (food production), Fonterra (dairy production) and Bulmers (cider production). Although, in theory, the efficiency gains of industrialized agriculture can be

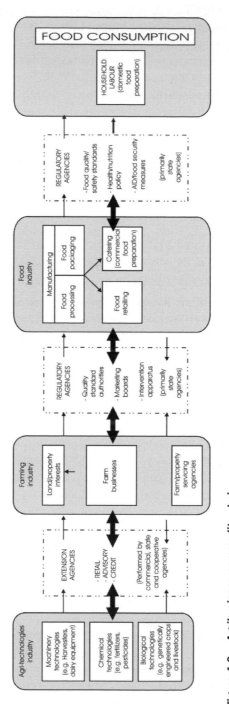

Figure 4.6   *Agribusiness commodity chain*

Source: Adapted from Whatmore (2002)

transferred to consumers as lower prices, generally the resulting non-competitive structures result in higher mark-up levels. Some also argue that food choice is eroded also.

Globalization of agribusiness

The agribusiness model has diffused globally over the past fifty years and is associated with the broader expansion of capitalism and modernization following the Second World War. Over the past two decades, facilitated by the liberalization of the world's economies, agribusiness TNCs have sought direct and indirect links into sectors outside their home countries. This may be conceptualized as an agricultural NIDL, with the aim of creating a mass food production system globally. Given relatively cheap land and labour costs combined with lenient environmental regulation, this has led to the diffusion of agribusiness to the Third World. This has been especially pronounced in fruit and horticultural complexes, for example, since these are labour-intensive at the production and packing stage, and because TNCs seek out localities where items can be produced counter-seasonally (i.e. harvested and supplied to affluent Northern Hemisphere markets in their winter). This has led to the production of what may be termed 'Fordist' fruit (McKenna and Murray, 2002). In general, neoliberal agriculture is drawing rural localities and communities in the Third World into the ambit of globalized circuits of capital as never before.

How is the globalized agro-food complex structured? Basically, different nodes of any given network are located in different nation-states linked through horizontal and vertical integration, and regulated by agents in between those nodes. Vertical integration may involve *ownership* of the different nodes of production by the same agribusiness TNC (direct vertical), or it may consist of the linking of different functions through *contract relations* (indirect vertical). Regulating agents – which may be state, quasi-state or privately owned – interact with commodity flows and circuits and intervene to govern the complex. For example, in New Zealand, the Ministry of Agriculture, Forestry and Fisheries (a state body) and Agri-quality (a commercial company) play a role in maintaining food safety and quality for retailers. Until recently the global marketing of New Zealand fruit and dairy products was undertaken by state boards that have now been privatized (See McKenna and Murray, 2002). Agricultural regulatory bodies may also be involved in: research and development of agro-technology; the extension of technology, knowledge and finance; health monitoring; and food security/aid issues. Given the move to open up economies it has been argued that the nation-states are less effective regulating agents than they once were. Thus, the rise of global agricultural networks raises the need for globalized regulation of supply. This is made more urgent by the rising concern in the West over food safety. The worldwide spread of diseases such as CJD alerts us to the vulnerabilities that are created by transnational food networks.

The globalization of agriculture also has a cultural globalization effect as populations are increasingly exposed to agro-produce from distant localities. This effect works in many directions. In the West, the new networks supply items that were hitherto unavailable (for example, fruit in the wintertime in the UK was relatively unheard of until the globalization of fruit began in the mid-1970s) (see Murray, 1998). On the other hand, non-Western countries are increasingly exposed to Western diets. McDonald's is of course the most notorious example of a TNC supplying relatively homogeneous Western products around the globe.

In order to make sense of the history of the global agro-food complex the idea of *food regimes* has been developed. This theorizes the transition of agriculture in the context of the evolution of global capitalism in general. Under this schema, which is influenced by the French regulation school of thought, the history of the agro-food complex is divided into three phases, separated by restructuring crises (see LeHeron, 1993):

- *First food regime* (1780s–1945) – This is based on the colonial division of labour whereby colonial territories supply imperial and former/neo-imperial powers with unprocessed and semi-processed items. The major trade during this period was in grains and meat, a classic example being the beef production system of Argentina in the 1800s. This phase corresponds to the industrial phase of the first wave of globalization identified in Chapter 3. The Great Depression and the First World War led to the end of this regulatory regime.
- *Second food regime* (1945–1980) – This is sometimes referred to as the 'productivist' phase in agriculture, and is characterized by the rise of intensive agriculture. Grain-fed livestock systems and the production of fats/durable foods played a central role, the UK dairy system being an example. The regime saw the rise of globalized capitalist agriculture in the periphery, as well as large-scale subsidy systems in the core (especially the EU and USA). This phase corresponds broadly to the modernization phase of the second wave

of globalization and created a NIDL in agriculture. The oil crises and the inability of nation-states to maintain subsidies, together with oversupply and safety concerns, are leading to the eclipsing of this regulatory regime.

- *Third food regime* (1980–) – This is referred to as 'post-productivist' agriculture, which is less intense and more diversified than the preceding regime (see section below). It involves the search for quality and safety by consumers. The central importance of biotechnology, as well as the rise of flexible and small-scale niche agriculture in the core, characterizes this phase. It corresponds with the neoliberal phase of the latest wave of globalization and more generally mirrors the rise of flexible accumulation. The rise of organic farming worldwide is one of the principal outcomes of this regime. The rise of counter-seasonal fruit and vegetable exports is also sometimes associated with this phase (see Box 4.6).

Overall, despite its analytical utility, global agriculture is more fragmented and uneven than this conceptual model can possibly capture.

Box 4.6

Fruitful globalization?

The global fruit sector is one of the most truly globalized of all the agricultural subsectors. Over the past 100 years the flows that constitute the complex have intensified and widened in geographical scope, creating a configuration that sees different nodes of supply networks spread across many continents (See Murray, 1998, 1999). Trade in fresh fruit has grown considerably in volume and value terms over the past four decades. In the years 1961 to 1963 the average total global export value of the ten most important fruits stood at US$1.56 million; by 1990 to 2001 this figure had risen to US$19.5 million (nominal figures). In 1969 the total volume of fruit exports was approximately 19 million metric tonnes per annum; over 40 million metric tonnes per annum are now traded. The role of fruit in the agricultural economy has also grown. In 1980, fruit and vegetable exports represented 11 per cent of total global agricultural exports; this figure had risen to 16.5 per cent by 2001. It is this evolution which allows citizens of the Northern Hemisphere to eat Chilean grapes in December, New Zealand apples in November and South African passion fruit in January. The fruit system illustrates many of the points made above concerning restructuring, spatiality and equity in global agriculture. Using a regimes framework the evolution of the sector may be traced as follows.

Box 4.6
continued

The first fruit regime

The origins of globalization in the fruit complex are found in the period of the first food regime, when the dominant form of fruit trade involved the export of a relatively small range of then 'exotic' products from colonies to imperial cores. Trade in bananas, a product which could not generally be grown in the core, and was relatively non-perishable, was most important. The fruit was produced in parts of Southeast Asia, Latin America, the Caribbean and some parts of Africa, shipped to Europe and subsequently to the USA. This system was almost fully extractive and based around plantation agriculture, leading to local enclaves where backward and forward linkages into the wider economy were limited.

Second fruit regime

It was during the intensive phase that the global fruit system, as we know it today, developed. An NIDL in fruit was forged, involving the 'going abroad' of fruit TNCs. The spatiality of this is illustrated in Figure 4.7, which conceptualizes the global fruit chain. Fruit TNCs, such as Del Monte, Standard Trading, Dole and Chiquita, played a central role in this globalization as they actively searched out cheap pools of labour and natural resources. Geopolitics also played a role in the diffusion of the complex as the USA supported its fruit TNCs in Central American countries such as Honduras and Nicaragua through economic and, during the Cold War, military

intervention (see section on geopolitics in Chapter 5). Many Third World countries, especially in Latin America, reoriented their agricultural production systems and became geared towards the satisfying of 'luxury' fruit demands in the core, often at the expense of domestic production of food. This threat to food security is one of the most damaging legacies of the rise of 'non-traditional agricultural exports'.

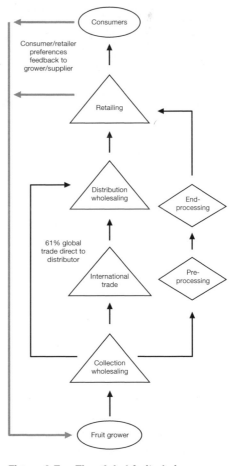

Figure 4.7   The global fruit chain
Source: McKenna and Murray (2002, p. 506)

Box 4.6
continued

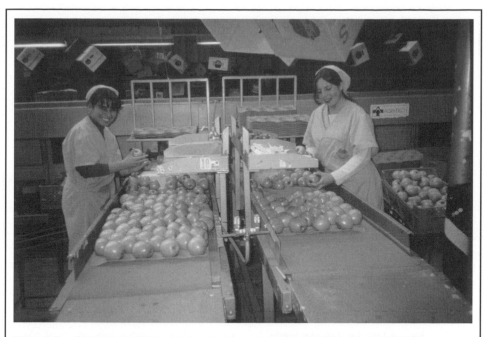

Plate 4.7 *Global fruit: Chilean* temporeras *pack apples for export to the EU. The evolution of the Chilean fruit export economy has created demand for female labour which, in many cases, is employed outside the home for the first time*

The factors driving globalization during the second fruit regime were complex and variable. They may be divided, somewhat artificially, into social-cultural, economic and technological factors with many of the impulses coming from the demand end of the chain.

1 Social factors

● *Evolution of middle class*. Higher disposable income and changes in lifestyles – especially travel – bring members of this class into contact with new fruits which are now affordable. Rising dietary concerns stimulate increased per capita fruit consumption. Furthermore, immigrants who arrive to take the place of the former working classes in Western economies bring new dietary habits with them.

● *Ageing populations*. Older populations accumulate greater knowledge, have more leisure time to inform themselves, use more resources on travel and thus spend a greater proportion of their income on fruit.

2 Technological factors

● *Development of cool chains*. These are integrated systems of supply from the orchard to the supermarket involving shipment in refrigerated vessels. They were developed essentially in the 1960s, and

Box 4.6
continued

facilitated the transport of perishable items over large distances.

● *Transfer of technology*. Fruit growing is a highly technical operation, with each locality needing special appraisal and adapted technology. The source of knowledge and hardware has been agribusiness TNCs, as Third World governments are often constrained in supplying such factors.

3 *Economic factors*

● *Rising disposable incomes*. Growth of the world economy leads to a more than proportional increase in demand for fruit given the 'income elasticity of demand' for this relatively luxury product. This may change, however, as fruit becomes a standard product and luxury status fades.

● *Increased mobility of capital*. Going abroad to establish foreign fruit supply began in the 1960s. In the neoliberal 1980s, foreign investment of this type increased dramatically.

Third fruit regime

According to regulation theorists we are now in the throes of a crisis in the global fruit system, involving oversupply in core countries, combined with safety issues and environmental concerns in general. It has been suggested that the second regime will be replaced by a third which bears much in common with post-Fordism in general (Whatmore, 2002). The possible nature of this regime in the context of fruit is contested. LeHeron (1993) has termed

Figure 4.8 *Different strategies in the global fruit complex*
Source: McKenna and Murray (2002, p. 506)

Box 4.6
continued

the period to which we are headed the 'integrated stage'. This will be characterized by a greater importance of biotechnology, employed in a flexible way to satisfy 'niche' demands. This will increase the possibility of research-intensive production in Western economies themselves and thus may reduce the need to go abroad. It may also be characterized by a renewed search and construction of 'exotics' – meaning that large-scale exploitation of vast amounts of stock fruits, such as grapes from Chile or apples from South Africa, may become less marketable. These trends are visible in some fruit-supplying localities – although the extent to which the basic Fordist model is becoming undermined is uncertain. In reality the future global fruit complex is likely to be characterized by multiple strategies by suppliers, with some following the standardized price/quantity route (the 'commodity' fruit), while others are more likely to aim for the quality end of the market, investing more in product differentiation and innovation (the premium or 'luxury' fruit). The Southern Hemisphere's two largest fruit exporters, New Zealand and Chile, are already differentiating their production on this basis (see Figure 4.8 for a model of this).

Although food regimes are useful concepts for organizing thought about the globalizing agro-economy they are macro-scale and require qualification at the local scale. There are many remaining controversies. For example, although some have seen the rise of the global fruit and vegetable system as symptomatic of the third food regime, much of the complex is intensive in its nature and standardized in terms of output. In reality, as is the case with many economic sectors, a mixture of Fordism and flexibility more accurately reflects the nature of the agro-food complex. Notwithstanding this, we look at some evidence for the rise of 'post-productivist' agriculture in the subsequent section.

Post-productivist agriculture – new food networks?

Although an increasing proportion of the world's agricultural production is undertaken by TNCs, localized non-corporate agriculture still remains very important. Millions of farmers across the world, the majority in terms of absolute numbers, are not linked explicitly into globalized networks and supply local markets only. There are also many millions of peasant farmers who are bypassed by commercialism entirely, and subsistence agriculture remains central in many societies in Africa, Asia, Latin America, and especially the Pacific Islands. This is not to say that global processes do not impact upon such groups of farmers. There is mounting evidence that such agriculture is being squeezed out by agribusiness as smaller farmers lose their means of production, including land, and are increasingly employed as labourers on commercial farms or migrate to large cities (Bryceson *et al.*, 2000; Murray, 2002a). Simultaneously, we are seeing the return to the practice of smaller scale agriculture in some places that is in direct resistance to globalization. In the core, it is possible to discern a partial shift to new alternative forms of flexible agriculture which some see as evidence for the third food regime. This has been labelled 'post-productivist' agriculture (Ilbery, 2001; Lowe *et al.*, 1993), and is characterized by a move away from intensive, concentrated and specialized agriculture towards diversified, extensive and niche-oriented production. The nature of this agriculture is outlined in Table 4.8 (based on Ilbery, 2001).

The transition to post-productive agriculture is explained by a range of factors linked to the perceived costs of the intensive globalized model of agriculture and the wider regulatory crisis. First, government intervention is being rolled back, meaning that large-scale productivist sectors can no longer be supported. Although agriculture is the last sector to receive the WTO-inspired deregulation there are signs that the largest intervention system of all, the EU's Common Agricultural Policy, will be dismantled. Second, increased concern for the environment has fuelled the promotion of less resource-intensive agriculture based on principles of permanence and sustainability. For example, the number of permaculture communities currently operating in New Zealand and Australia has increased of late

Table 4.8 *Post-productivist agriculture*

Major characteristics

- A reduction in food output and greater emphasis on food quality
- Progressive withdrawal of state subsidies for agriculture
- The production of food within an increasingly competitive and liberalized international market
- The growing international regulation of agriculture
- The creation of a more sustainable agricultural system

Farm restructuring

- Diversification of farm resources in to new agricultural products
- Diversification of farm resources into new non-agricultural products (e.g. tourism, retailing)
- Off-farm employment
- Hobby and semi-retired farming

Source: Adapted from Ilbery (2001)

as some people choose to drop out of the industrial agricultural system
for reasons of both lifestyle choice and food integrity. Finally, consumer
demand has been transformed, and consumers in the West, and elites in
developing countries, are increasingly motivated by quality over price.
This has led to a worldwide boom in organic farming, for example,
and safety and equity concerns have led to the development of sectors
where consumers are able to trace the exact chain that delivers their food.
Post-productive agriculture is generally thought to be applicable in richer
economies only, which have markets where consumers are sufficiently
affluent to trade price for other product traits. Ironically, in countries
that have been largely bypassed by the globalization of agriculture,
the implementation of such production may be easier to establish, as it
broadly mirrors practices that have existed for centuries. For example,
virtually all fruit production in the Pacific Islands region is 'organic'.
However, the problem for peripheral societies is gaining access to
networks of finance and marketing which will allow them to trade on
their new comparative advantage in the context of post-productivist
demand shifts in the affluent North.

Rather like flexible technology concepts, it may be argued that the
notion of post-productivism is exaggerated. The bulk of the world's
agricultural production is, and is likely to remain, intensively produced
and standardized. In New Zealand agriculture – one of the world's
most 'cutting edge' agronomies – the majority of production is still
'productivist'. In reality we are likely to see a mixture of the two in any
given situation. Broadly supporting these findings, Ilbery (2001, p. 266)
argues that:

> [on] the one hand, an intensive, high-input, high-output system
> emphasising food *quantity* and continued integration into the agro-food
> system may continue in 'hot spot' regions in both developed and
> developing countries. On the other hand, a more extensive, low-input,
> low-output system which emphasises *quality* and sustainability may
> develop in more marginal agricultural areas in developed countries.

Added to this there is some, albeit limited, evidence that such agriculture
is being evolved in emerging market economies, such as Argentina
and Malaysia, and it may be that with advances in transport technology
products will be supplied to Northern markets in a 'just-in-time' way.
In theory however, post-productivist agriculture has the potential to
partly reverse the agricultural NIDL. Whether this is seen as a positive or
negative outcome depends on one's opinion of the impact of agribusiness
TNC investment in the periphery (see Chapter 7 for some case studies of

Map 4.7 *GM varieties across the world*

Source: Adapted from Potter *et al.* (2004, p. 231)

this). A further complication is that post-productivist agriculture is arising at a time when genetically modified varieties are spreading across the planet in a 'second green revolution' (See Map 4.7). This has the potential to revive the agricultural NIDL, and some Third World countries, such as India, are rapidly adopting such technology in order to capture export markets. What this points to is that global agriculture remains highly uneven and, despite penetration of national and local agricultures by 'global' capital, the patterns on the ground have increasingly diverse geographies.

Conclusion – a transformed economy

Circuits of capital and spatial divisions of labour have been stretched through the process of economic globalization, leading to increasingly complex and hybrid outcomes that result in the co-existence of Fordist and flexible modes of production. This evolution is tied up with the shift from the colonial division of labour, through the NIDL to one based on flexible accumulation and the ascendancy of neoliberalism. TNCs have been among the most important, and controversial, driving forces of this transition. The evolution of the globalizing economy is summarized in Table 4.9.

Table 4.9 *Summary of the transforming global economy*

- Time–space compression, and the desire to decrease the turnover rate of capital, has led to a rapid expansion of the global economy over the past two centuries.
- The colonial division of labour faced a crisis sometime following the Second World War.
- It was replaced in 1960/1970 by a NIDL which saw increased FDI to the periphery. This created a global semi-periphery.
- TNCs were the main actors in the forging of the NIDL – they have bought costs and benefits.
- The expansion of TNCs has been facilitated by the shift to outward-oriented, free-market (neoliberal) development policy in the Third World.
- More recently there has been a further complication in the processes that created the NIDL – of note is the rise of NIC investment elsewhere in the periphery and to the core.
- The crisis of Fordism has led to a more flexible regime of accumulation.
- Flexibility has spatial implications which may lead to an exacerbation of the already concentrated pattern of global economic activity.
- However, in reality, the global economy is characterized by a mixture of Fordism and flexibility. The only truly supportable fact is that the global economy is far more extensive and complicated than it has ever been before.

The transformations outlined in Table 4.3 have led to increased unevenness across the global economy. Patterns of exclusion/inclusion are more complex today than they were in the past as networks of power have replaced simple core–periphery dichotomies. Understanding these new patterns requires incorporating both chain and network concepts. At the same time as 'global' flows have increased, powerful regional economic blocs have evolved which, to some, undermine the thesis of economic globalization. Notwithstanding this, as the case study of globalized agriculture illustrates, economic flows have intensified, become more extensive, move at increased velocity and have more profound impacts for those on the networks. Human geography has produced some critical work on the nature of economic globalization. In this context, Yeung (2002, p. 302) argues:

> The empirical orientation of geography as an intellectual discipline allows us to bridge the still largely unfilled gap between overwhelmingly stylised abstractions in most globalization studies and the contradictory experiences of most people who are living with economic globalization as an everyday fact.

However, notwithstanding this promise:

> [D]espite – or perhaps because of – the immense and fast-growing literature on globalization, we remain a long way from really understanding what is happening in the global economy.
>
> (Dicken *et al.*, 2001, p. 89)

Further reading

Angel (2002): This Editorial introduces a range of useful papers collected in one edition of the journal *Economic Geography* that deal with global economic change.

Bryson and Henry (2001): This is the entrance point for those wishing to grasp the distinctiveness of new economic geographies.

Daniels and Lever (1996): This collection of chapters deals with transformations in the most important sectors of the global economy from a geographical point of view.

Dicken (2003): This is by far the most comprehensive book on the geography of the global economy and is used across the social sciences. A measure of its globalization is that it is translated into Japanese.

Gwynne *et al.*, (2003): This text looks at the concept of 'alternative' capitalisms and adopts a regional geography perspective in comparing and contrasting experiences from Russia, East Asia and Latin America.

Knox and Agnew (1998): This work provides an excellent introduction to the spatial nature of the world economy and useful examples from across the world.

LeHeron (1993): This is the classic book on the globalization of agriculture and employs a regulation theory perspective to tease out the transformations taking place in the sector.

Globalizing political geographies

- New political geography and globalization
- The rise of the nation-state
- Trans-state governance from above and below
- Reflections on the nation-state
- Geopolitics and world orders
- The war on terror – new world order?
- New social movements and anti-globalization
- Conclusion – a multi-polar world

New political geography and globalization

The nature of political geography has changed markedly over the past fifty years. Early political geography was dominated by traditional geopolitics, which focused on the formation of, and interaction between, nation-states. Geopolitical ideas were used to justify the expansionist ambitions of some states, and as such often informed a colonial mindset (see e.g. Barton (1997) on Pinochet's geopolitics in Chile). During the quantitative revolution in geography in the late 1960s to early 1970s positivist tools were used to describe and analyse the playing out of formal politics in space – electoral geographies, political parties, the business of government were modelled in a spatial manner. This work played an important role in our understanding of the spatiality of the polity. With post-positivist shift, the rise of radical and critical geography, and the cultural turn, the nature of political geography has been transformed. This is reflected in two main developments in the subdiscipline. First, the impact of formal politics (e.g. government, democracy, elections, transnational governmental organization) on communities and individuals has been analysed in a more critical way. Furthermore, a critical geopolitics school has emerged which seeks to challenge accepted ways of understanding governance and governability in world politics (see Ó' Tuathail, 1996, 2002; Painter, 1995; Sidaway, 2001a). Second, informal politics, sometimes referred to as grassroots politics, have been increasingly researched. By focusing on the politics of the home, the body, the mind, the street and the community as practiced outside of the formal sphere, the idea that politics is personal has entered into contemporary political geography (Cloke *et al.*, 1999). Thus geographers have been active researching themes such as citizenship, national identity, nationalism and new social movements, focusing on both the concrete and imagined outcomes of political discourse at various scales.

The above changes are related to the evolution of globalization in important ways. The broadening of the subdiscipline comes in response to the increased complexity of political geographies in a globalizing world. It is apparent that the stretching of social relations across space is giving rise to new networks through which political demands and power are transmitted, and these may be both formal (for example, transnational governmental institutions such as the UN) and informal (for example, grassroots political communities such as the anti-globalization movement). Further, the critical perspective that has evolved in political geography enables researchers in the field to grapple with moral questions concerning how best to manage, regulate, respond to and resist globalization. The world is at a crossroads and political geography is in a strong position to analyse the political choices ahead.

The rise of the nation-state

The role of the nation-state is being contested. Critics from across the political spectrum have come to question the relevance and durability of the concept as the principle organizing unit of globalizing society. Authors such as Ohmae (1995) and Castells (1996) suggest, albeit in very different ways, that it is being eroded both from 'above' and 'below'. The argument underlying each of these analyses is a 'scalar' one: the nation-state is viewed as too small to cope with global economic, political and cultural flows and too large to represent effectively the needs and desires of local and trans-local communities. On this basis, hyperglobalists have predicted the *end* of the nation-state. Geographers have, on the whole, responded to such arguments in a balanced way; while acknowledging the transformed nature of political governance, arguments predicting the death of the state are seen as exaggerated.

Nation-states are such a central part of most of our lives, albeit to different degrees, since we have grown up in a world divided by neat, seemingly impervious, borders. Behind these borders we are often led to believe, exist *inherent* national socio-cultural traits. Questioning the role and survival of the state rocks the very foundations of what many have been led to believe – through discourses of nationalism – is their very identity and place in the world. The nation-state, however, is a relatively new concept in the history of global society, and is the product of the European political imagination. The concept diffused across the world through the processes of globalization itself, during the colonial wave when Enlightenment notions of the state as the 'natural' unit of society were proliferated. This was consolidated during the postcolonial wave after the Second World War, which saw the creation of many newly independent modernist nation-states. The fact that the nation-state is threatened by the very thing that propagated it is a paradox (Waters, 2001). If we understand the roots of this *discourse* we can better understand how and why notions of the state are shifting in the light of contemporary globalization, and we are better placed to accept that changes in governance practices might actually be *required*.

The concept of the nation-state

The nation-state may be defined as:

> A complex array of modern institutions involved in governance over a spatially bounded territory. It claims sovereignty over that territory and generally protects that sovereignty through a monopoly on the means of violence (military).
>
> (Adapted from Johnston *et al.*, (2000) and Daniels *et al.*, (2001))

There are two processes involved in nation-state formation. First, *state-building* sees the state centralize power over its territory, penetrating civil society in order to construct the political and social infrastructure to implement decisions. Second, *nation-building* involves elites using power gained through state-building to create a national identity and/or culture, and thus a notion of *citizenship*. Nation-states are not to be confused with nations, where the latter refers to a community of people that shares a common sense of 'togetherness', sometimes expressed as a common culture. As such, nations can transcend nation-state borders, and there are few nation-states, if any, which contain one homogeneous nation. For example, the Kurdish nation stretches across the borders of Iraq, Turkey and Syria. Under contemporary globalization the diffusion of nations through migration can lead to *diasporas*. For example, the Tongan nation exists largely outside of Tonga itself and comprises populations in Australia, New Zealand, the UK and the USA (Lee, 2004). As such most nation-states are home to multiple nations. The UK, for example, is home to English, Scots, Welsh and Irish, other smaller groups (such as Cornish and Manx) as well as a multiplicity of immigrant diasporas which may or may not continue to practise their national culture (see Chapter 6).

The evolution of the nation-state

The concept of the nation-state is only around 300 years old. Prior to this there existed a range of organizational forms governing territory. Compared to today's system these were much less rigidly organized in space. Across Europe in the sixteenth century the feudal state dominated, where aristocratic landholdings were loosely integrated in a relationship to barons and a superior monarch, sometimes bounded in empires (see Map 5.1). By the early seventeenth century in Europe, monarchical power had been consolidated, power centralized in bureaucracies, and professional armies were established to impose the will of the state. Such states were often absolutist and 'depended on the monopolisation of control over a particular territory in which it could raise both armies and taxes' (Waters, 2001, p 97). Some have argued that England and Holland represented the first examples of constitutional states where the powers of the monarch were limited through the collective interest of the emergent capitalist bourgeoisie (Mann, 1988).

The emergence of the modern nation-state and a system of international relations is often argued to have been established by the Treaty of Westphalia in 1713. This was signed at the culmination of the Thirty Years War in which the Austria–Hungary Catholic Empire sought to impose its religion on the protestant states of Northern Europe. The Treaty established three major points:

1 *The king is emperor in his own realm* – that states should not be subject to external control.
2 *The ruler determines religion* – the state has internal sovereignty.
3 *Hegemons should not be allowed to arise* – that a balance of power should be struck between different states.

The Treaty enshrined territorial sovereignty, borders and citizenship. This ushered in a period where diplomacy replaced war as the principal form of interstate relations. By the late nineteenth century the modern industrial state, based on models from Northwestern Europe, had been

Map 5.1 *Early states – Europe in 1500*

firmly established and had been exported to North America through colonization. This transformation culminated in a set of interrelated processes (Waters, 2001):

1 *The centralization* of state power to overcome internal resistance.
2 The *autonomy* of state power and sovereignty to redefine tradition and resist external dominance.
3 The *broadening* of the political community through constitutional reform and the extension of suffrage.
4 The development of the *symbols of nationhood* – flags, anthems and national languages.
5 The *activization* of the state to move beyond military function into areas of daily regulation.

Underlying the formation of the modern state lay the rise of the transnational industrial bourgeoisie which, based on Adam Smith's neoclassical ideas, sought *laissez-faire* economics combined with an enabling state. This class achieved its economic and social goals through the constitutionalization process and the consequent shift away from absolutist control. The bourgeoisie also used the state to expand control over new colonies and to control resource flows to its benefit, beginning the eventual globalization of the nation-state concept.

Trans-state governance from above and below

Until recently, the global political economy was not much more than the sum total of its nation-states, which were the regulating agents of change at virtually all scales. Today's polity is more like a 'web of interdependence' as Dicken (1998) calls it. This involves three basic types of institution:

- National governments
- International organizations (e.g. UN, WTO)
- Transnational organizations (e.g. NGOs including TNCs, environmental groups, international protest groups).

The contemporary political system is far more complex than the old system and it involves states 'as merely one level in a complex system of overlapping and often competing agencies of governance' (Hirst and Thompson, 1999, p. 183). In this 'world system of differentiated power relationships which operate at different levels of geographical analysis' (Dicken, 1998, p. 80), states are *permeable* and global governance is *polycentric*. As such we are living in a less predictable 'multi-polar' world where power is exerted from above and below the scale of the state (Waters, 2001). It seems paradoxical that today's world has more nation-states than ever (over 170), and this number is likely to rise further. It is possible, however, to see this as part of a broader dialectical reaction to political globalization as discussed further into this chapter.

The rise of global governance

A factor that has influenced theorization of the demise of state power is the rise of global 'regulation' practised above the level of the state. Of late, this issue has become one of enormous public interest as the nature and practices of global institutions such as the WTO, the IMF, the World Bank and the UN have been widely criticized. Later in this chapter we look at one of the most important manifestations of this – the 'anti-globalization' movement. Here we trace the rise of institutions of global governance, which have proliferated since the end of the Second World War (1939 to 1945). It is argued that, to date, such entities in fact represent the will of the individual nation-states involved in their construction, particularly a small handful of very powerful states. It is this latter point that has given rise to concern over the 'undemocratic' nature of such bodies. Indeed, it has been argued that current malcontent is so overwhelming that we are possibly 'witnessing and participating in the establishment of a new order of global regulation' (Roberts, 2002, p. 143).

'Global regulation' is a term that needs unpacking before we can understand the rise of its formal institutions. The French Regulation School sees capitalist society defined at any given time by an economic 'regime of accumulation' within a broader socio-cultural 'mode of regulation'. According to Lipeitz (1987, p. 33), a mode of regulation includes 'institutional forms, procedures and habits, which either coerce or persuade private agents to conform to (the regime of accumulation's) schemas'. Thus a mode of regulation includes not only a system of rules, but also social and cultural norms that sustain and reproduce the regime of accumulation. When we consider the evolution of global governance over the past fifty years it is apparent that it operates as a 'system of rules' which forms part of a wider capitalist mode of regulation designed to uphold successive regimes of accumulation (we reviewed the Fordist and flexible regimes in Chapter 4).

The current mode of regulation has evolved in contested and often uneven ways over the past six decades and its path has been historically contingent. Two points are worth noting in this regard (adapted from Roberts, 2002):

1 *Trans-state organizations did not begin after the Second World War –* although they certainly intensified. The initiation of the current global governance regime may be dated back to 1920 and two factors were especially important as a result of the end of the First World War. The first was the creation of new nation-states as the Austro-Hungarian, Ottoman, Prussian and Russian empires dissolved, and the second was the establishment of the League of Nations (1920). The system set up after the Second World War built on the failures of its precursors. As such, global regulation reflects both contemporary and historical institutions and power relations.

2 *Global regulation continues to evolve* – and each of the major
   institutions has been significantly contested since the Second World
   War. Roberts identifies three major developments that have shaped this
   evolution: (1) the internationalization of capital; (2) geopolitical shifts,
   notably post-war independence of former colonies, and transition of
   formerly socialist countries; (3) the recent protest against major global
   institutions. Some thinkers argue that the terrorist attacks against the
   USA of 11 September 2001 (9/11) represent a further watershed in the
   evolution of global regulation as discussed in the geopolitics section.

The eventual victors in the Second World War were instrumental in
establishing the post-World War order. This was first discussed in 1944
when representatives from forty-five countries met at Bretton Woods,
USA, in order to negotiate the regulation of the post-war political
economy. This conference confirmed the role of the USA as the dominant
power given its overwhelming economic size, political influence, and
the devastating impacts of the Second World War on its European allies.
By 1945, two Bretton Woods institutions had had been established – the
IMF and the IBRD (the World Bank). These institutions would attempt
to stabilize the global economy and prevent another catastrophic war
by assisting in the reconstruction of war-torn economies and the
modernization of newly independent nation-states. They were designed
in a way that favoured already powerful nations who played the major
role in deciding what shape they would take.

Arguably, the most important post-war global institution has been
the UN, which came into existence in 1945 (see box 5.1). There were
further key moments in the evolution of the post-war regulatory system.
For example, the General Agreement of Tariffs and Trade (see Chapter 4)
was established in 1947 to promote global free trade. GATT was
intended as a precursor to the establishment of the International Trade
Organization, but this was blocked by the USA's continued protectionist
stance. It was not until 1995 that GATT was replaced by the WTO (see
Chapter 4). In 1960, the Organization for Economic Cooperation and
Development (OECD) was formed by the twenty-five most powerful
Western countries building on the success of the Marshall Plan (see
Chapter 7).

The withdrawal of two of the Second World War victors, the USSR and
China, from the capitalist system in favour of socialist development
signalled the start of the Cold War (see section on geopolitics below for
details) which was deeply influential in the further development of global

regulatory systems both in the capitalist and the socialist world. Following the collapse of the socialist states post-1989, the institutions that govern the world system are effectively those established at Bretton Woods and Dumbarton Oaks – the World Bank, IMF, WTO and UN. These institutions have great influence in the design of economic and social policy in their member countries. To some radical thinkers, the overt support of capitalist accumulation practised by these bodies represents a form of neo-imperialism visited on politically independent, but economically subordinate, postcolonial states.

Box 5.1

The United Nations

The United Nations is one of the most powerful and important global institutions. The seeds of the formation of the UN were laid after the end of the First World War (1914 to 1918) when through the Treaty of Versailles (1920) the League of Nations was established. This sought to restore global stability in order to administer the transition of the new nation-states carved out of the Ottoman, Prussian, Russian and Austro-Hungarian Empires. The League failed, partly because the USA turned inwards, and partly because the policy of shaming Germany laid the seeds for a nationalist-fascist movement there. After the Second World War it was clear that the system needed reform to achieve the twin goals of stability and peace, and the fostering to independence of nation-states from the former French, British, German and Dutch empires. It was clear that the Westphalian state system, which enshrined competition between states and the use of force (see above), had failed and developments in nuclear weapons meant

that there was a real possibility that a third world war could end civilization. In October 1943, the allied countries (the UK, USA, USSR and China) met in Moscow and agreed to create an improved version of the League of Nations following the war. In 1945, fifty-two countries – mainly the war victors – met at Dumbaton Oaks, USA, and, based on the Atlantic Treaty signed by Churchill and Roosevelt in 1942, established the UN Charter. This was signed in San Francisco on 26 June 1945. The major principles were:

- The use of force and war between states was prohibited except in special cases.
- The Security Council was to have a monopoly on violence and to use this to maintain peace.
- States could use force only as a means of protection.

A Secretary General heads the UN, a post currently filled by Kofi Annan, who is the first African to hold the position. Its membership currently stands at about 180 and is

Box 5.1
continued

growing. The two central bodies are the Security Council and the General Assembly. The *Security Council* consists of five permanent members (France, Russia, UK, USA and China – the victors of the Second World War) and ten non-permanent members elected every two years. It is the more powerful of the two major bodies: it is charged with maintaining peace and security, and can call on members to send troops for 'peacekeeping' operations. Each of the members of the Security Council has a veto, and thus during the Cold War in particular it became relatively ineffective. The membership of the Security Council remains contested. Some nations protest the composition of permanent members as not reflective of current global political realities, for example. The *General Assembly* consists of all members of the UN, each of which has one vote. The assembly debates a wide range of issues related to the UN Charter and holds sessions for matters of special concern (such as the 2002 meeting to discuss the then-impending war in Iraq). For constitutional matters a two-thirds majority is required, whereas for non-constitutional matters a simple majority is sufficient. Its resolutions are binding (in theory), although its recommendations are not.

Other principal organs of the UN include the *Economic and Social Council* which deals with issues related to human welfare and development; and the *International Court of Justice* located in the Hague, Holland. This court deals with disputes between nation-states, as opposed to individual prosecutions (a factor behind recent pushes

for an *International Criminal Court*; see Roberts, 2002). Beyond the principal organs there is a host of other specialized agencies including the UNDP (United Nations Development Program); FAO (Food and Agriculture Organization); UNESCO (Educational, Scientific and Cultural Organization); ILO (International Labour Office); WHO (World Health Organization); and the International Monetary Fund (IMF). The WTO is independent of the UN, while the World Bank is linked informally.

The UN's record in international conflicts has been one of mixed success. The original aim of the Charter was to create an independent military force, but this has never come to fruition. Instead the UN relies on calling up troops to enforce 'peace'. The greatest successes have been in the arbitration of conflicts such as Kashmir (1949), Cyprus (since 1964) and the Iran–Iraq War (1980 to 1988). In larger scale conflicts the results have been less clear-cut, as the UN has relied on the will of the more powerful nation-states providing troops. During the Cold War, the USA used the UN to further its own cause. For example, it cited self-defence as the rationale for the invasion of Korea (1951). The UN has failed to prevent ongoing conflict in places as diverse as Israel/Palestine, the former Yugoslavia, Rwanda, East Timor and Sudan.

In 2002, the USA and the 'coalition of the willing' flouted UN resolutions and invaded Iraq on the basis of the supposed existence of weapons of mass destruction which they argued posed a threat to Europe, the USA and allied Middle Eastern neighbours. The

Box 5.1
continued

most important allied country in the latter region is Israel which was created by the UN following the Second World War to formalize Zionist-inspired Jewish resettlement on their historical holy lands – a move which has led to nearly five decades of Arab–Israeli conflict. The authority of the UN was seriously compromised by the invasion of Iraq. In reality, some have argued, the UN has always been at the whim of the USA – the major budget contributor and hegemonic power. Thus the impartiality of the institution is called into question constantly. The role of the UN in other areas has not escaped controversy; in the case of the IMF the institution is charged with forcing politically unpopular structural adjustment reforms on poorer countries while rarely subjecting richer countries to the same discipline (the UK in 1976 being a notable exception) (see Chapter 7).

The end of the Cold War had an important impact on the UN in that the transition from superpower geopolitics reduced the incentives for liberal-capitalist states to provide financial and military support. Despite its recent undermining however, the UN has made great advances in many areas since it began. In the development arena the UNDP, while remaining overtly modernist, often provides a relatively holistic counterbalance to the purer neoliberal stance of the other global institutions. Overall, it could be argued that:

> the UN system represents a very clear advance on the Westphalian system and is clear evidence that peace and security is a shared global problem that can neither be left to private treaties between states nor to the dubious intentions of any hegemon.
>
> (Waters, 2001, p. 119)

Global regulation from below

The perception of global institutions as undemocratic, combined with frustrations over the capability of the nation-state, is leading to a global civil society movement that seeks to reconstruct and regulate the system from below (Glasius *et al.*, 2002; Goldsmith and Mander, 2001). This has seen the rise of transnational non-governmental organizations (NGOs) representing common interests, shared needs and new globalized identities (see S. Amin, 1997; Bello, 2002). The number of transnational NGOs has risen sharply since the 1960s, and generally much more rapidly than the rise in intergovernmental organizations or nation-states (see Figure 5.1). A number of broad dissenting alliances have driven this process, including the global environmental movement (see Chapter 8) and the anti-globalization movement (see below), as well as more contested entities such as the global women's movement (see below). Individual examples include NGOs as diverse as Greenpeace, The Fair Trade Network, Stop the War Coalition, Globalize Resistance, Médecin

Sans Frontières, Amnesty International and Oxfam. Although diverse
in their perspectives, such bodies generally argue that powerful Western
associations such as the G8 and the World Economic Forum are
undemocratic, and that Bretton Woods institutions are conduits for
the agenda of the dominant elites in rich industrialized countries. These
bodies have propagated, it is argued, damaging neoliberal perspectives
that have become increasingly deterministic in the daily lives of billions
of people across the planet. These NGO networks are organized
transnationally, and as such are substate in their constituents but
globalized in their extent (Roberts, 2002).

New forms of representation and lobbying have been developed, most
visibly in the street protests of the anti-globalization movement, but
also in the day-to-day resistance of individuals (for example, those who
boycott GM crops in the USA). Elements of this general movement

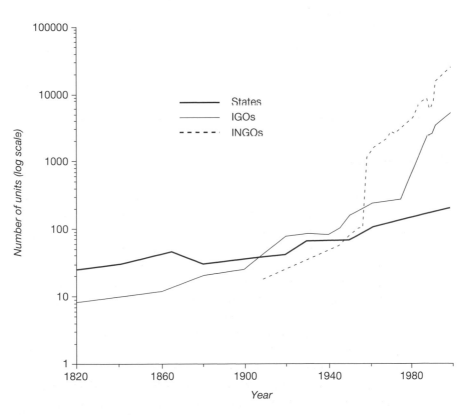

Figure 5.1 *TNGOs, IGOs and nation-states*
Source: Waters (2001, p. 150)

are found in both the rich and the poorer worlds, which in many ways transcends the traditional delimitations between First and Third Worlds and East and West. In this context, Brecher *et al.* (1993, p. xv) argue that the emergent 'globalization from below':

> in contrast to globalization from above, aims to restore to communities the power to nurture their environments; to enhance the access of ordinary people to the resources they need; to democratise local, national and transitional political institutions, and impose pacification on conflicting power centres.

This resistance from below represents a new force that challenges the post-Second World War mode of regulation. We should expect this sector to become increasingly strident in the future as the global income polarizes further and environmental issues come to the fore. Whether such organizations have sufficient power to change the global order is a debatable point; however, the enhanced role of global civil society does at the very least signify a watershed in world politics (see Brecher *et al.*, 1993).

Reflections on the nation-state

Broadly speaking, there are two schools of thought with respect to the impact of globalization on the nation-state. The first school, which Waters (2001) terms the 'modernizers', argues that the state is becoming increasingly incapable of coping with the demands of globalization, and that its sovereignty is being eroded from above and below. Moderate modernists such as Held (1995) and Waters (2001) argue that although the state is unlikely to disappear as a political form it will become less powerful as transnational social and political flows and grassroots demands multiply. More radical modernists argue that the state is just about finished as a meaningful political and economic force (Castells, 1996; Kindleberger, 1967; Ohmae, 1995).

In contrast, the 'realist' school is much more optimistic about the role and weight of the state in global affairs. Authors such as Albrow (1996), Dicken (2003), Giddens (1985, 1999), Gilpin (1987), Harvey (2001), McGrew and Lewis (1992) and Rosenau (1980) assert that while the state is being transformed, and governance is being played out at a range of scales from the local to the global, it still fulfils a central regulatory role. Some thinkers point to the fact that the global regulatory regime itself is a product of nation-states and that globalization processes have been facilitated by the existence of states. For example, Robertson (1992) argues that: 'the diffusion of the idea of the national society as a form of institutionalised societalism was central to the accelerated globalization which began just over a hundred years ago' (p. 58).

Although the 'modernizer' versus 'realist' framework is imperfect – there are many overlaps – it does provide a useful way of thinking about the relationship between globalization and the nation-state. The broad arguments of these two camps are used as springboards for the discussion below.

The powerless state?

Held (1991) argues that nation-states will eventually be replaced by a world government. The steps in the argument are as follows:

- Cultural and economic flows are reducing the power of governments at the nation-state scale and flows are becoming more difficult to regulate at their borders. Transnational societal and cultural links are flowering.
- Transnational processes are growing rapidly – this includes TNCs who are increasingly powerful; sometimes being larger in economic size than nation-states.
- Given the above, nation-states are forced to increasingly cede many areas of their traditional sovereignty to international, intergovernmental and regionalized institutions. These powers include defence, security in general, economic management and communications.
- Linked to the above, larger territorial units have evolved to which states are increasingly ceding power: these may be political (EU), economic (NAFTA), military (NATO) or even broader (WTO, UN, IMF).
- The above leads to a system of global governance which is emerging and which states must increasingly abide by.
- This system provides the seeds of a world government with coercive and legislative power – and this is already beginning.

In this sense then, Held sees the major threat to the state as coming from 'above'. Waters (2001) identifies a number of what he calls external 'threats' to the sovereignty of the state, the most important of which is the *global human rights* agenda. This agenda, established, ironically, by nation-states after the Second World War during the prosecution of war criminals, has provided an important rallying cry for many across the

world. It has created a sense that there is an authority higher than
the state, together with the institutions to enforce this view. Waters
argues that all the cross-societal social movements of the second half
of the twentieth century are ultimately rooted in the human rights agenda.

There are those who see processes from 'below' as more troubling for
the future of the state (Routledge, 1999; Spybey, 1996; Waters 2001).
It is possible to be critical of the idealism surrounding the celebration
of the rise of globalized NGOs however, since it tends to overstate the
importance of cross-societal struggles. Intra-national differences in
welfare are as great as ever, and a wide gulf between Western and Third
World populations in terms of needs and wants remains. Castells argues
in the *Power of Identity* that the state is actually sandwiched between
'internal' and 'external' threats (1997, p. 8):

> The growing challenge to states' sovereignty around the world seems
> to originate from the inability of the modern nation-state to navigate
> uncharted, stormy waters between the power of global networks and the
> challenge of singular identities.

To 'modernizers', then, the state is in crisis and its powers have peaked,
and this decline was clearly visible in the last quarter of the twentieth
century. In Western nation-states there was a marked shift away from
corporate welfare states to more liberalized and 'downsized' government,
often referred to as the 'rolling back' of the state. This 'state weakening'
included, and continues to include, privatization, welfare budget
reductions, liberalization, and the ceding of sovereignty to transnational
institutions. Underlying this crisis, claim some sociologists, is the rise of
'post-materialist' society (see Box 5.2).

The powerful but transformed state

Many argue that the nation-state remains a powerful entity in a
globalizing world, simultaneously transforming and being transformed
by globalization processes. Some of the arguments put forward by social
scientists since the Second World War on the transformed state are
reviewed below.

Traditional international relations arguments conceptualized global
politics in terms of a snooker ball analogy. Each ball has autonomy but
is impacted by the movement of other balls as the game progresses.
Although arguably accurate until the end of the First World War, this
'dynamic mosaic'-type approach is unsuitable for the contemporary

Box 5.2

Post-materialist society and the nation-state

Post-materialist society is a term coined by Inglehart (1990) to describe the rise of a new culture with distinctive political values. In traditional politics, the dividing lines are determined by material values – that is to say around issues involving redistribution of wealth. Typically, this involves political differentiation along 'leftist' and 'rightist' lines, with the former favouring redistribution, state regulation and intervention, and the latter favouring private ownership, individualism and economic market freedom. The values of post-materialist society are different however, and revolve around community, quality of life and identity. In general this shift is most pronounced in richer Western countries but in poorer countries too such values have diffused as they globalize. We have seen the evolution of what have become known as the 'New Left' and the 'New Right', where the former emphasizes minority empowerment, human rights and the environment, while the latter stresses consumption, individual expression and

small government. This shift has led to the transformation of traditional conservative and labour parties in many democracies as the formal political system attempts to keep up. In the UK, the New Labour Party led by Tony Blair epitomizes the transformation of traditional labour values. In the USA, the Democrats under Bill Clinton underwent a similar shift. Both the New Left and the New Right are sceptical of the role of the state in different ways. The New Right regards the state as undermining the market system and the freedoms associated with this, while the New Left sees the state as a potential oppressor and the agent of powerful corporations and materialistic elites. The shift to post-materialism challenges the state and deepens cross-societal value systems across borders, including on concepts such as human rights, and environmentalism. According to Waters (2001) post-materialism lies at the core of the formation of new social movements and international non-governmental organizations.

world. It fails to capture the political networks of power which bypass formal government and underestimates the possibility of alliances of interest. International relations theory has evolved to try and capture the new complexity. Burton (1972), for example, proposed the rise of 'world society' characterized by a 'layering of inter-state relationships with networks or systems relationships between individuals and collectivites that transcend or subvert state boundaries' (Waters, 2001, p. 99). Burton talks of the need to incorporate a concept of 'effective distance' which describes the mesh of relations between places more effectively than territorial space. At the core of his argument however the nation-state retains the power to control this process.

Rosenau has been especially influential in terms of the theorization of the transformed nation-state, introducing the concept of 'transnationalization'. This conceptualized a system where new actors and groups were brought into the arena of global politics facilitated by technological advances. This would render the global system an increasingly complex web of interactions between governments and non-governmental organizations. The effect of this in terms of the propulsion of inter-societal groups on to the world arena would have a disintegrative effect, causing a: 'transformation, even a break down, of the nation-state system as it has existed throughout the last four centuries' (Rosenau, 1980, p. 2). However, Rosenau concludes that despite the existence of such corroding processes the nation-state would remain central as governments generally have the power to adapt, design and implement policy.

More recently, Rosenau (1990) suggested the 'two worlds' of world politics conceptualization. This argued that we are entering a 'post-international' system where chaos and unpredictability rule, and formal nation-state governance is unable to cope with the emergence of truly global problems (such as environmental change), and the demands of an increasingly informed population that is able to articulate its distrust of traditional state power. This has led to a bifurcation of the world system into the 'state-centred' world, where the USA, Japan, the EU and the USSR are dominant (the thesis was written prior to the collapse of communism), and the 'multi-centric' world comprised of interactions between a diverse range of groups at various scales cutting across national boundaries, striving for autonomy from the state. Although attacked, and possibly eroded, the state remains a powerful agent in this conceptualization however.

Other authors who have written supporting the continued relevance of the state from various viewpoints include Gilpin (1987), Robertson (1992) and Giddens (1985, 1990). Geographers have also entered into these debates. Dicken (2004) sees states performing three important roles with respect to the globalization of the world economy, for example:

- *States as containers* of specific business practices – all economic activity is embedded in broader cultural patterns and processes.
- *States as regulators* of economic activity within and across jurisdictions.

● *States as competitors* where national differences can determine the extent to which nation-states have a competitive advantage.

Numerous geographers have argued that the state continues to play a central role in reproducing the 'mode of production' in the socio-cultural sphere (Peck and Tickell, 2002). Nation-states remain active in the reproduction of cultural values and norms, and perhaps even more so in the era of accelerated globalization as they seek a distinctive place on the global stage. Geographical work on nation-state building, discourses of nationalism (see Gruffudd, 1999) and citizenship (Painter and Philo, 1995) has revealed that nation-states are extremely active in the area of identity formation (Goodwin, 1999). For other geographical perspectives on this debate see Taylor and Flint (2000) and Smith (1994).

The future of the state

By the end of the twentieth century there were more nation-states in existence than ever before in the history of humanity. The remnants of the European empires had disappeared, leaving only very small dependencies such as the Pitcairn Islands, Tokelau and the Cayman Islands scattered across the world. In this sense 'the institution of the nation-state focused on unified sovereignty over a single territory and population had become globalised' (Waters, p. 2001, p. 120). The state system, Waters argues, is no longer Westphalian given that nation-states increasingly surrender sovereignty in military alliances and hegemons have been allowed to arise throughout the century. Does the proliferation of nation-states contradict the political globalization thesis? Decolonization and independence, as well as devolution and decentralization, are part of the *dialectical* process of political globalization which creates new space for the assertion of localized identities, ethnicities and cultures struggling against global forces and traditional national ideologies. It is this process which is partly behind the rise of nationalist movements we have witnessed over the past twenty years as nations (not nation-states) seek their place in the globalizing polity. It is likely that the number of nation-states will increase further, probably along ethnic/cultural lines (as is the case within the context of the European Union: see Box 5.3), leading to a proliferation of less powerful states. In the absence of enlightened global political regulation, this multiplication will make the world more chaotic.

Across existing nation-states we see a drift to the political Centre-right where legislation from the economic to the environmental, from welfare to workers' rights, is made in the light of the demands of neoliberalizing global institutions. The evolving nature of nation-states allows globalizing agents, such as TNCs, to penetrate boundaries far more easily than ever. But, the global governance system does not supersede nation-states – although it changes the way they interact. As Roberts (2002, p. 157) argues:

> trans-state organisations are not extra-state organisations. They do not mark the eclipse of the state. The formal structures of post-1945 world have been creations of states and reflect the unequal power relations between states.

The current global 'mode of regulation' is under deep scrutiny, since it was designed for a very different world. Global regulation from above is being challenged from below as civil society groups increasingly globalize their concern in an attempt to forge a more democratic and equitable system of global governance.

In this context, Waters (2001, p. 117) argues:

> INGOs [international non-governmental organizations] might be regarded as more important in globalization terms than IGOs [international governmental organizations] because they outflank nation-states and threaten borders. They are unruly because their complexity defies command and their capacity to link diverse people in relation to common causes and interests undermines the saliency of the state.

Whether the forces of transnational civil society can reform the deeply ingrained system remains to be seen, especially in the light of pro-neoliberal and pro-military discourses which have proliferated in the USA since 9/11. The importance of such movements, however, is attested to by the fact that many nation-states have been co-opting some aspects of the movement's arguments. There has been a widespread mainstream adoption of environmentalist ideas, some feminist ideas, and other more specific concepts such as the participatory budget model emanating from urban areas of Brazil.

The human geographical view has tended to fall somewhere in between the 'realists' and the 'modernizers'. For example, Smith (2000a, p. 535) argues:

> The process of globalization, in the form of both the internationalisation of capital and the growth of global a regionalised forms of governance,

challenge the ability of the nation-state effectively to practise its claim to a sovereign monopoly over its bounded space and to protect its citizens from external incursion. . . . There is however, general consensus among political scientists that while the powers of the state have been eroded as a consequence, it is a myth to claim that the state has no influence over the impact of such globalising processes.

What we are left with, then, is a multi-polar world, where nearly all states are forced to act in concert in order to pursue their will. Multinational military efforts in Europe, the Middle East and Asia, as well as regionalism (see Box 5.3), are testimony to this. At the same time, global institutions are heavily influenced by the desires of powerful nations and seek to co-opt other nation-states to perpetuate the current mode of production, while transnational civil society seeks to resist, transform or revolutionize this process. We are moving away from a state-centred politics to what Held *et al.*, (1999) call 'multi-layered governance'.

Box 5.3

The European Union

The seemingly paradoxical process of the surrendering of sovereignty by nation-states to transnational institutions combined with the devolution of power and increasing salience of localized politics is best illustrated through the case of the European Union. Some have seen the rise of the EU as evidence that refutes the political globalization thesis (Hirst and Thompson, 1999). Others see it as a product of globalization, providing a model of what a 'global government' might eventually look like (Held, 1991; Held *et al.*, 1999). Regional integration, focusing on economic processes, was discussed in Chapter 4. The EU has gone beyond the economic, and represents a political, social and cultural phenomenon however.

Attempts at European integration rose out of the ashes of the Second World War and the desire among Western European nations to prevent such a war from occurring again. In 1951, the European Coal and Steel Community (ECSC) was founded by Belgium, West Germany, France, Italy, Luxembourg and the Netherlands. Focusing on the industries that were crucial to the production of arms, it was intended that the unilateral rearmament of Germany and the escalation of arms production in other nation-states should be discouraged. Buoyed up by the success of this institution, the six members went on to form the European Economic Community (EEC), enshrined in the Treaty of Rome of 1957 and brought into effect in January 1958.

Box 5.3
continued

This created a common industrial and agricultural market and laid the ground for economic harmonization in other sectors. At the same time, EURATOM was established, focusing on the research and development of nuclear power. These various bodies functioned as separate institutions but were overseen by a European Parliament. In 1966, the Common Agricultural Policy was introduced, which has proven to be the most controversial, and yet binding, of the agreements in force to date.

In 1967 the three institutions – the ECSC, EURATOM and the EEC – were combined to form the European Community. Other European nation-states considered joining at this point, notably the UK, which had two applications (1961, 1967) turned down under the veto of French President de Gaulle. Following the latter's retirement, Denmark, Ireland and the UK joined in 1973. In the early 1980s Greece (1981), Spain (1986) and Portugal (1986) joined the EC. In 1986, the *Single European Act* introduced a rule of majority decisions and thus greatly enhanced the powers of the Council of Ministers and the Parliament. It also introduced the goal of removing all barriers to trade by 1992. The Maastrict Treaty, which came into force in November 1991, was the most far-reaching agreement hitherto produced. It confirmed the agenda for the removal of trade barriers and set in motion the establishment of a European Central Bank, a single currency and a range of social regulations. The Maastrict Treaty established the *European Union*, which was joined in 1995 by Austria, Finland and Sweden, augmenting the total number of member states to fifteen at this point (see Map 5.2). On 1 January 2002 the *Euro* single currency was introduced in twelve of the fifteen member states of the EU. Some, such as the UK and Denmark, opted out of this system with the possibility of joining at a later date. The largest single enlargement of the EU took place On 1 May 2004 when ten further nation-states, mostly from Central and Eastern Europe, joined. Currently, Turkey and Russia are in discussions for further enlargement.

Both the reality and the concept of the EU have been enormously influential in domestic politics across Europe. The desire to join, or not, and issues surrounding the integration of social and economic policy have been influential in national parliamentary elections and shaped political parties. In the late 1980s, the integrationist aspirations of France (under Mitterand) and Germany (under Kohl) clashed with the more isolationist stance of the UK's Margaret Thatcher who argued aggressively against the perceived loss of sovereignty. In general, the UK's Conservative Party has traditionally been 'euro-sceptic' (despite the fact that a Conservative Prime Minister took the UK into the community), while the Labour Party is broadly supportive – although both have recently hit the middle ground. Blair's New Labour has promised a referendum on joining the single currency for two successive general elections but has not yet delivered, since defeat would bring considerable embarrassment. Euro-scepticism in the UK is not just an economic or political issue; it is also a cultural one. The recent formation of

Box 5.3
continued

The European Union
Year of accession

■ 1957
■ 1973
□ 1981
▨ 1986
▨ 1995
□ 2004

Map 5.2 *Membership of the EU over time*

the UK Independence Party, whose main policy platform is to retain Sterling, has struck something of a chord with parts of middle-class England who see the potential loss of the pound as a form of identity erosion.

Across Europe in the late 1990s there was a general retreat into euro-scepticism, both at the level of (some) national governments and the general electorate. The increased enlargement of the EU, and the issue of immigration from new member states, became political 'hot potatoes'. Some argued that this relative disinterest was due to the fact that the majority of the population had no direct recollection of the Second World War, as well as the end of the Cold War, which removed the external

Box 5.3
continued

'threat' posed by communism. The fact that the recent turnout for European Parliament elections was very low signifies a continued lack of concern, or possibly exasperation, with the EU among the general population. As nation-states have increasingly surrendered sovereignty to the EU (albeit with a host of opt-out and veto clauses), we have seen a rise in local autonomy and independence movements across the Continent. These processes are linked, as the new EU creates space for the assertion of such aspirations given the erosion of state power that formerly thwarted such desires. In the UK, political power was devolved in the 1990s to Wales, Scotland and Northern Ireland, where Assemblies and Parliaments have been variously established. In Spain, Catalonia and Galicia have achieved significant autonomy (while the Basque Country has not, largely in response to the violent tactics of the armed independence-seeking faction ETA). There are autonomy movements in parts of France, Italy and Germany. In this sense the evolution of the EU appears to confirm the dialectic process of centralization and devolution that is unleashed by globalization. In some ways, the EU is a laboratory for the potential impacts of globalized politics.

Geopolitics and world orders

Geopolitics is 'a long-established area of geographical enquiry which considers space to be important in understanding the constitution of international relations' (Smith, 2000b, p. 309). Some might argue that in the era of globalization focusing on international relations in this way is not compelling. The following discussion of the evolution of world orders contradicts that view however. World orders, or geopolitical worlds, as O'Tuathail (2002) calls them, 'come together after a general war and subsequently develop until a new crisis or war changes the order of power' (p. 177). During any given order the 'distribution of power and configuration of alliances remains relatively constant' (p. 177). The evolution of geopolitics and 'world orders' from 1945 onwards illustrates the increasingly important role of globalization.

The Cold War

Following the end of the Second World War many former European colonies gained independence (see Chapter 3). The overstretch of empires, combined with the rise of the USA, led to a watershed in global geopolitics which eventually led to the Cold War. The USA emerged as the richest and most powerful nation on Earth, not least because it had been relatively undamaged during the Second World War. This led to a process of economic expansionism which replaced the political expansionism of the former colonial powers. The USA's post-war foreign policy was squarely aimed at maintaining its dominant position. As Rennie-Short (1993, p. 38) argues:

> The USA has a key role in the world economy and is intimately associated with overseas economies, which it needs for supplies of raw materials, as sites for capital investment and as markets for agricultural and industrial produce. Foreign policy objectives are primarily aimed at maintaining and servicing these links.

In the immediate post-war years the evolution of a bipolar world, dominated by the USA and the USSR, emerged. The USA attempted, largely through the Bretton Woods institutions, to promote capitalism, wishing to take a leading role in the ideological formation of the post-war order. The USSR, a communist state since 1917, had different ideas, forbidding countries under its influence to accept the Marshall Plan (see discussion below). It occupied much of Eastern Europe and installed satellite polities there, seeking *buffer-states* to provide resources for a recovery from its devastating war. The West interpreted this as an aggressive move, summed up by Churchill's argument of 1946 that an 'iron curtain' had descended across the continent. Soon the Western powers argued that communism should be arrested lest it knock over successive countries like 'dominoes'. The resultant Cold War proceeded through a number of phases (Rennie-Short, 1993).

In the *First Cold War* (1947 to 1964) US policy was organized around two pillars – both of which were intended to promote capitalism, contain communism, and thus promote US interests. The first was the *Truman Doctrine* developed in 1947, which saw the USA effectively appoint itself as the global 'policeman' of freedom, modernization and stability. The speech that launched the doctrine also saw the introduction of the concept of 'development' (see Chapter 7). The doctrine resulted in the development of alliances for the purposes of security – the three most important being NATO (Western Europe), SEATO (Southeast Asia) and CENTO (Central Asia). The policy of the containment of the USSR was explicitly pursued, resulting most explicitly in wars in Korea (1950 to 1953) and Vietnam (1964 to 1975). Smaller scale interventions – military, political and economic – also took place during this period. These included many interventions in Latin America and the Middle East based on the 'domino theory' (see Map 5.3). As Rennie-Short argues,

Map 5.3 *US interventions in twentieth-century Latin America*

'The recorded interventions are only the tip of the iceberg. Submerged beneath the waves of secrecy lie the endeavours of the CIA to bolster friendly governments, topple unfriendly ones, get rid of troublesome politicians and advance the cause of US interests' (1993, p. 42). Subsequent presidents including Eisenhower, Kennedy, Johnson and Nixon echoed the Truman Doctrine. Notably, President Carter said in 1980 that 'an attempt by any outside force to gain control of the Persian Gulf region will be regarded as an assault on the vital interests of the United States. It will be repelled by the use of any means necessary, including force' (quoted in Rennie-Short, 1993, p. 43). Indeed, the Truman Doctrine set the tone for US foreign policy until the fall of communism, was absent during the 1990s, and resurfaced – arguably – following 9/11, although the enemy was no longer communism. Its spirit was especially visible in the subsequent invasions of Afghanistan (2001) and Iraq (2003), and George W. Bush's articulation of the 'axis of evil' in 2002, which included Iran, Iraq and North Korea (see below).

The second major pillar of US foreign policy was the *Marshall Plan* which saw the USA invest nearly $17 billion dollars of aid in Western Europe's post-war reconstruction. A similar approach was taken in East Asia, although not as extensively, under the Colombo Plan. These aid regimes were explicitly intended to bolster capitalism and prevent the spread of the communist 'threat'. Meanwhile, the USSR had set up the Communist Information Bureau in 1947, COMECON in 1949 (Council for Mutual Economic Assistance) supporting communist governments in satellite states, and the Warsaw Pact in 1955 (the equivalent of NATO). Although Stalin's successor, Khrushchev, was more intent on co-existence with the West, and more tolerant of different forms of socialism in the periphery, this period was still marked by irreconcilable conflict.

In the *Period of Détente* (1964 to the late 1970s) there was a heightened realization by the superpowers that the Cold War was *enormously* risky. This was epitomized by the Cuban Missile crisis of 1962, which almost led to the outbreak of nuclear war due to superpower conflict over the issue of Fidel Castro's Cuban revolution. In 1963, however, a treaty of non-nuclear proliferation was signed. The military-industrial complex in both countries had vested interests and remained a powerful lobby group, stimulating further clashes, most notably in Vietnam. Most US interventions during this period were covert (e.g. Chile in 1973) or economic in nature. The investment role of the USA in East Asia during the late 1960s to late 1970s underpinned subsequently spectacular economic growth there. Both countries increased their global scope and drew in more countries, especially from what was then termed the Third World, into their influence.

During the *Second Cold War* (late 1970s to mid-1980s), relations between the two superpowers again deteriorated. Tension in the US administration between more 'hawkish' elements and those who wished to reduce the USA's role in global affairs was high. Meanwhile the buildup of the USSR's arms was seen as a threat to the USA. With the election of Ronald Reagan in 1980 there was a return to the more explicitly anti-communist rhetoric of the post-Second World War years. This resulted in direct interventions in Grenada and Libya, and covert activity in Nicaragua (the Iran-contra affair), for example. In the USSR, the military-industrial complex played an increasing role in determining policy there. People began to talk of MAD geopolitics (mutually assured destruction) as an unprecedented arms race took off. The level of hysteria in some Western countries was tangible: in the UK in the 1980s the prospect of nuclear war and the resultant winter were images which hung heavy over the population. In the popular media Hollywood films, such as *Rambo*, portraying the 'commies' as evil and non-human, were propagated. It was a period of intense and dangerous propaganda, and genuine fear.

The bipolar world order was destroyed by the fall of the USSR in 1989. By the late 1980s, partly because of the costly arms race, the USSR was devastated by economic decline, political dissent and tension. Gorbachev's *perestroika* (restructuring) led to the undoing of the union as centrifugal forces resulted in independence movements across it (Bradshaw, 2004a). The USA emerged as *the* global superpower, and the world order, it seemed, shifted from a focus on geopolitical to economic forces.

Post-Cold War – the rise of liberal democracy

The 1990s saw the dispersion of policies of neoliberalism and democratic governance across the former Third World, to former communist satellite states, and to those formerly subsumed by the USSR itself. It seemed the West had won. It is no coincidence that this was the time of greatest rhetoric with respect to the liberating potential of the processes of globalization. Capitalism and liberal democracy, hyperglobalists argued, had succeeded in removing the threat of global conflict. As such, globalization at this point was used extensively as a *discourse* to provide rationale for the extension of neoliberal democracy, but it also intensified as a *process* as an increasing array of economies, especially those in Africa, Latin America and Asia, were opened up to the world's markets.

Democratization is 'political change moving in a direction from less accountable to more accountable government' (Potter, 2000, p. 368). The problem with measuring democratization is that it is subject to competing definitions. The aid and loan powers certainly have fairly fixed ideas with respect to what democratization should look like. For some commentators the enthusiasm with which Western governments have embraced the democratization of the rest of the world has more to do with its complementarity with free-market economics than with any *moral* superiority. 'Good governance', a mixture of neoliberal economics and liberal democracy, formed the backbone of the so-called 'Washington Consensus', which provided an organizing framework for the post-Cold War order.

There has been a major shift towards democratic governance over the recent past. In particular, Latin America, Sub-Saharan Africa and the former USSR have moved towards electoral politics and away from dictatorships. The most explicit outlining of this shift is found in the work of writers such as Fukuyama (1992) and Huntington (1991), who herald the arrival of a 'new global political culture' based on liberal democracy founded in the globalization of Enlightenment values. As discussed previously, the shift towards liberal democracy began in the seventeenth and eighteenth centuries in Western Europe and North America, and intensified through waves of decolonization following the Second World War with the explicit support of the USA and its allies. However, both fascist and socialist dictatorships persisted, the latter supported in many cases by the USSR or China. In the 1970s fascist dictatorships fell in Spain, Portugal, Greece and Turkey, and in the 1980s there was a widespread transition to democracy in Latin America and some parts of Asia. The end of the Cold War led to a great acceleration in liberal democracy across the world. The vigour for democratic governance is illustrated by the proliferation of secessionist movements across the former Third World – for example, in Indonesia, Morocco, Sudan and Ethiopia, and also in the West (see Box 5.3). The rise in the number of liberal democratic states is shown in Figure 5.2.

Does this mean that the globe has become more democratic for ordinary people? There are many interpretations of 'democracy', and this is

Figure 5.2 *Democratic nation-states over time 1790–1999*
Source: Waters (2001, p. 155)

greatly complicated by the fact that political impacts on people's lives now come from above and below the scale of the state (see above). There are broad swathes of global society that are not democratic by *any* measure – for example, China, and some Islamic states in the Middle East, such as Saudi Arabia. However, what is for sure is that, as an ostensible ideal, liberal democracy has become the goal for the majority of states across the world. Fukuyama (1992, p. 45) ascribes this shift to the evolution of post-materialist values and neoliberal economics. He is triumphant in terms of his prognosis:

> What is emerging victorious . . . is not so much liberal practice, as the liberal idea. That is to say, for a very large part of the world, there is now no ideology with pretensions to universality that is in a position to challenge liberal democracy.

The rise of democracy is not irreversible however, and rests on the tacit and real support of international institutions. There are regions, however, which are going in the other direction (for example, the Pacific Islands).

Were the USA, its allies and linked institutions to drop the favour
which they accord democracy and 'good governance' then a major
global shift would be likely. The era of 'good governance', it may be
argued, has come at a time when people's real choice is more constrained
than ever. For example, how do we measure the gradual erosion of
power away from nation-states and local populations towards TNCs
and supra-national institutions that back big business?

The war on terror – new world order?

The war on terror instigated by the first George W. Bush administration
(2000 to 2004) following the attacks of 9/11 is, arguably, shaping a new
global geopolitical order. The USA and its major allies, including the
UK, Spain, Pakistan and Australia, have undertaken both military and
'diplomatic' offensives in countries including Afghanistan, Iraq, Iran,
Yemen, the Philippines, Indonesia, Syria, Sudan and Lebanon, in an
attempt to neutralize the terrorist networks that were behind 9/11.
Intelligence sharing at the nation-state level has also formed a central
part of this 'war'. The principal target is al-Qaeda, an alliance of radical
Islamic groups that have used terrorist and military tactics for over two
decades to defend what they perceive as the oppression of Muslims. The
group was formed in 1988 by Osama Bin Laden in order to expand
the *mujahideen* resistance to the Soviet occupation in Afghanistan.
Despite US backing for this movement in the context of the Cold
War, the USA and its Western allies (including Saudi Arabia) became
the group's principal targets in the 1990s, motivated principally by the
West's intervention in the Gulf War of 1990 to 1991 and the ongoing
Israeli–Palestinian conflict. The events of 9/11 represented the climax
of this conflict and unleashed US efforts designed to remove the threat,
although attempts to uproot the network, and to invade Iraq, were
ongoing during the Clinton years and before. The rhetoric of
al-Qaeda is that the West, led by the USA, is waging a new crusade in
the Middle East designed to impose its will and culture, and as such the
conflict has far deeper historical roots. The terrorist networks that have
evolved, partly in *response* to the war, are thought to be truly worldwide
in their extent and represent an amorphous and globalized target for the
USA and its allies. In this sense the war is the first of its kind – fought
against a group that is not spatially bounded and that does not identify
itself with any one nation-state. The International Institute for Strategic
Studies has estimated that al-Qaeda has over 18,000 militants at its

disposal spread across sixty countries. Critics have argued that this greatly overestimates the true extent of the network and that such claims should be seen as part of the USA's wartime propaganda.

The rhetoric of the USA during this 'war' has been 'you are either with us or against us', an approach which seeks to draw new geopolitical lines in ways which echo the Truman Doctrine. Those who clearly demonstrate that they are with the USA can expect grand rewards, as the issuing of reconstruction contracts in Iraq following the 'end' of the war there in 2004 illustrated. Some have argued that the current US administration is using the war to re-inject life into the industrial-military complex and protect fossil fuel reserves in the wake of declining supplies at home (see Box 5.4). This may explain the administration's great leap from Afghanistan into Iraq, citing unsubstantiated terrorist links and weapons threats as part of the rationale. The invasion of Iraq, for example, was predicated on the existence of weapons of mass destruction (WMD) that could be used to attack the West and its allies in the Middle East. In January 2005, the USA finally admitted that there were no such weapons in Iraq. The invasion of Iraq was also represented as an expansion of

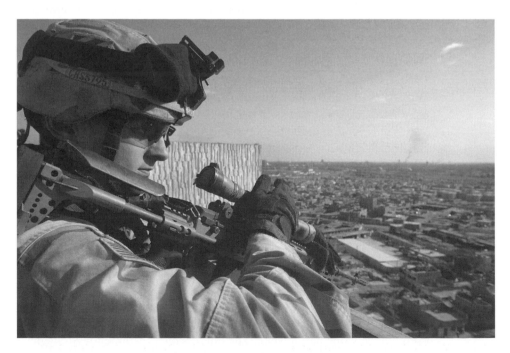

Plate 5.1   *Global war on terror: a US soldier surveys an Iraqi city*

Box 5.4

The geopolitics of oil

The supply of oil plays a major role in global geopolitics and, since the modernist industrial model of development has been diffused, it has become increasingly important. Many recent conflicts, such as the Gulf War following Iraq's invasion of Kuwait (1990 to 1991), or indirect interventions such as the recent US intervention in Venezuela, have an 'oil link' of some description. The West's tolerance of the deeply undemocratic and anti-humanitarian regime in Saudi Arabia is also illustrative of the power oil has. Demand for oil currently stands at unprecedented levels, and rapid economic growth in large 'emerging' economies in South, Southeast and East Asia lies behind some of this increase. In China alone the International Energy Agency estimates that demand will rise from the current 5.5 million barrels per day (mbpd), to 11 mbpd by 2025. The rise in demand in 2004 was the largest single annual rise in history and pushed OPEC (Organization of Petroleum Exporting Countries) to 96 per cent of its capacity. By far the world's largest consumer both in an absolute and per capita sense, however, is the USA, and this consumption is set to rise.

These developments are having a notable impact on the price of oil which rose to $50 per barrel for the first time in history in 2004. Some commentators have argued that this is a temporary blip caused by supply pressures in Russia, uncertainty in Venezuela, and insecurity in the Middle East/Iraq given the recent escalation of conflict there. Others argue that the rise in prices represents a more long-term scarcity-induced trend that is unlikely to be reversed. The IEA estimates that the world will require 120 million barrels per day by 2025 – which is double today's demand.

There is thus a scramble for new supply locations. The largest reserves exist in Saudi Arabia (36 billion tonnes) followed by Iraq (15 billion tonnes). As the wars in Iraq and Afghanistan illustrate, there is a campaign to install Western-aligned liberal democracy in the Middle East, arguably motivated in part by the desire to secure oil supplies. However, the security situation in the region means that the oil companies are turning their attention elsewhere. The USA has become more directly involved in Venezuelan politics and the Mexican sector, for example. There is a major diplomatic offensive underway in the Gulf of Guinea of Africa, which currently supplies 15 per cent of the USA's oil demand. There are ongoing efforts in countries such as Angola, Equatorial Guinea, Nigeria and Chad.

It is in Central Asia however where some of the greatest reserves lie – centred around Kazakhstan, Turkmenistan and Azerbaijan. Control of oil in this region is likely to cause a future stand-off between the USA and Russia, the latter seeing it as a natural area of influence. Post-9/11 the USA has been

Box 5.4
continued

expanding its military presence in the region as well as its economic links – encouraging the formation of a new association without links to Moscow named GUUAM (Georgia, Ukraine, Uzbekistan, Azerbaijan and Moldova). Moscow has sought to increase its influence through the Collective Security Treaty Organization as a counter to this.

These shifts are indicative of the role that energy, and more particularly oil, play in the evolution of geopolitics. In the absence of alternative energy, the continued expansion of the Asian economies and rising demand in the West is likely to result in complex and dangerous geopolitical shifts over the coming decades.

the war on terrorism as it was claimed that Saddam Hussein harboured Islamic militants. Again there is little evidence of this, although it is certainly true that terrorism in Iraq has been inspired by the invasion. The core motivating rationale for the war, radicals argue, is a desire to protect economic prosperity and the spoils of Western-directed globalization (see Mike Moore's film *Fahrenheit 9/11* for a radical exposition of this idea).

Is it possible that we are moving into other defining moments in global geopolitics in the post-Cold War world? The Washington Consensus of neoliberalism and good governance is certainly being eclipsed by simplistic notions of 'with us or against us'. If this were not the case, why has the West increasingly turned a blind eye to the most illiberal and undemocratic politics of Pakistan, Saudi Arabia and Indonesia, for example? What is clear is that, as always, the lines around which imagined global geopolitical divisions are drawn are contingent on the desires and needs of those who are most powerful.

What is for sure is that the broader war on terror has already been used as a means to promote the diffusion of neoliberalism. At a recent Pacific Island Forum meeting in 2003, Australia cited the war as the rationale for an increased Australian security presence in the region and for the further extension of neoliberal economic reform – which it is argued would reduce the threat of terrorism by increasing prosperity (Murray and Storey, 2003). The challenge for the USA, as the hegemonic global force, is that it must have legitimacy in what it does – lest it fall prey to accusations of isolationism and self-interest. That the will of the United Nations was flouted in assembling the invasion of Iraq, that WMDs were not found, and that some terrorist suspects have been held without trial in the US base Guantanamo Bay for over three years, has spurred on the global resistance movement which seeks to counter the unregulated and illegitimate actions of the US-led coalition (see also discussion below). As it stands, however, it is probably most accurate to see the war on terror as an expansion of the post-Cold War order propagated by the world's only superpower in its efforts to spread neoliberalism and the political models that best support this.

New social movements and anti-globalization

The past three decades have witnessed the evolution of many *new social movements* including, for example, feminism (see Box 5.5), environmentalism (see Chapter 8), gay rights (see Box 6.3), indigenous rights and anti-racism. Geographers have argued that global economic networks have disembedded people from their cultures, histories and places, and civil society is reacting against the perceived loss of control over collective and individual destinies (Routledge, 1999). Others have linked the rise of transnational new social movements to the evolution of post-materialist society and associated dissatisfaction with the nation-state and its inability to cope with a multifaceted society (Box 5.2). There is no single theory of new social movements but an analysis of the literature suggests that they must contain the following to warrant that name (based on McCarthy, 2000; Mercier, 2003):

- A collectivity of people
- A level of organization
- A shared ideology
- Cross-class and cross-societal linkages
- Self-identification to that collectivity
- An alternative or transformative social vision
- A variety of conventional and unconventional tactics.

Whether any new social movements fulfil all of these characteristics is open to debate, and the distinction between 'old' and 'new' movements is deeply contested. These movements are considered *new* because they appeal to values which are cross-societal rather than focusing their struggle in the arena of the nation-state, as do traditional social movements such as organized labour (see Box 5.6). Furthermore, the techniques employed by such groups differ from those of the past and are focused away from traditional channels of representation such as established political parties. These are the forces that are seeking to regulate and resist the current mode of production from below (see above). People are losing faith, it seems, in formal politics.

The anti-globalization 'movement' – and it is debateable whether it should be termed a movement at all – is now a major phenomenon in world politics which has come to the fore over the past decade and peaked very recently. We explore the roots, evolution, nature and impact of this phenomenon below and ask whether it represents a new, or possibly even *newer* social movement.

Box 5.5

Feminism, geography and globalization

Feminism is one of the most important examples of a new social movement. An amorphous and contested concept, it has been defined as:

> [A]n awareness of women's oppression on domestic, social, economic and political levels, accompanied by a willingness to struggle against such oppression.
> (Wieringa, 1995, p. 3)

There are, however, multiple feminisms which vary, among other things, according to the particularity of place and moment in history (McDowell, 1999). Is it feasible to argue that there has been a globalization of feminism and that this represents a new social movement? Women's movements actually have an extensive history, with formal activism dating back to the suffragette movement in the UK in the 1800s. It was subsequent to the 1970s, however, that a *global feminist struggle* was articulated. This perspective crystallized in the 1980s with the publication of the book *Sisterhood is Global* (Morgan, 1984). A sense of commonality evolved in terms of analysing and acting against patriarchy across the planet. Arguably, the rise of

globalization has provided a space for all women to act from within, while the nation-state had proven adept at excluding and marginalizing women. Even in those democracies where female suffrage had existed for some time, the participation of women in politics was much lower than that of men. It is this reaction to the state which lies behind the formation of the various local, national and transnational NGOs that comprise the broader feminist movement.

The UN, with its framework of universal human rights, has been a particularly important vehicle for the above aspirations. In 1975 it declared the Decade for Women and held a number of conferences subsequently in Mexico (1975), Copenhagen (1980), Nairobi (1985) and Beijing (1995). The latter conference was instructive in terms of the divisions that have evolved in this supposed 'global' movement, as representatives were often explicit about their perspectives on women's issues from their own national/cultural perspective. The globalization of feminism was seen, by some, as a reflection of the agendas of middle-class feminists from Western countries. The result of this

Box 5.5
continued

meeting, however, was the *Platform for Action* signed by 189 delegates. Thus culturally relativist arguments, which, for example, might allow or deem necessary acts such as polygamy, purdah and genital mutilation in some societies, were rejected in favour of universal values. What was especially notable about this conference also was the high representation from NGOs (over 4000), far outnumbering the nation-state representatives.

While globalization creates opportunities for the promotion of universalized women's rights, it also presents threats to the condition of women. Feminist geography is engaging increasingly with the nature and impact of globalization. At the broad level globalization has a differential impact on women and men according to a host of contingencies in any given case. This work has included more traditional feminist analysis with respect to gender inequality and oppression of women. Christopherson (2002) argues that globalization has had the effect of differentiating women in new ways and that 'class has arguably increased in importance as the variable determining women's economic position' (p. 246). Further research has revealed the continued ignorance of the value-adding and social reproductive role of women in capitalism and the 'hidden ingredient' that is female labour in many new flexible industries and global commodity flows (see Barrientos *et al.*, 1999). Christopherson (2002, pp. 236–237) argues for example that:

> [W]hile women continue to be disadvantaged relative to men with respect to wages, total work hours, and access to societal resources, they are playing a critical role in the processes structuring a global economy. Their continued invisibility within these processes is a result of the measures we use to gauge economic and social contributions; of ideologies that marginalise women as either peripheral to, or victims of, economic change; and of the personal nature of many women's responses to exploitation.

Women's relative position is also greatly impacted by their location in regions that have been restructured in the neoliberal era. African and Latin American women have suffered greatly through the imposition of structural adjustment policies, especially in increasingly prevalent female-headed households (Chant, 1999).

There has also been a shift to adopt feminist *perspectives* (epistemologies and methods) on globalization that seek to complement positivistic empirical studies (Nagar *et al.*, 2002). A more reflexive perspective built on feminist epistemology has been called for, as well as pleas for the mainstreaming of feminist perspectives on global change, which are 'missing a key explanatory variable' (Christopherson, 2002, p. 247). Partly because feminism is itself broad and diverse there has been no single feminist response to globalization. The aim of building a global grassroots feminism which delivers gender equality and strengthens civil society is a core aim of the movement however.

Box 5.6

Globalization and the labour movement

Herod (2002) argues that there are two tendencies associated with globalization that are changing the conditions under which labour operates. First, the increased mobility of TNCs, given developments in transport and communications technology, means that any worker, or group of workers, has to compete with counterparts in other areas of the world. This has led to an increasing 'race to the bottom' as employers seek to push down wages to remain competitive at the global scale. Second, the end of the Cold War and the triumph of neoliberalism have led to a rewriting of the rules with respect to organized labour. In particular, the deregulation of labour markets across the world has seen labour unions undermined. In the West, regulatory changes that weigh against collective bargaining were brought into force with the surge to the political Right in the 1980s. In Thatcher's Britain and Reagan's USA, for example, there were concerted efforts to 'break the unions' in the supposed search for efficiency. The EU is unique in the current era in that it has regulated to encourage international labour interaction. Under the Maastricht Treaty of 1993 each company employing over 1000 people has to meet the costs of bringing labour representatives together on an annual basis where they can share information with counterparts. This grants rights to workers in around 1500 TNCs across the continent (Dicken, 2003). In the Third World, labour is often organized weakly

or regulated heavily by the state, and the shift to neoliberalism has compounded this trend. During the Latin American dictatorships of the 1970s and 1980s, for example, unions were outlawed in many countries, as they were seen as a source of economic inefficiency which did not sit well with the free-market model, and were considered hotbeds of radicalism in the context of the Cold War.

The rise of TNCs (see Chapter 4) presents special problems for the labour movement. Labour unions are traditionally nation-state-based, although there were some examples of international solidarity movements in the nineteenth and early twentieth centuries (see Herod, 2002). TNCs by their very nature are not constrained by borders. Thus organized labour in host countries is often concerned that decisions made by TNCs are 'distant' and do not take into account the specificity of location. Although it is the case that the level of centralization of TNCs varies greatly, there has generally been a tendency to shift important decisions that affect labour to corporate headquarters. Given this trend, it has been challenging for labour to organize in order to face policies such as branch plant closures and lay-offs. Organized labour has had to transform itself rapidly in order to face the demands of globalized business. Two developments are notable in this respect. The first is the increased action of global networks of organized labour, such as the International Confederation of Free Trade Unions. In

Box 5.6
continued

some cases networks of unions within single large TNCs have evolved. Furthermore, international labour movements from the past, such as the various International Trade Secretariats (ITS) developed in the late nineteenth century (e.g. in mining, transport and metalwork) have been revitalized. The goal of labour solidarity in this case has been to prevent TNCs from playing off groups of workers in one country against those in another, a process referred to a 'whipsawing'. Herod (2002) provides a number of examples of successful international cooperation of this nature. For instance, in 1990, metalworkers in a plant in Virginia were locked out in a dispute with the Swiss owners over work rules. The United Steelworkers of America networked with the International Metalworkers Federation (the industry's ITS) and evolved contacts with unions across the world where the same company had operations. Protests over the sacking of the Virginia workers took place in twenty-eight countries, and they were rehired. A second transformation in organized labour has seen it foster the technologies of globalization to conduct its operations. Internet technology, for example, enables the diffusion of information about strikes and other protests at rapid speed between distant territories. Thus, among other things, unions have developed web pages, chatroom and email listserves to divulge their message (Herod, 2002). The Association of University Staff of New Zealand receives regular email bulletins from its counterpart in the UK, for example, allowing a sharing of information and potential responses to labour issues. The way unions structure their business has also altered due to technological change; paid 'organisers', whose job was previously to develop contacts, have been replaced by less formal 'networkers' whose goal it is to promote 'network activism'. Herod (2002) concludes that 'through these activities unions are not just responding to processes taking place at the global scale but are also actively shaping those processes' (p. 87).

The anti-globalization movement

The anti-globalization movement (AGM) is a nebulous term used to describe a wide range of protest, lobby and interest groups. Although the movement broadcasts its message through a variety of channels it is the proliferation of street protests, often planned to take place at the same time as important capitalist summits or events, that has brought the AGM to the attention of the world. Actions have been staged in places as diverse as Seattle, Genoa, London, Hyderbad and Wellington. Some see this 'globalized resistance' as a unified reaction/resistance to transnational regulatory processes from above (Routledge, 1999; N. Smith, 2000; Spybey, 1996). For others, the range of the AGM is so broad and diffuse as to render the terms 'movement' and 'anti-globalization' inappropriate. The news media, which coined the term, jumped on the growing tide of protest against global capitalism and for the sake of simplicity gave it an all-embracing name (see Cartoon 5.1). In the *New York Times*, Friedman (1 December 1999) was particularly scathing with respect to its diversity, calling the WTO protests in Seattle 'a Noah's ark of flat-earth advocates, protectionist trade unions and yuppies looking for the 1960s fix' (Quoted in Mercier, 2003, p. 35). It is certainly true that the protests which have taken place have involved the participation of a vast range of groups including environmentalists, peasant farmers, anarchists, feminists, consumers, unionists and workers – to name but a few.

Cartoon 5.1 *Tell us your complex message in a thirty-second soundbite*
Source: Kirk Anderson

Pieterse (2001) argues that the AGM has crystallized around a number of linked concerns including the inherent instability of the post-Fordist capitalist system; the cycles of crises this has tended to create; increased global inequality since the dawn of the era of development; the perception that the USA is seeking to control the cultures and economies of the world; and, the environmental degradation associated with the increased industrialization and modernization of the planet. These concerns, of course, existed before the dawn of the 'anti-globalization' movement, and indeed before the term 'globalization' itself appeared. However, globalization provides a 'catch-all' set of elements against which protest can be waged in a collaborative form. At its root, the AGM articulates a geographical concern – that the power of locality and individuals within localities has been usurped by forces which exist at 'higher scales' (Hertz, 2001). At the core of the AGM is the conviction that neoliberal policies are further marginalizing the already impoverished and disempowered, and that the globalization agenda is actually *designed* to exacerbate the concentration of wealth (Klein, 2001).

The evolution of the anti-globalization movement

As noted previously, resistance to neoliberal policies and corporate capitalism is hardly new. It has been continuing in various forms since the creation of industrial capitalism itself. More recently, across Latin America, in particular in the 1980s, there were widespread protests against the structural adjustment policies of the World Bank and IMF which imposed neoliberalism on the nation-states of the continent. But protest was not just evident in the Third World; in countries such as the USA, UK and New Zealand protests against Reaganomics, Thatcherism and Rogernomics respectively were mounted throughout the 1980s. Routledge (1999) argues that the first example of a 'global' protest against neoliberalism came in 1994, following the actions of the Zapatista National Liberation Army (*Ejército Zapatista Liberacion Nacional*) in Chiapas, Mexico (see Box 5.7). This group, then, is often credited as being the spark of the current AGM. In reality, there were a number of large-scale protests and other actions that took place in the years preceding the uprising; some of these are outlined in Box 5.8.

After the second Zapatista *encuentro* (see Box 5.7), around fifty NGOs from across the world created a network called People's Global Action Against Free Trade (PGA) with the intention of establishing coordinated protests. The first 'global days of action' occurred in late May 1998, to coincide with the WTO's second ministerial conference in Geneva. Protests took place in over twenty-eight countries. In Hyderbad, India, 500,000 people took to the streets with the rallying cry; 'we the people of India declare the WTO our brutal enemy.' These groups also linked with movements from the West such as 'Reclaim the Streets' from the UK. By 1999, the movement had gained significant momentum and was beginning to identify a 'common enemy'. This concluded in a meeting at

Box 5.7

The Zapatistas

On 1 January 1994, when the NAFTA came into force (see Chapter 4), about 3000 lightly armed men and women took control of the main towns in Chiapas – a poor southern Mexican state. Most were peasants who had lost their land to large landowners who prospered in the restructured Latin American economy of the 1990s. Economic liberalization badly damaged local economies as protectionism for coffee, forestry and cattle were removed. People were killed in the ensuing struggle, a factor leading to widespread sympathy for the Zapatistas both at home and abroad. A ceasefire was announced in late January and negotiations over land reform and social change were initiated.

The Zapatistas were protesting for democratic reform to fight perceived domination by political elites and foreign TNCs. They explicitly pinpointed neoliberalism as the main adversary. In 1996 and 1997 they organized two 'encuentros' (meetings or conferences) 'for humanity against neoliberalism', bringing together a wide range of social movements from around the world – especially the Third World. The intention was to discuss the impacts of neoliberalism and how it might best be resisted. A major declaration was communicated as a result. Some credit this as the founding document of the anti-globalization movement. What is especially notable is how the Zapatistas used globalizing technology as a means of spreading their message across the globe, appealing to social networks with common interests (see www.ezln.org).

the 'Other' Davos by five large protest groups, running in parallel with the World Economic Forum in Davos, Switzerland. On 18 June 1999, the day of the G8 economic summit in Germany, PGA co-ordinated a day of protest against financial centres called 'Carnival against Capital' and action was taken in over 100 cities worldwide. On 19 August 1999 PGA met in Bangalore India to coordinate the next global day of action which would coincide with the opening of the 'Millennium Round' of WTO trade talks in Seattle (see Box 5.8).

The Battle of Seattle

The protests in Seattle in late 1999 are seen as a watershed in the evolution of the AGM and new social movements in general. Approximately 60,000 people descended on the city – the home of

Box 5.8

A selected chronology of the pre-Zapatista uprising 'anti-globalization' movement

1982. Eighth World Economic Summit (G7) in *Versailles*. Attack on the Paris representative of the World Bank and International Monetary Fund as well as protests for peace and against Ronald Reagan.

1984. Tenth World Economic Summit (G7) in *London*. An alternative World Economic Summit was staged at the same time under the title of 'Global Challenge: Out of the Crisis, Out of Poverty'.

1985. Eleventh World Economic Summit (G7) in *Bonn*. Several protests, including one of 25,000 participants and a 'Tribunal against the World Economic Summit' held in the town hall of *St Godesberg*.

1986. Twelfth World Economic Summit in *Tokyo*. Critical flyers distributed and rockets fired over the guesthouse of the Japanese government exploded near the Canadian embassy. The rockets did not cause any damage.

1987. Thirteenth World Economic Summit in *Venice*. Several thousand demonstrators protested. Bombs exploded in front of the British and US Embassies in *Rome* on the second day of the negotiations. No one was hurt.

1988. Fourteenth World Economic Summit in *Toronto*. Around 500 representatives of

the international Third World and economic groups participated in the 'Other Economic Summit'. Protests in several German cities later that year (26–29 September) as the International Monetary Fund (IMF) and representatives of the World Bank met in *West Berlin*.

1990. The IMF and World Bank meet in *Washington, DC*. Small demonstration and an alternative forum. March against child labour later that year in New Delhi (29–30 September) to coincide with the UN summit of children in *New York*. GATT World Trade talks in *Brussels* prompt a protest conference entitled 'GATTastrophe'. Demonstration by thousands of farmers in Europe as well as the USA, Japan and South Korea. Protests also took place in Argentina, Japan and Switzerland.

1991. World Economic Summit in *Munich*. Largely peaceful large-scale demonstration with about 17,000 participants. About 500 people were later arrested at the opening of the conference and 6000 anti-globalization demonstrators protested, demanding the resignation of the Interior Minister and Munich Chief of Police.

1992. In *Bonn*, the autumn conference of the World Bank blockaded to protest the construction of dams in the Indian Narmada River, funded by the World Bank. On the occasion of the planned GATT Compromise

Box 5.8
continued

between the EC and USA, 40,000 Farmers demonstrated, arriving primarily from Germany and France but also from Japan, South Korea and Canada.

1993. Mass protests throughout the year against the GATT agreements, particularly of farmers: in *New Delhi* (particularly seed-patent provisions); in *Seoul* (against the opening of the South Korean

rice market); in *Geneva* (including farmers from Switzerland, France, Spain, Japan, India, Canada and the USA); and throughout *Asia*. Leading Indian and Pakistani politicians complained that the USA was not willing to reduce its tax on textiles far enough.

Source: Compiled by
Anton Griffith and author

Microsoft and Starbucks – to take part in what has become known as the 'Battle of Seattle' between 29 November and 2 December. Simultaneously, protests in world cities around the globe, such as London, Paris and Bombay, took place. In Seattle itself, over 700 different groups took part. Steger (2002) argues that the truly transnational character of the participating groups set it apart from earlier anti-globalization protests. What was especially notable was that the protestors had come to the 'hearth' of global capitalism and corporate globalization, and a large contingent was from the USA itself.

Due to the protestors' efforts, the trade talks were postponed. Rioting took place and damage was relatively widespread – although the majority of participants used peaceful methods. However, early disagreements in the AGM with respect to tactics quickly became apparent, as some groups demanded more direct, and sometimes violent, action. Less radical contingents claimed that such tactics would be futile and counterproductive. The US military intervened and declared martial law. The media emphasized the anarchy and chaos of the protest, a trend that has continued in mainstream coverage of such events. Bill Clinton, then President of the USA, decried the method of protest – but acknowledged that the protesters had some worthwhile points to make.

Since Seattle, almost every major international meeting of this type has been targeted and the number of protestors has remained high. Although the exact focus of the protests might vary, the same range of groups would often take part. In 2000, for example, 250,000 people took part in a Paris anti-fascist demonstration in reaction to the rise of the right wing in France; in Sao Paulo, Brazil, 100,000 people demonstrated against neoliberalism in the home of anti-liberal protests; and an incredible

Plate 5.2 *Anti-globalization protests in Genoa, 2001*

7,200,000 took part in a general strike to protest against structural adjustment programmes in Argentina. In 2001, the protests continued with, for example, 80,000 people in Quebec protesting the proposed Free Trade Agreement of the Americas. A notable development during this year was the establishment of the World Social Forum in Porto Alegre, Brazil, involving tens of thousands of attendees from across the political left united, broadly speaking, by their opposition to globalization as currently practised. This conference – which runs in parallel with the meeting of economic leaders at the World Economic Forum held in Davos, Switzerland – was founded by a coalition of trade unions, NGOs and environmental groups from Brazil. Also in 2001, the 'Siege of Genoa' – involving an estimated 300,000 people – resulted in the first death of a protestor. In many ways, the scale and nature of this particular protest represented the pinnacle of the AGM (Bygrave, 2002).

9/11 – where did all the protestors go?

Following 9/11, the mainstream media suggested that the AGM would lose its head of steam. It was argued that the global war on terror (see above) waged by the Bush administration would seek to classify such protestors as 'terrorists' – raising the stakes for participants considerably. The *Guardian* newspaper from the UK asked: 'Where have all the protestors gone?' (Bygrave, 2002). The movement responded with its own question: 'Where have all the journalists gone?' (Taylor, 2002). It appeared that the new tactic of Western governments and allied media was to ignore the protests to the extent that this was possible, denying the global publicity given to the AGM up to that point. For example, the protest of 250,000 in the streets of Barcelona in March 2002 was covered far less extensively than were previous demonstrations on the news networks over which Western governments hold influence (Bygrave, 2002).

In reality, the attacks of 9/11 provided a renewed stimulus for some in the movement as the belief that they came about precisely because of the historical injustices created by the capitalist West became one of the central explanatory discourses adopted by the political left. At this point, however, the movement began to fractionate even more than it had hitherto. The invasion of Afghanistan and the rising possibility of war in Iraq provided new focal points for protest. The AGM partly metamorphasized into an anti-war movement, and this culminated in 350,000 protestors marching on London on 28 September 2002. Many protestors seemed to interpret the protest against a possible war as part of the anti-globalization ethic. Others did not see it this way and a questioning of the coherence of the AGM, which seemed able to reshape itself to any global issue at hand, permeated media discussion. The World Social Forum, which epitomizes the broad range of issues that unites the global left wing, has grown in importance and scale, with over 100,000 people attending the 2004 and 2005 meetings in Mumbai and Porto Alegre respectively. The issues dealt with at the 2005 forum – such as GM crops, the global environment, debt and the continued conflict in the Middle East – illustrated that it had become a general meeting of minds for alternative thinkers. Strictly speaking, however, it is more accurate to see the World Social Forum as part of the evolving alter-globalization movement (see discussion below). It seems unlikely that the fervour surrounding the AGM in the early 2000s will be captured again. To claim dissolution, of course, implies a former unity – and this is controversial, as is discussed below.

Coherent movement or ineffective hybrid?

The term 'anti-globalization movement' was imposed externally, and there are a number of problems raised in using the term. First, the movement, given its inherent diversity in tactics and causes, has no one alternative model which it proposes. Second, the movement is itself facilitated and made possible by the processes of globalization, which is seemingly paradoxical. Third, leading from this last point, it is important to note that many within the movement are not opposed to globalization per se. Hardt and Negri (2001, p. 102) argue that 'the vast majority of the protestors are not against globalising currents and forces as such; they are not isolationist, separatist or even nationalist'.

Alternative titles for the movement have been suggested. Examples include the 'anti-capitalist movement', 'anti-imperialism' and 'anti-neoliberalism', all of which emphasize protest at the expansion of the capitalist nucleus (Petras and Vettmeyer, 2001). Other suggestions have included the 'movement against global corporatism' or the 'anti-corporate globalization movement', which emphasize the opposition to TNCs and supranational bodies (George, 2001; Starr, 2000). Further examples highlight the 'governance' aspects of the movement, and include the 'Civil Society Movement', 'Citizens' Movement for World Democracy' (Bygrave, 2002), the 'New Democracy Movement' (Barlow and Clarke, 2001) and the 'Global Justice Movement' (Lefrancois, 2002, cited in Mercier, 2003). Finally, and most nebulously, the term the 'Alternative Movement' has also been proposed. Naming is more than 'mere' semantics; the title chosen reflects the nature of the adversary, something of the politics of dealing with it, and helps sustain the identity required to operate as a movement. The diversity of the movement as reflected in virtually all the suggested titles here is at once strength and weakness. It captures many of the pertinent issues in today's global society, while at the same time threatening to dissolve the concerns to the point of ineffectiveness.

The AGM, in a broad sense, claims a number of victories. Two of the largest and most publicized were the halting of the construction of the World Bank-funded Narmada dam hydroelectric project in India and the blocking of the Multilateral Agreement on Investment (Brecher et al., 2000). There have been others, but perhaps the main accomplishment has been symbolic and ideological – and thus very difficult to measure in the short term. Globalizing the issues of concern and bringing them to the attention of millions of people is a significant feat, although it could be argued that mounting large-scale media events have caused the institutions of global governance to retreat further into themselves, thereby lowering transparency further. However, despite the fact that much of the media tried, at first, to portray the movement as violent and anarchistic, and later ignored it, the AGM has undoubtedly captured the public's imagination. There have even been a number of 'defections' by former insiders of neoliberal institutions (see Box 5.9).

Box 5.9

Defections from the inside

One factor which makes the anti-globalization movement so hard to pin down is that former 'insiders' have joined its ranks. While such people have not necessarily abandoned their adherence to market principles they have criticized the manner in which prescriptions are made. Two notable examples are reported in Gwynne *et al.*, (2003).

Jeffery Sachs was the World Bank's most prominent consultant in the 1990s. Despite being part of the neoliberal apparatus he reflected critically on the imposition of neoliberal reform on the Third World in 1999:

> The Washington consensus listed 10 or 12 steps – the recipe for economic development. When you look at those, they're all pretty reasonable. But it's a kind of bland list of commandments, rather than a real blueprint of how to get from A to B. . . . It became a substitute for assistance, because the idea was 'you don't need us. You don't need any help. . . . You just have

to follow the magic rules'. The actions of the IMF and World Bank became very stylised . . . at that level of simplicity, it just doesn't work.
> (Sachs, 1999, in Gwynne *et al.*, 2003, p. 5)

Joseph Stiglitz was President Clinton's chief economic adviser between 1993 and 1997 and became the World Bank's chief economist and vice-president from 1997 to 2000. He resigned under protest and was especially critical of the IMF:

> When the IMF decides to assist a country, it dispatches a 'mission' of economists. These economists frequently lack extensive experience of the country; they are more likely to have firsthand knowledge of its five-star hotels than of the villages. . . . Critics accuse the institution of taking the cookie cutter approach to economics, and they're right. Country teams have been known to compose draft reports before visiting.
> (Stiglitz, 2002, in Gwynne *et al.*, 2003, p. 5)

Not surprisingly, the alternative or transformative vision of the AGM is not well defined. It has, at various times, included diverse suggestions such as: debt write-off; the end to unjust war; environmental sustainability; reform of global institutions; and a global welfare system (Mercier, 2003). Through the work of authors such as Aguiton *et al.* (2001) in *The Other Davos*, and Bello (2002) in *Deglobalization*, we are seeing the formation of some concrete policy alternatives, which have at their core a global regulatory system that seeks to address inequality within a sustainable framework. In many ways these suggestions bear much in common with earlier work on policies for 'sustainable development' (see Chapter 7) (see also Hines, 2000; Houtart, 2001).

Alter-globalization from below?

From a theoretical standpoint the problem with the AGM, while powerful in its scope and force, is that it has tended to commit the error of *metanarrative* – ascribing all of the world's woes to the inexorable rise of globalization. In this sense the movement has become over-global and has lost sight of the localized forms of resistance from whence it was born. It is partly due to this that the movement has fragmented, as some seek to distance themselves from what they see as a relatively ineffective marriage of concerns. This has also involved the evolution of the *alter-globalization* movement, which is linked to the idea of *progressive globalization* (see Chapter 9). This perspective argues that it is not globalization per se which is the problem but the way it is currently practised. There is nothing inherently bad about the time–space compression of global society, but it does need to be regulated in order to prevent growing inequality and non-sustainability. These debates hinge crucially on the definition of the term 'globalization'. If we accept that globalization is synonymous with the corporatization of the world then the AGM has a natural enemy (Hardt and Negri, 2000). If we accept, in contrast, that the corporatization outcome is but one possible outcome, then we can reconstruct the concept of globalization in a more progressive way.

This perspective has been especially influential in France where a large alter-globalization contingent gathered at the European Social Forum in September 2003. An editorial in *Le Monde* (*Guardian Weekly*, 20–26 November 2003, p. 33) argued that this group 'which is as heterogeneous as it is dynamic', had succeeded in creating a 'new political dynamic' in the country. In particular, in transcending accusations of being influenced by 'an antiquated form of socialism, they have now acquired a vigour which makes them capable of mobilising seasoned and inexperienced activists alike'. Furthermore, they have distanced themselves from the 'anti-globalizationists' who, according to the newspaper, 'want to barricade the nation-state and return to a policy of state control and protectionism'. However, the article points towards some of the ambiguities which still remain, the main one being the:

> lack of any convincing counter-project, while the alter-globalizationists can muster solid arguments against 'neoliberal globalization' . . . the alternatives they propose are lacking in clarity and coherence. . . .
> From the defence of 'free' software, the proposed Tobin Tax on capital movements and the defence of environment, to the denunciation of brand marketing, the struggle against poverty and the campaign against genetically modified foods, they peddle a host of totally uncoordinated notions.
>
> (*Guardian Weekly*, 20–26 November 2003, p. 33)

Another strand in the evolving alter-globalization movement, based on a more radical stance, is that of 'globalization from below'. Brecher *et al.* (2000, pp. 38–39) identify a number of common goals within this camp including: policies to strengthen the ability of national governments to counteract global economic forces; support for the empowerment of local people to take control of their natural resources; the organization of repressed groups in civil society; and the construction of a transnational social movement of organizations across borders. Mittelman (2000) argues that globalization from below should involve self-determination and economic self-governance by the majority. There are some groups that have been established specifically to promote grassroots globalization, as it is alternatively called, including the *International Forum on Globalization* and *People's Global Action*. Undoubtedly, the crowning achievement of this part of the broader movement to date is the establishment of the *World Social Forum*, which seeks concrete routes to world reform.

Anti-globalization: movement of movements?

Callinicos has likened what he refers to as the 'anticorporate' movement (the AGM) to the new social movements of the 1960s and argues: 'there has not been such a resurgence of activist activity since the Vietnam War' (2001, p. 387). As globalization comprises a complex set of processes, agendas and outcomes, anything defined in opposition to it will be similarly complicated. The common enemy is not as easy to define as for, say, the feminist movement or the anti-nuclear movement. It is perhaps best described as a 'movement of movements' – one which captures the globalization era and its dialectal processes.

Overall, it is uncertain to what extent the anti-globalization movement incorporates the elements thought necessary by commentators to define a new social movement (McCarthy, 2000; Mittelman, 2000). Mercier's primary research on the anti-globalization movement in London illustrated a diverse, fractured and internally incoherent set of movements operating under a banner which many felt had been imposed upon them externally. It may be that current terminology is not able to capture what is happening. It is perhaps time to develop a theory which addresses the rise of 'newer social movements', 'global social movements', or some other term which captures globalized diversity within unity and the constellation of varied, shifting and dynamic interests which characterize the AGM.

Conclusion – a multi-polar world

Globalization is altering the way politics is practised. The nature of the nation-state is being transformed as a result, civil society is evolving rapidly, and governance practised at regional and global scales is flourishing. Political geography needs to move quickly to keep up with these changes. The argument presented in this chapter is that nation-states remain central players in global affairs. However, they will become less powerful as they proliferate due to the devolutionary pressures created by globalization and as their borders are transcended by cross-societal NGOs and NSMs that appeal to universal values. At the same time, global institutions will need to be reformed and strengthened in appropriate ways if they are to deal with the transnational flows of the contemporary world in an effective and equitable way. In this context, the reinvigoration of the United Nations is absolutely essential for global stability.

Geopolitics has evolved rapidly since the Second World War, moving through various 'world orders' around which global politics has been organized. Arguably, the Cold War was replaced by the neoliberal-democratic world order. More recently, and contentiously, the war on terror might be seen as part of an evolving order that confirms the rise of the USA as the global hegemon. This chapter argued, however, that the war on terror is best seen as an extension of the neoliberal-democratic world order, which – despite the overwhelming power of the USA – is in fact multi-polar and multi-layered, and involves the practice of governance at a range of scales. Underpinning the design of the successive 'world orders' and regimes of accumulation has been the capitalist cultural and economic mode of regulation. It is this imperative, and the globalization to which it gives rise, that requires rethinking if political geography is to contribute to a just and sustainable world.

Further reading

Brecher, Costello and Smith (2000): This is an excellent review of the anti-globalization movement in all its complexity.

O'Tuathail (2002): This chapter provides a solid introduction to geopolitics and the concept of world orders, or 'geopolitical worlds', as the author terms them.

Rosenau (1990): This classic book presents some early arguments concerning the impacts of globalization on political governance and the state that remain salient.

Sidaway (2001a, 2001b): The first chapter is another useful introduction to geopolitics and its relationship to geography. The second chapter provides a geographical interpretation of the role and rise of the nation-state.

Taylor and Flint (2000): This is a classic book on political geographies that adopts a World Systems perspective and is thus relatively sceptical of globalization.

Waters (2001) ch. 7: In this chapter the author discusses the rise of the nation-state and the trends that currently threaten it.

6 Globalizing cultural geographies

- Global cultural shifts and geography
- Defining cultures – landscapes, change and interaction
- Historical cultural globalization
- Contemporary cultural globalization
- Cultural consumption – commodification and cosmoplitanism
- The global music industry case study
- Conclusion – towards progressive cultural spaces?

Global cultural shifts and geography

Advances in telecommunications technology in the twentieth century meant that cultural symbols – such as images, music or texts – flowed across the world as never before, and capitalism commodified these symbols and perpetuated their globalization. Cultures became less spatially bounded than they once were, leading to more intense cultural traffic. Some argue that this has given rise to a new 'global consumer society' based on the diffusion of Western culture, building around culturally imbued commodities marketed by companies such as MTV, McDonald's and Coca-Cola. While it is true that such symbols have become increasingly pervasive, the idea that their propagation is forging a homogeneous global culture has been rigorously debated, and this chapter pursues some of these arguments.

Although culture is seen as central in the *postmodern* epoch, in reality it has *always* been central to all forms of political, economic and social change. It is impossible to separate cultural, economic and political processes – especially in the context of globalization. We do this in order to be able to deal with what would otherwise be an overwhelming complexity. There can be little doubt, however, that culture has become more *visible*. This new visibility is intimately linked to the evolution of globalization, as we will explore below.

The cultural turn in social science and human geography

Over the past two decades, much of social science and human geography has been rewritten with a new emphasis on the way culture permeates other spheres. In the 1980s a cultural turn took place in the social sciences in general, influenced by developments within academia and society as a whole. In the academy, the rise of postmodernism, based partly on the humanist tradition of the 1970s, challenged 'grand' theories or metanarratives that sought to reduce social complexity to singular explanations (such as Marxism or World Systems Theory). This involved a resounding critique of positivism – an epistemology which views the world as existing independently of human perceptions of it and reducible to laws derived through empirical analysis. In contrast, postmodernism argues that 'reality' is a social construct and absolute truth is therefore impossible. This general critique gave rise to a range of perspectives on

Plate 6.1 *Postmodern architecture – Te Papa, New Zealand/Aotearoa's national museum, represents an example of postmodern architecture with its juxtaposition of styles, patterns and shapes*

social change that is still unfolding today, including post-structuralism and postcolonialism. These critical perspectives often place 'culture' at the centre of their explanations and/or interpretations of the world. Postmodernism, argue Cloke *et al.*, (1991), is not only a way of thinking; it may also be seen as an 'epoch' or time in world history and a set of associated material outcomes in which culture plays a more visible role than hitherto. Thus postmodernism has been theorized in the economic (e.g. flexible accumulation) and political spheres (e.g. postmaterialist society) (see Chapters 4 and 5 respectively). The cultural turn was as much influenced by the changing world around the academy, and its shift to a postmodern condition, as by theorization from within.

In the late 1980s there was a 'cultural turn' in human geography informed by postmodernism (see Crang and Thrift, 2000; Jackson, 1989; Philo, 1991). This endeavour sought to elucidate the links between culture and space and emphasized the role of culture in social change. The shift involved the widespread use of *interpretive* techniques using qualitative methods, as opposed to empirical techniques (Johnston and Sidaway, 2004). At the core of the transformation lay a critique of the 'quantitative revolution' of the 1960s and 1970s – often by the very people who led that revolution. Geographical models and perspectives that sought to explain the world in terms of economic, political or environmental determinism were challenged – such as industrial location theory or core–periphery models. Geographers dealt with issues such as identity, race, gender, ethnicity, tradition, indigeneity, and the cultural roots of economic and political change in more sophisticated ways than previously (Johnston and Sidaway, 2004; McDowell, 1997). In contrast, traditional cultural geography had often described exotic 'mosaiced' cultures from an anthropological viewpoint. Some new cultural geographers committed the same error of which they accused positivists – seeing culture as deterministic in social change. Notwithstanding this, a new vitality pervaded the discipline. As is characteristic of a healthy discipline, the cultural turn in human geography has not been without its detractors, however (see Box 6.1).

A central question in the study of cultural globalization from a human geographical viewpoint is: In what direction does the heightened cultural traffic referred to in the opening of this chapter flow? Does it move from the global to the local, the local to the global, or is it more complex and multiple? This takes us back to the discussion in Chapter 2 concerning the definition of, and interaction between, scales. The reality may be that the world is so complex that the traffic flows in all directions at the same

Box 6.1

The 'cultural turn' and its relevance in human geography

The 'cultural turn', informed by postmodernism, sent shock waves through human geography that still resonate today. What are its implications for the relevance of human geography? This depends much on how one defines relevance. Two opposing views are represented by Johnston (2000) and Pacione (1999). For Johnston, the turn helps human geography in its quest for relevance. He argues (2000, p. 696) that work in this area:

> is promoting awareness of how all knowledge is produced and situated in particular contexts and then used to promote privileged positions for some forms over others . . . Such work is highly relevant to the creation and maintenance of humane societies.

Pacione (1999, p. 4) is more critical:

> The failure to address real issues would seem to suggest that the

advent of postmodernism in radical scholarship has done little to advance the cause of social justice. Discussion of relevant issues is abstracted into consideration of how particular discourses of power are produced and reproduced. Responsibility for bringing theory to bear on real-world circumstances is largely abdicated in favour of the intellectually sound but morally bankrupt premise that there is no such thing as reality.

To Pacione then, the cultural turn is promoting conservatism and paralysing progressive collective action. New cultural geographers would argue that Pacione dismisses postmodernism because his conception of the term 'relevance' is too narrow however.

time. Human geographers have engaged in the theorization of global cultural change at all levels. Global scale work has focused on 'the movement of people, objects and images around the world through telecommunications, language, the media industries, radio and music, cinema, television and tourism' (McEwan, 2001, p. 157). This work has critically engaged with theses of homogenization, McDonalization and Coca-Colonization, for example. At lower levels there has been particular interest in the reassertion of local cultures in the face of 'global' processes; how such processes are interpreted; and how they are resisted, regulated or reconstituted from below. A great deal of the work in this area has argued that global and local cultural processes interact in many different ways in different places to produce 'structurationized', 'hybridized' or 'glocalized' outcomes. For example, in the UK, CNN has presenters with UK accents, and in Australia, TV actors with Australian accents overdub adverts made in the USA in order to appeal to the local market. Central to new cultural geography work in this area is how globalization is affecting cultural production and consumption, and how this in turn is affecting people's sense of identity, sense of place and cultural politics (Massey, 1994). This chapter seeks to throw light upon simultaneous global, local and glocal processes which make up cultural globalization. We can only understand how cultures change and interact in the current era by first defining what we mean by culture, however.

Defining cultures – landscapes, change and interaction

Defining 'culture' is a difficult task. There are many popular misconceptions of the term. McEwan argues that 'students of geography often imagine studying cultures to be either about looking at exotic peoples in other places, or about high culture, such as the arts' (2001, p. 155). These aspects are certainly part of studying culture, but do not present the full picture. Two definitions of culture are offered below:

> Cultures are maps of meaning through which the world is made intelligible.
>
> (Jackson, 1989, p. 1)

> Cultures are part of everyday life, and are systems of shared meanings that can exist on a number of different spatial scales (local, regional, national, global, among communities, groups or nations). They are embodied in the material and social world, and are dynamic rather than static.
>
> (McEwan, 2001, p. 155)

Culture provides direction, guidance, rules and limits concerning how humans interpret habitats and environments and make decisions on how to exploit resources, including each other. These elements are reproduced through formal and informal means. Cultures are thus constructed by society and are subject to the operation of power relations. There is often tension between different cultures in any one given space or place. Often, a dominant culture may seek to impose meaning of what 'culture' is or should be, and subcultures – some of which arise in resistance to dominant culture – may challenge this. An important contribution in thinking about culture in recent years is to conceptualize it as a *process* rather than as a fixed entity. Culture involves the practising of identities

– be they political, sexual, ethnic or national – around sets of shared beliefs and values.

Cultural traits are individual components of a cultural complex. A group of people that is distinguishable by its possession of a unique assortment of traits may be called a cultural group. Traits are the building blocks of culture and may be divided into three categories:

● Sociological – interaction forms, laws, regulations
● Ideological – religion, politics and non-secular belief systems
● Technological – technical systems and abilities resulting in infrastructures.

In tracing the geographies of the globalization of culture we need to examine the stretching of these traits across space and time through social networks and the resultant impacts on cultural 'landscapes'.

Cultural landscapes

Cultural landscapes were one of the central components of 'traditional' cultural geography, developed largely by Carl Sauer (1925). He argued that 'landscape' – and this meant far more than the 'natural' landscape – reflects the traits of the past and/or currently occupant culture. It was argued that humans as an agent of change are constrained by their environment and culture. Landscapes thus differ greatly according to local geographical conditions, defined in the broadest sense. Material culture consists of the physical objects created by cultures as they carry out their way of life. For example, urban-industrial society creates landscapes of concrete and steel. Material culture reflects the non-material aspects of culture such as religion, and non-secular values. An obvious example would be the building of Christian churches or Muslim mosques; a less obvious example might be the clear felling of forest to make way for agro-industrial development.

Ideas concerning cultural landscapes have shifted of late as cultural globalization has made cultural landscapes much more difficult to discern, and bounded cultural spaces have 'melted into air' (Harvey, 1989). The reality is that even in the supposedly 'mosaiced' past, few entirely isolated cultures existed and the majority have been subject to outside influence of some form. This disrupts notions of 'indigeneity', 'tradition' and 'authenticity'. Isolation is a relative concept of course. Many still expect cultures, such as those in the islands of the South

Pacific, to be relatively untouched, or 'authentic', by virtue of the large absolute distance between populated land. There is no necessary correlation between absolute distance and cultural isolation, however. One of the most physically remote islands on the planet, Tahiti, as part of French Polynesia, is far more culturally connected to the West, and exhibits a cultural landscape constructed accordingly, than many other Pacific Island nations at closer proximity to Western countries such as the Solomon Islands or Papua New Guinea. This arises because of the territory's intimate connections to the colonial power, France. Conversely, there are large cities in densely populated regions such as Tehran (Iran), Baghdad and Kinshasa (Democratic Republic of the Congo) that, for historical reasons, are largely bypassed by globalized cultural flows becoming what Rennie-Short (2004) refers to as 'black holes'. In short, the influence of globalization on cultural landscapes is a function of the connectedness of that landscape to global networks together with the geographical and historical contingencies which have hitherto shaped that place. All cultural landscapes remain unique, but they are more complex and hybridized than they once were. It is these new globalized cultural landscapes, and to a lesser extent the 'black holes' in the new

Plate 6.2   *Cultural landscapes – Kuching, Sarawak, Malaysian Borneo*

networks, that form the basis of much investigation in contemporary cultural geography.

Cultural change and interaction

Before moving on to look at the arguments surrounding the 'globalization of culture' in more detail, we need to understand a little more about concepts of cultural change. Cultures change at different paces across space and time. Both 'internal' and 'external' forces influence them – and there is a range of different 'modes of interaction', as we will see below. Internally, innovations such as the development of new technology or new philosophical understanding can alter cultures. External influences may include colonialism, global marketing and tourism. In reality, there is a blurring between what comes from 'outside' and what comes from 'inside'. The problem with understanding cultural change today is that distinguishing between the 'internal' and the 'external' has become increasingly problematic given the processes of deterritorialization.

Human history may be traced in terms of a series of *cultural revolutions* (e.g. agricultural, industrial, democratic, technological). A cultural revolution occurs when a culture changes dramatically and the roots of this change may be multiple. It may occur subsequent to the invention of a new groundbreaking technology such as the wheel, settled agriculture, the combustion engine or the internet, but deeper historical forces often drive technological innovation. For example, the industrial revolution was inspired in a technological sense by the invention of the combustion engine, which resulted in mechanization and possibilities in terms of the division of labour and economies of scale, all of which facilitated capitalist accumulation. However, the industrial revolution would not have occurred or been sustained without the earlier agricultural revolution and, later, industrial colonialism. Although cultural revolutions, by definition, had shattering consequences at the time, a more important process over the long run is cultural diffusion.

Cultural diffusion is the process of the spreading of cultural traits from one place to another. A locality or region that has been a rich source of traits is known as a cultural hearth. In the past, hearths were bounded spaces; for example, Mesopotamia (the area roughly corresponding to Iraq today) became a cultural hearth from which agriculture diffused. Diffusion arises through various processes of personal and impersonal interaction. In the past, migrations and empires were the most important diffusing conduits, whereas today shared meanings, ideas and cultural norms are diffused through the broader spaces of flows. This leads to the possibility of deterritorialized hearths. Arguably, neoliberal globalization's cultural hearth is the world city network, dominated by the USA and Western Europe.

As cultures move and/or expand, either in real space or cyberspace, they will inevitably interact with others. *Cultural interaction* may be conceptualized in three main ways, which exist on a continuum and are subject to overlap:

- *Assimilation* occurs when a culture changes completely and loses its original traits due to subsumption by a dominant one. The geographical outcome of this process, if it is extended globally, is universalization and spatial homogeneity.
- *Acculturation* takes place when one group assumes the traits of others while retaining some of its own. The geographical outcome of this is glocalization or hybridity, characterized by heterogeneous patterns.
- *Autarkism* occurs where a culture reasserts its authenticity and specificity in the face of a perceived or actual threat from another culture. The geographical outcome of this is localization and 'mosaiced' heterogeneity.

As in every other sphere, commentators are divided on the extent and nature of cultural globalization. Broadly speaking, opinions may be divided into the three camps as outlined in Chapter 2. The arguments of the three schools correspond, in part, to the models of cultural interaction outlined above.

- *Hyperglobalists* – argue that globalization is leading to the homogenization of world culture, towards a Western – especially US – form. Radical hyperglobalists (i.e. from the left wing) agree and see it as cultural imperialism. This corresponds to the *assimilation* model that leads to universalization.
- *Sceptics* – argue that hyperglobalists' claims are greatly exaggerated and that national culture in particular continues to exert the overwhelming influence on cultural change. This corresponds, in part, to the *autarkic* model, although less radical interpretations might favour a mild *acculturation* model. In both cases, heterogeneity is the outcome, although this may be localized or glocalized in form, depending on the exact nature of cultural interaction that is envisaged.

- *Transformationalists* – argue that the processes of globalization are leading to an intermingling of cultures creating new hybrids and networks. This corresponds to the *acculturation* model that leads to glocalized heterogeneity.

Historical cultural globalization

Held *et al.*, (1999) argue that, to date, much of the discussion in this area has focused exclusively on the impacts of contemporary processes and that the enduring legacies of earlier historical periods on current patterns of cultural globalization have been underestimated. Cultural globalization, arguably, pre-dates significant globalization in other spheres. Global migrations and the populating of the Earth, the diffusion of language families (see Map 6.1) and the initiation of trade routes were the first manifestations of this. Following this, arguably, the first true examples of global networks evolved in the form of *world religions*. There are six religions that are conventionally described as world religions – Christianity, Islam, Confucianism, Hinduism, Judaism and Buddhism. The latter four, although significant in numbers both in the past and today, are essentially regionalized (see Map 6.2). It is the first two, however, which can lay claim to being global in their extent. In the

AU Australian	Ha Hamitic	MP Malayo-Polynesian	Su Sumerian
Ba Bantu	Ho Holophrastic	ST Sino-Tibetan	UA Ural-Altaic
Dr Dravidian	IE Indo-European	Se Semitic	

Map 6.1 *Cultural diffusion – early spread of language families*

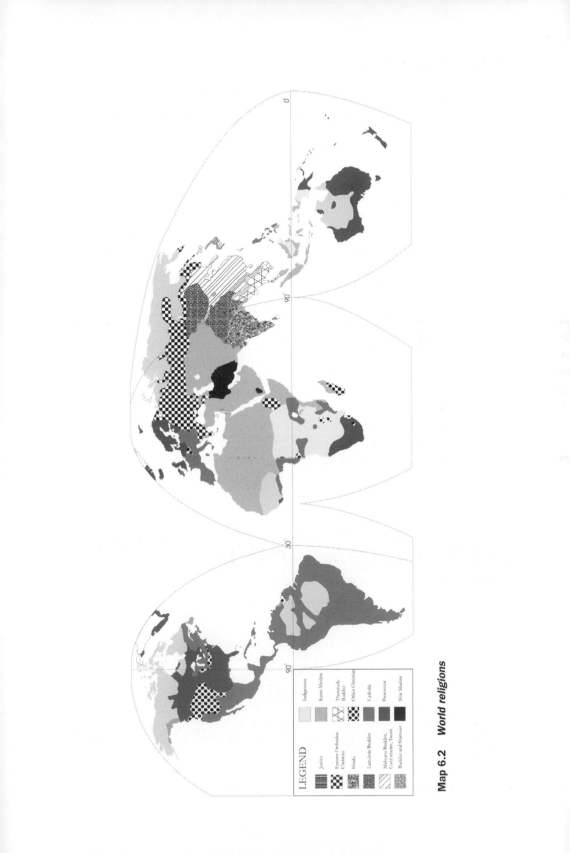

Map 6.2   World religions

case of Islam, but especially Christianity, political networks and military force were combined to propagate the respective religions, which were used as both justification and means of conquering territory. Christianity was especially efficient in this respect, and during the sixteenth and seventeenth centuries began a major diffusion across the globe linked to the expansion of the European empires. This has left an enormous legacy in terms of the commonality in belief systems across Latin America, North America, much of Africa and the Pacific Islands. It should be noted that Christianity was often acculturated according to existing belief structures (see Plate 6.3).

Linked to the rise of religion was the rise of *global empires*, which have left major legacies. The most successful empires, such as the Roman and British Empire, are the ones where 'the extensive reach of military and political power has been reinforced by the reach of cultural power' (Held *et al.*, 1999, p. 334). In this way, global empires were able to overcome the inherent dialectic of imperialism whereby pressure is placed on centralized control by the inherent decentralization of political

Plate 6.3 *Tongan cultural landscape – Christian graves in Tonga are often adorned with tapestries with pre-Christian motifs*

networks. Thus, in the case of the British Empire, for example, extensive work in communications infrastructure, including the development of the telegraph and undersea cable system, gave the British a distinct advantage. This allowed the transmission of cultural information intended to create a sense of shared identity. Education policy was also instrumental in holding the empire together through the production of standard textbooks, curricula and the Oxbridge education of colonial elites.

In the above way, European imperialism laid the cultural infrastructure for more recent cultural globalization. Throughout the nineteenth and twentieth centuries the emphasis shifted to two areas which remain salient today – the rise of *national culture*, and *secular ideologies* of liberalism, socialism and science. As described in Chapter 5, nation-states came to prominence in the nineteenth century. An essential part of this rise was the construction of nationalism and national cultures, which were only sometimes based on pre-existing cultural commonalities. The concept of the nation-state was a Western one, and diffused across the world largely through the cultural infrastructures of colonialism. Crucial to the establishment of the nation-state system was the forging of identities that cut across class. Nation-building involved a variety of tactics including the pursuit of a common language, an education system, national communication networks, national symbols (e.g. flags) and a standing military. Paradoxically, empires sowed their own seeds of destruction through the diffusion of national self-determination.

Alongside the development of national cultures, various secular ideologies or discourses were also evolved and dispersed. *Socialism*, based on early work by Marx subsequent to the industrial revolution in Britain, was one example of this. *Liberalism*, which remains of central importance in the current era, also arose during this period. The role of *science* as a rationalizing element in society was also propagated at this time. It was through the nation-state system that the culture of *capitalism* was diffused across the world, and this remains the dominant force in global society and culture today. On this basis, it could be argued that cultural globalization reached a climax during the colonial wave. It is certainly the case that the global networks of contemporary culture were laid over many centuries. World religions, capitalism and nationalism are arguably more important today than they have ever been, with the latter making a recent comeback in the form of the reassertion of local identity in the face of globalizing processes.

Contemporary cultural globalization

The cultural globalization of the past fifty years differs both qualitatively and quantitatively from that of the past. Held *et al.*, (1999, p. 327) argue:

> Despite the complexity of cultural interactions between societies over the last three thousand years, the intensifying movement of images and symbols and the extraordinary stretch of modes of thought and modes of communication are unique and unparalleled. . . . There is no historical equivalent of the global reach and volume of cultural traffic through contemporary telecommunication, broadcasting and transport infrastructures.

Current cultural globalization departs from that of the past in terms of six trends and processes:

1 New global cultural infrastructures of technology which operate at unprecedented scale and efficiency.
2 The consequent rise in the velocity of cultural exchanges across borders and between societies.
3 The unparalleled rise of Western culture as the core marker of global cultural interaction.
4 The rise of TNCs in the culture industries which create and perpetuate the infrastructure necessary for heightened diffusion.
5 The rise of business culture as the main driver of cultural exchange linked to point 4.
6 The shift in the 'geography' of cultural interaction compared to the pre- Second World War world.

In the case of the latter, Held *et al.*, argue that although flows from the West to the rest of the world are far greater in an absolute sense we are seeing a partial reversal of this through increasingly complex patterns of migration (see Box 6.2), the evolution of mass tourism, and the rise of new sectors, such as the world music industry (see case study on global music). Some have argued that globalization inexorably leads to cultural homogenization, however. James Petras (1993), for example, claims that a global culture is emerging which is directly linked to the USA. This has both political and economic goals and is part of the geopolitics of capitalism. We consider the arguments for homogeneous global culture below.

Box 6.2

The 'age of migration' and cultural change

International migration is playing a more central role in demographic and cultural change than ever. Castles and Miller (1993) refer to the current era as the 'Age of Migration'. They observe four key trends in international flows:

1 Migration is globalizing in the sense that the range of origin and destination countries has greatly diversified over the recent past.
2 The volume of migratory flows is increasing in all regions.
3 The nature of migrants, and their reasons for migration, is becoming increasingly differentiated. Today, no one type of movement dominates with flows characterized by combinations of economic, environmental, political and social migrants.
4 Migration is being feminized as women move increasingly for economic reasons and dominate refugee flows.

Migration, of course, has always been important in demographic evolution, and has played a major role in historical cultural globalization. Flows of note include the Atlantic slave-trade and the mass migrations from Europe to the 'New World' between 1850 and 1950. The latter example saw 55 million people relocate; 30 million people immigrated into the USA between 1860 and 1920 alone. However, Champion (2001) argues that migration has become more important in contemporary demographic

transformations because of the decline in natural increase across the world, but especially in the West. Thus migration accounts for an increasing proportion of population growth, and this is especially the case in richer countries. Indeed, in Europe over the past twenty years, migration has offset natural decrease and has been the *only* source of population growth. Overall, migration accounts for three-fifths of population growth in Western countries as whole. In the Third World, where natural increase remains relatively high, migratory movements are less important in demographic change.

At the global scale, international migration over the past fifty years has exhibited some clear spatial patterns. There has been a general shift from the Third World to the West, with Europe and North America receiving the bulk of migrants. This has led to the establishment of significant diaspora populations in the West and also increased cultural hybridity. In Europe this has been fuelled particularly by the end of the Cold War and the Balkan crises of the 1990s which saw large-scale movements from the east to the west of the Continent. In North America the two major source regions of recent migration have been Central America (especially Mexico) and East and Southeast Asia (especially the Philippines, Vietnam and China). The USA has also received a steady flow of Cuban migrants since the revolution of 1959. Mexican immigration

Box 6.2
continued

dominates, however, and the 1990s saw the legal shift of over two million people across the border into the USA, most of these in search of better economic opportunity. As a consequence of Hispanic immigration the culture of the USA is being transformed; for example, linguistically there has been a major shift to Spanish over the past twenty years.

Three groups of migrants dominate the flows that are currently taking place. The first is 'professional' migrants who move for career prospects. The rise of TNCs has seen an increase in postings abroad and the shortage of skilled labour in some Western economies has led to a 'brain drain' from some Third World countries. The second major group is unskilled labour, which often moves without specific offers of work. This has led primarily to increased flows from the Third World to the West as employers search for workers who are willing to labour for lower wages. A third major group is refugees and asylum seekers, which was larger in magnitude in the 1990s than at any other time in history. In the early 1990s the

number of such migrants stood at around twenty million and this has continued to rise. This does not count the large numbers of people 'displaced' through conflict inside their own countries (such as the Palestinians and Kurds). Although the refugee 'crisis' is highly visible in the public sphere in Western countries – it has dominated recent election politics in Australia, Germany and the UK, for example – in absolute terms the number of refugees entering Third World countries is far higher. Because immigrants into Western countries have a significant impact on population composition, migration has become a highly politicized issue. This has led to the revival in some European states, for example, of nationalistic political groups who bemoan the economic and cultural impacts of immigration. In terms of the latter, migrant groups are often painted as undermining the position of 'indigenous' non-skilled labour, and in the case of the latter the arrival of people with different traditions, religions and languages is seen by the right wing as undermining the integrity of the national identity and the state.

Cultural imperialism

Why is it that some aboriginals in Australia send each other Christmas cards featuring snow? Why do most maps of the world show Britain in the centre and at the top? Why is the Middle East called so? The answers to these questions are found in the history of imperialism and spread of Westernization. In this context Pieterse (1995, p. 45) argues:

> The world is becoming more uniform and standardised, through a technological, commercial and cultural synchronisation emanating from the West, and globalization is tied up with modernity.

Proponents of the cultural imperialism thesis date its inception to the industrial colonialism phase. As discussed in Chapter 3, it was during this phase that colonialism reached its zenith, peaking just prior to the First World War, when the British Empire reached its maximum territorial extent. Through colonialism Western culture was diffused, driven by and driving the penetration of capitalism. The end of formal colonialism in the second half of the twentieth century did not spell the end of cultural imperialism. According to this argument, cultural imperialism is no longer played out explicitly by nation-states and it has become an economic process as well as a political one. It is constructed by TNCs that represent the interests of the elite, especially those of the USA. Powerful Western governments facilitate cultural imperialism, and it is bolstered further by various transnational institutions (IMF, World Bank and WTO especially) that are designed to serve capitalist needs. The powerful companies, institutions and countries that have the greatest say in the governance of the contemporary system have thus built directly on the spoils of colonialism. As the most powerful country with the largest corporations, the USA plays the

Plate 6.4   *Cultural imperialism – a Tahitian Santa Claus in most unsuitable attire given the climate*

Plate 6.5   *McChurrasco – a Chileanized McDonald's product*

major role in the diffusion of Westernization and thus has great influence over its nature.

In 1993, Ritzer wrote an influential book concerning the McDonald's franchise. He defined *McDonaldization* as the process through which the organizational, productive and representational principles of McDonald's were redefining globalization. At the core of the new model is the efficiency and standardization of the McDonald's approach; 'Drive-thru' windows and assembly line production impose a new rationality on the process of production and consumption. A key point is the predictability of the McDonald's menu, virtually standardized across the world, imposing uniformity on the customer and the workforce alike. Authors such as John F. Love (1995) celebrate McDonald's, and see it as playing a positive role in 'exporting Americana'. It is notable in this supposed age of flexibility and niche marketing that standardized production epitomized by McDonald's has become so prominent. This thesis has been criticized, however, by those who argue that McDonald's products often reflect the locality in which they are marketed (see Plate 6.5).

Those who subscribe to the cultural imperialism thesis worry that the world's cultural diversity will be threatened by the diffusion of Westernization/Americanization (Waters, 2001). There are around 300 million people who identify as members of an indigenous culture and there are many examples of threatened survival and/or erosion of these in regions such as the Amazon Basin, North America and Australia. Aside from the immorality of culture death, with the decline of such entities much 'traditional' wisdom is lost, rather as reduced biodiversity leaves a depleted ecosystem. Given the rise of universal human rights, however, these transitions have taken on a political form.

Box 6.3 presents some arguments that support the thesis of cultural imperialism. Evidence in five areas is discussed: language, tourism, consumerism, media and democracy. The global music industry is considered separately in a subsequent case study.

Box 6.3

Global cultural imperialism?

Language – There are around 6000 languages in the world and this may drop to 3000 by 2100. Approximately 60 per cent of these languages have less than 10,000 speakers; a quarter have less than 1000. English is becoming *the* world language.

Box 6.3
continued

Although Mandarin is more widely spoken as a first language, if second-language speakers are taken into account the total number of English speakers is close to one billion. This number increases significantly if those who have a working knowledge of the language are included. English is the medium of communication in many important fields including air travel, finance and the internet. Two-thirds of all scientists write in English; 80 per cent of the information stored in electronic retrieval systems is in English; 120 countries receive radio programmes in English; and at any given time over 200 million students are studying English as an additional language. It is an official language in much of Africa, the Pacific, and South and Southeast Asia. This dominance may be explained by the cultural imperialism of the British and the current hegemonic role of the USA, and new technologies are perpetuating this trend (Crystal, 1997).

Tourism – Tourism is now the world's largest industry. Mass tourism has often been compared to Fordist methods of production. The standard package holiday seeking sea and sun, often involving accommodation in a global hotel chain such as the Holiday Inn or the Sheraton, has proliferated as a desirable form of relaxation. The journey of many British to the Costa Del Sol, Spain, where they practise cultural traits such as drinking beer and eating fish and chips while lying on crowded beaches surrounded by tall buildings, is a stereotype which captures the essence of this type of standardization.

Global brands – The growth in the influence of TNCs is the rise of global consumer culture built around world brands. McDonald's, for example, operates in over 26,000 outlets in 119 countries. In 1997 it opened one outlet every four hours. The queues to get into Suva's (Fiji) new store when it opened in 1997 were over 400 metres long. Coca-Cola is sold in nearly every country. It is a transcultural item yet it is very much linked with US culture. This type of standardized consumption is mirrored across many sectors and is perpetuated by the capitalist system it supports and is borne out of.

Media – National media systems are being superseded by global media complexes. With global deregulation during the neoliberal era, and WTO regulations since 1997, the media have become more expansionist and consumerist, rather than aiming coverage at citizens of a particular nation-state. The industry has become increasingly concentrated, privatized, vertically and horizontally integrated, and predatory. Around twenty to thirty large TNCs dominate the global entertainment and media industry, all of which are from the West, and most of which are from the USA. These include giants such as Time-Warner, Disney, New Corporation, Universal and the BBC. In terms of the news media there has been a standardization of presentation, with increased focus on 'human interest' stories, and a generally unquestioned bias towards the West and its allies – powerfully illustrated during the CNN, BBC and ABC coverage of the War in Iraq (2003–). Herman and Chomsky (1988) see a powerful intentionality at work in this trend which they term 'manufacturing consent'.

Box 6.3
continued

Democracy – As outlined in Chapter 5, the spread of liberal democracy has been profound and is now practised in the vast majority of nation-states across the planet. Underlying this diffusion is the Western enlightenment belief that it is the most desirable form of governance, and this is evidenced by its defeat of socialism. Parliamentary systems have spread to all corners of the planet, diffused through Western empires and successor states and more recently through the conditionality of loans from the IMF to poorer countries. This is eclipsing the diversity of governance methods that other cultures exhibit, which in some cases have deep historical roots.

A variation on the cultural imperialism argument sees the creation of a universalized hybrid culture. This type of culture is homogeneous but not entirely Western in nature. Thus we are still left with the impact of Westernization as something that is very powerful – but it is not unidirectional. The influence of non-Western cultures on dominant Western culture has been marked. For example, the British drink tea because of the British imperial connection with India, and a number of words in the English language, such as *bungalow*, *shampoo*, *thug* and *pyjamas*, are borrowed from languages of the subcontinent. These traits are acculturated into the dominant culture and form part of the subsequent round of Westernization as they disseminated through neo-imperial networks. The influence of Black American and Hispanic dialects on rap, the most popular music globally at present, and the fact that football (soccer), which diffused through the British Empire, is thought to have been invented in China, are further examples of universalized hybrids.

Criticisms of the homogeneous global culture arguments

So what are we make of the two main global culture theses? Both the idea of a cultural imperialism and a universalized hybrid culture have been criticized. This is particularly the case in the context of the former. It has been argued that the concept of cultural imperialism ascribes globalization with too much determining power and is too broad a generalization of the processes that are going on on the ground (Tomlinson, 1999). The power of locality, and of local culture, is thus overlooked. The result is a metanarrative or grand theory – the kind that geographers have been rallying against since the mid-1980s at least (Johnston and Sidaway, 2004) (see also Box 6.4). In particular Tomlinson has argued that grand theories of homogenization have misrepresented, even patronized, the supposed recipients of dominant cultural signals and symbols emanating from the West (Tomlinson, 1991). It is naively assumed that masses of the population of the culturally marginalized will passively accept these transformations. This is eurocentric and invokes a one-way flow of cultural symbols and meanings without recognizing the countervailing possibilities of localized hybridization or localization. Furthermore, the rise of 'cosmopolitan consumption' in the West casts further doubt on the homogenization thesis (see below). In sum, McEwan (2001, p. 160) argues that:

> in reality both of these processes are flawed explanations for what is happening today. If global culture exists, it is far from a product of unidirectional 'Westernisation'. However, alternative ideas about cultures mixing to produce a universal global culture are also problematic. Cultures are mixing, but this mix does not necessarily mean we are all becoming the same.

Box 6.4

A critical view of the 'global gay identity' thesis

Bell and Binnie (2000) are highly critical of what they see as 'the export of western definitions of sexual practices, identities and cultures around the world' (p. 116). This has led to the proposition, in some circles, that there exists a 'global gay identity', with critics such as Herdt (1997) associated particularly with this idea. In the book *Same Sex, Different Cultures* the latter argues for a model of gay behaviour which is diffusing from the West to marginalized regions and localities. An integral part of this evolution is the development of what Sinfield calls a 'queer diaspora' which cuts across traditional boundaries of state and citizenship. Watney (1995, quoted in Bell and Binnie, 2000, p. 117) argues that what defines this:

is the sense of relief and safety which a gay man or lesbian finds in a gay bar or a dyke bar in a strange city in a foreign country. Even if one cannot speak the local language, we feel a sense of identification.

Bell and Binnie's critique of the globalization of such identities is multiple. They argue that the global gay culture thesis presupposes a common form of homophobia across the world – rather as the globalized feminist movement assumed a single form of patriarchy. In reality, homophobia is constructed in vastly different ways in different local and national contexts. Second, the construction of what Cant calls a McPink economy does not take into account those who are marginalized

Box 6.4
continued

from the global gay capitalist circuit. As such, 'there is a danger in exporting a mode of "being gay" from late-modern westernised liberal democracies that does in fact "McDonaldise sexual minorities"' (Bell and Binnie, 2000, pp. 116–117). In criticizing this point they echo Cant (1996, in Bell and Binnie, 2000, p. 117), who argues that globally speaking:

> The romantic myth of homosexual identity cutting across class, race, and so on doesn't work in practice any more than it does in the West. The experience of sexuality in everyday life is shaped by such variables as the gap between city and country; ethnic and religious differences; and hierarchies of wealth, education and age.

Overall, the globalized model privileges Western sexual identities and emphasizes 'sameness over difference' between and among lesbians and gay men. The overemphasis on the role of the affluent, middle-class, jet-set gay subject masks 'a multitude of local differences' (p. 119) whose reality is far more embedded in local practices and regulatory policies at the level of the nation-state. In general, Bell and Binnie argue that a postcolonial analysis is required in order to more adequately capture the diversity of experience, claiming that:

> The complexities of national, transnational and global identifications and disidentifications thus reproduce their own disjunctive flows around the world, mapping out the complex implications that globalization has upon sexual citizenship.
>
> (p. 122)

Samir Amin (1997) argues that globalization has actually led to a reassertion of ethnic and cultural difference, and that this has come about as a form of resistance to homogenizing tendencies. Another term for this is 'reterritorialization', as used in post-structural political geographies. Paradoxically, globalization makes space for such assertions – which mirrors change in the political sphere. In the field of language, for example, there has been a revival in small European languages (such as Welsh, Irish Scots, Catalan and Basque) over recent years. Potentially, communications technologies can play an important role, informing us of other cultures, and providing the impetus and infrastructure necessary for the protection of cultural diversity.

Localized hybridity

The idea of localized hybridity allows us to move beyond what McEwan terms 'mosaic' (multiculturalism) and 'melting-pot' (assimilation) concepts of cultural interaction. Multiculturalism is a term which is often considered 'progressive' but which can ghettoize and segregate cultures. For example, in a migratory setting arrival cultures may be accepted and tolerated but aspects of identity (be they racial, social and so on) are essentialized and used as a means of differentiation. Localized hybridity refers to a situation where cultures mix to produce unique outcomes. There are many instances of this (see Box 6.5 and global music case study). For example, 'Australian cuisine' is a mixture of traditional British cooking and exclusively Australian ingredients (such as kangaroo), together with the fare of the many waves of migrants that have arrived in the country including Greeks, Italians and Spanish. The essential point is that these forms do not just exist in a multi-culinary form – they are often combined in new, and tasty, ways. Indeed, Australia is becoming known for what is sometimes referred to light-heartedly as *con*fusion cooking. A second example also relates to cuisine. Curry is the most popular food in Britain and central to the identity of millions of Brits. British curry, and derivatives such as balti, are actually constructed foods based on Indian cuisine, which Indians from India would probably not recognize if they dined at a restaurant in Small Heath, Birmingham – the self-proclaimed 'balti capital of the world'. Another example of localized hybridity comes from the field of language. There are many words in Chilean Spanish that differ from Mainland Spanish because of the influence of Mapudungun, the language of the Mapuche (Southern Chilean indigenous). Indeed, as is argued in Box 6.5, there are many local varieties of 'global' languages that have developed based on contingent factors.

In the context of the above, the concept of *glocalization* is particularly important – whereby the global and the local interact to produce hybridized outcomes. As such, local agents are no longer seen as victims of global cultural shifts but as interpreters of such processes (Allen and Hamnett, 1995). When we accept this proposition we are able to more accurately analyse the influence of non-Western cultures on other Western cultures, the influence of Western cultures on Western cultures, and the influence of non-Western cultures on other non-Western cultures. In other words, this perspective gives us more subtle tools for interpreting complex cultural change. Lash and Urry (1994) take this argument a stage further, arguing that diversity and uniqueness become part of cultural capital and may be used in some cases in order to promote livelihoods in the new postmodern economy (see section on cultural commodification below). An extension of this idea leads to the concept of renewed localism (Levitt, 2001). This process might entail the rise of ethnic fundamentalism (Waters, 2001) – where a group attempts to re-create and rediscover its fundamental and radical roots, for example. Although the evidence points towards increased hybridity, there can be little doubt that there are some very real examples of renewed localism in all spheres (for example, in the form of de-linked local economic trading schemes (LETS) or devolution movements across Europe).

In some ways cultures have always been hybridized, since no culture is static and cultural interaction is constant. But there can be little doubt that the speed of cultural 'traffic' in the contemporary world is such that interactions – or processes of transculturation, as McEwan calls them – are speeding up. To many commentators localized hybridity is liberating and exciting; it provides a means of resisting and subverting dominant and hegemonic cultures and disrupting homogenizing processes. It can create a richer and more 'progressive' sense of place (Massey, 1991).

Diaspora

The fixed relationship between place and culture is disrupted by the concept of diaspora. A diaspora refers to the scattering of a culture from its 'homeland' and was originally used to refer to the Jewish peoples. With the high contemporary incidence of migratory movements (see Box 6.2) such 'communities' have become increasingly common. Cyberspace provides particularly fertile space for diasporic cultural interactions (see Box 6.5). The Indian diaspora, for example, stretches across all continents with significant 'Indian' populations found in places as diverse as South Africa, Brazil, the UK and Fiji. The diasporas of the Pacific Island nations are, relative to the size of 'hearth' populations, among the largest in the world (see Box 6.5 on the Niuean diaspora). McEwan (2001, p. 171) argues that:

> Diasporic identities are important because they are at once local and global, and are based on transnational identifications encompassing both 'imagined' and 'encountered' communities.

Within 'diasporic space' processes of hybridization are paramount as migrants interact with 'host' cultures to create new forms. Britishness, for example, can take on many forms – African-Caribbean, Indian, Jewish and so forth – and these diasporic identities recondition the original concept of 'Britishness'. Diasporas stretch across traditional boundaries, forging new connections between places, creating cultural forms that challenge accepted hierarchies and structures. This challenges many accepted geographical notions – such as 'home' and 'belonging' (Dwyer, 1999). As such, diaspora are profoundly transforming the spatiality of nation-state cultures.

Box 6.5

The Niuean diaspora

Niue is a country located in the Western South Pacific Ocean occupying one island of approximately 250 square kilometres with a population of approximately 2000. It is one of the smallest independent territories on Earth and is isolated at over 200 km from its nearest neighbour, Tonga, with only one wharf and a small airport. Populated by Polynesians beginning some 1000 years ago, it was colonized first by Great Britain in 1901 and handed over to New Zealand jurisdiction from 1904 onwards. It was not a settled colony and has retained a strong indigenous identity. In 1974 it gained independence as a 'territory in free association' with New Zealand, with the latter responsible for its defence and foreign policy. Niue receives substantial amounts of aid, sometimes totalling 50 per cent of GNP, the rest being made up by a mixture of remittances, tourism receipts and agro-exports (Murray and Terry, 2004).

Since independence Niue has been losing its population rapidly, falling from 5000 in 1966 to the current level (see Plate 6.6). Today, there are far more Niueans living in Pacific Rim countries than on the island

itself. There are some 20,000 living in New Zealand, and thousands more in Australia and other nations on the Pacific Rim. The Niuean diaspora has close ties with 'home' – extensive remittances in cash and kind are sent by first, second and third generation migrants. Cultural practices are relatively well maintained in Niuean population centres, such as Auckland, where ear-piercing and other rituals regularly take place. Since independence Niueans have had the option of dual New Zealand/Niue citizenship which has fuelled outmigration. Friction can be experienced when expatriate Niueans return home, as some of their cultural traits have been transformed. The adoption by young Polynesians of New Zealand-mediated US hip-hop culture is widespread, for example, and this is alien to those who have stayed on the island. Cultural values are thus changing rapidly, and this can lead to conflict. One salient example is the issue of land. All male Niueans are entitled to land as part of their *magafoa* (family group). As most Niueans are away from the island, this has left much land idle. Those who still remain have become increasingly frustrated

Box 6.5
continued

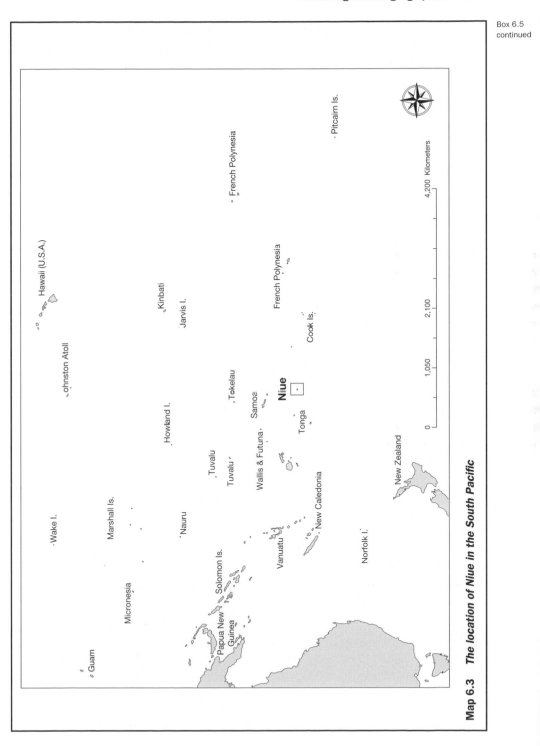

Map 6.3 *The location of Niue in the South Pacific*

Box 6.5
continued

with the desires of the diaspora population to retain their titles. The latter group cite the right to land as part of the *Moie Faka Niue*, or Niuean Way, however (McIntyre and Soulsby, 2004).

From isolation comes the need to develop links however, and Niue is the only country in the world where the entire population is provided with free access to the internet. As a consequence we are seeing the emergence of a 'cyberdiaspora' as Niueans communicate over the internet through email, a dedicated chat room, and numerous Niue-oriented sites (Gibson, 2004). The realities of isolation, and vulnerability to climate, were highlighted in January 2004 however, when Hurricane Heta destroyed much of the island's

infrastructure and agricultural crops. It is predicted that this, together with pressures on the country's economy, will lead to further depopulation. Economic problems have been heightened by the insistence of development 'trustees', such as AUSAID and NZAID, that Niue undertake neoliberal-like reform. The resultant restructuring led to a collapse of the state employment sector in the mid-1990s (Murray, 2002b). Is it possible that we have witnessed, through a mixture of voluntary and involuntary outmigration, a relocation of the cultural hearth of this society? Or is the culture so changed as to be unrecognizable? These are the kinds of political questions that cultural geographers often grapple with in the context of marginalized peoples.

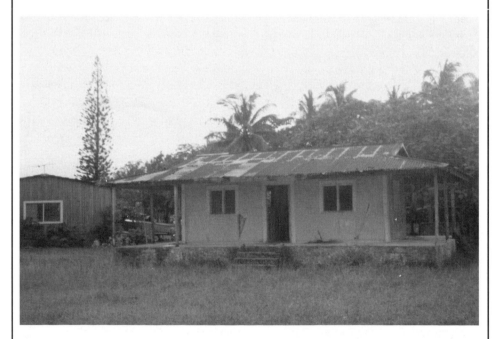

Plate 6.6 *Abandoned housing arising from outmigration from Niue is a common sight on the island*

Overall, it is easy to run away with the idea that globalization is destroying cultural diversity. It is certainly altering the cultural map of the world profoundly. However, as illustrated above, there are signs that the process is not always characterized by one-way cultural traffic. To most geographers the conditioning or deterministic impacts of cultural globalization have been greatly overemphasized. Five examples, corresponding to those explored in the context of the global cultural homogenization thesis in Box 6.3, are presented in Box 6.6.

Box 6.6

Global cultural heterogeneity

Language – Although English is becoming the dominant world language, enormous variety remains. Important world languages include Mandarin, Spanish, Hindi, Arabic, Portuguese, Bengali and Russian – all of which have over 100 million first-language speakers. There is no reason to suppose that these numbers will decline. It is likely, however, that an increasing number of people will learn English as a second language, leading to increased bilingualism. In parts of the USA, such as southern Florida, given the high Latino population, English and Spanish are equally visible. *Dora the Explorer*, a bilingual US children's television character with Latin American cultural roots, embodies the shifting linguistic landscape of the USA. Within English itself, there are many existing and evolving dialects and communalects that contain elements – including grammar, intonation and vocabulary – of both the indigenous language and received English. Jamaican English and Fijian English are unintelligible to some English speakers from England, for example.

Tourism – For a growing number of people 'otherness' is what makes a destination attractive. It is the very process of the globalization of the media which has made people more aware of otherness and increased the demand for it. It has been argued that some places are incorporated into global capitalism in ways which reinforce rather than erode their distinctiveness. The rise of eco-tourism, and even poverty tourism, seems to point to a shift away from the 'Fordist' tourism outlined in Box 6.3 (Scheyvens, 2002).

Global brands – Despite the undisputable rise of global brands, products are often adapted to local conditions. For example, mutton is used in Indian Big Macs and vegetable oil is used for frying. MTV, one of the archetypal examples of cultural pop globalization, is adapted for audiences across the world, and MTV Asia and MTV Latin America play a mixture of Western and regional videos. The Barbie Doll comes in thirty different national varieties.

Box 6.6
continued

Media – Although the global media is increasingly concentrated and standardized both within and beyond major corporations, product differentiation is actively pursued. Local newspapers, radio stations and television stations remain important information providers and offer their own particular slant on the news, although many of them are ultimately owned by large media conglomerates. During the Afghanistan and Iraq wars the Qatar-based Arabic news network *Al Jazeera* came to the fore as an alternative voice, which was often highly critical of US and UK actions in the Middle East. Ironically however, rumours circulating in January 2005 suggested that partly due to pressure from the US administration, the station was to be sold to a Western conglomerate. Small entertainment operations, such as independent film-makers, exist, although it is increasingly difficult for them to find distribution channels.

Democracy – Just as there are alternative capitalisms there are alternative democratic structures. Democracy in many Pacific Island nations is grafted on top of traditional and other forms of governance. In New Zealand, for example, seats are reserved for ethnic Maori in Parliament. Outside of the democratic hegemon there exist many other formations, which account for a significant proportion of the global population. The transition in China, for example, is highly nonconformist, since it attempts to combine liberalized economics with authoritarian politics. Socialism persists in a range of other countries such as North Korea, Cuba and Vietnam – albeit in very different ways with very different outcomes.

Cultural consumption – commodification and cosmopolitanism

Recent work by geographers has dealt with the evolving nature of consumption in the face of cultural globalization (see Crang, 1998; Mansvelt, 2005). There has been considerable debate concerning the rise of 'consumer culture' over the past three decades in particular. Consumption has come to be understood as far more than a material transaction process. Items are constituted of symbolic as well as material value and their consumption is tied up with the creation and expression of identities (Waters, 2001). In particular, 'brands' have become increasingly important markers of identity. Adidas shoes and clothing, for example, are linked with the hip-hop culture of the USA, and many purchase the clothing in order to associate themselves with this particular youth movement. In this sense then, consumption has become a key source of social and cultural differentiation. This is linked to globalization in a number of ways – not least as such 'signifiers' are rapidly diffused across the planet, constructing global cultures that transcend borders. Some have argued that this emergent culture is now beyond the control of any one group (Baudrillard, 1988). That such cultures are evolving is beyond question – whether they are creating one global culture is another matter however, as the rise of the 'cosmopolitan consumer' discussed later in this section illustrates (see also previous discussion).

When the objects, ideas and traits of a culture become part of the capitalist system of exchange and are bought and sold, a process of cultural commodification is said to be occurring. Linked to the rise of the cosmopolitan consumer has been the increased demand for and supply of 'authentic' and 'exotic' cultural experiences. This is most often associated with the tourist sector where, as previously noted, the dominant mode of tourism provision has shifted from the Fordist/mass market 'sun and sea' model to a more differentiated sector based around 'new' cultural experiences (see Box 6.7). Cultural commodification is an increasing feature of global society. In supermarkets across the West, for example, new rows of produce from non-Western countries are conflated under the label 'ethnic' foods – a process not unlike the evolution of 'world' music (see case study below). In urban areas, especially world cities, a prominent marketing strategy of local administrations of late has been to re-create and promote 'ethnic' districts such as Chinatown in Soho, London, or the Arab Quarter in Singapore (Chang, 2005). Ironically, in the past such areas may well have been the 'ethnic ghettos' that tourists were encouraged to avoid. Cultural commodification works both ways, involving flows from the Western to the non-Western world. Indeed, in many respects this direction of flow pre-dates the flow from poorer to richer countries. Symbols such as Coca-Cola and McDonald's are marketed as Western ideals, implying that consumption confers status. In both the Western and the non-Western world experiencing the 'other' can confer status and 'cosmopolitanism' on the consumer.

Bell and Valentine (1994) define 'cosmopolitan consumers' as a consumerist elite which seeks cultural signifiers that are not touched by homogenizing forces and actively explores transcultural diversity (Thompson and Tambyah, 1999). This group accumulates cultural capital – such as music, clothing, food, beverages or furnishings – as a means of disidentifying themselves from the mainstream (Shurmer-Smith and Hannam, 1994). Increasingly, such consumers do not have to leave their localities in order to achieve their goals, and thus heightened cultural flows have the dialectical impact of escalating the potential for

Plate 6.7 *Cultural commodification – tourists in La Boca, Buenos Aires, Argentina, where the old has been preserved and enhanced for visitors*

heterogeneity. Cosmopolitan consumers are not only found in richer countries, although they are relatively more common in such contexts, since practising such accumulation requires financial capability and access to global networks. In poorer countries such elites do exist, and they are often highly visible. In Chile, for example, a cosmopolitan consumer might take his or her holidays in Miami, drink Chivas Regal whiskey and listen to rock music in English. The fact that the same markers of identity would not confer cosmopolitanism on consumers in the USA, for example, highlights the fact that cultural symbols are interpreted in very different ways across space according to local contingencies.

Baudrillard (1988) is a post-structuralist who has written extensively on the process of cultural commodification and has influenced a number of human geographers in this area. He argues that the 'postmodern' culture is one of signs and symbols. Today's world is characterized by the proliferation of simulacra where the symbols or signifiers of an event or item replace direct experience. As such, 'reality' is experienced chiefly through representations of it – meaning that distinguishing between the concrete and the imaginary becomes increasingly difficult and redundant. This creates a social environment of 'hyper-reality' which disembeds cultures and their component traits, and creates new spaces where the former contingencies of history become less important. Geographers and others have extended these ideas to the analysis of capitalism and economics and services such as tourism (Lash and Urry, 1994).

It may be argued that cultural commodification, if managed appropriately and sensitively, offers the prospect of improved livelihoods for marginalized cultures as well as helping to preserve some cultural 'traditions' (see Box 6.7). In contrast, some see such commodification as tantamount to the 'prostitution' of culture (Britton and Clarke, 1987). In this latter argument, cultures move out of the control of those that inhabit them and into the hands of market forces. Such forces are characteristically fickle and cyclical – and thus actually threaten sustainable livelihoods. Commodification can also compound stereotypes of what 'authentic' and 'indigenous' mean. In this way, creating marketable cultures may 'strait-jacket' cultures so that they do not evolve as they would have under less 'exploited' circumstances. It could be argued that this last perspective patronizes and simplifies the marginalized who are not necessarily passive receivers of market signals. The new cultural politics is a highly contested terrain.

Box 6.7

Tourism and the invention of traditional culture in Fiji

Culture is used to sell things, and tourism – the largest single economic sector in the global economy – is one of the best examples. During the late 1980s the mass consumption beach and sun ideal was challenged by the growth in tourism to more 'exotic' locations. In the South Pacific, for example, eco-tourism and cultural tourism have become increasingly important within the general context of tourist industry growth in the region. In Fiji, in relatively mainstream resort holidays and cruises, the 'Fijian experience' is sold to attract visitors. This has seen the reproduction and adaptation of certain cultural practices, such as the *kava* drinking welcome ceremony, and the invention of others such as new forms of *meke* (dance) to cater for those seeking an 'authentic' experience (see Plate 6.8). The Fiji Visitors Bureau website dedicates considerable space to explaining traditional Fijian social structures, beliefs

Box 6.7
continued

and ceremonies. In the world outside of the tourist resorts such 'traditions' are rapidly evolving. Most Fijians now live in urban areas and are increasingly Westernized in both their tastes and attitudes. 'Traditional' practices still exist in some interior and coastal villages, but they are largely the lot of beach-side resorts, which have become commodified cultural enclaves. Tourism, trading on both invented and authentic Fijian traditions mixed with modern resort-based provision, is a truly glocal product. Some have argued that globalized tourism in places such as Fiji relies on a damaging 'orientalism' of the Pacific (see Nicole, 2000). Others have argued that commodification of this nature actually helps preserve culture and ethnicity in ways that would be otherwise challenged.

Plate 6.8 *Fijian tourism: villagers perform a 'traditional' dance for tourists in the Yasawa island group of Fiji*

The global music industry case study

It is a well-worn cliché that music is a universal language. Although transferral between cultures is not unproblematic – due to vast variations in terms of both musical style and structure and lyrical subject matter – it does not depend on written or spoken communication explicitly for its major impact. It is therefore a prime candidate for cultural interpenetration. The geography of the production and consumption of music is highly uneven however, with the bulk of musical flows moving from the West to elsewhere, reflecting economic and political relations of power that are rooted in colonial history. The industry is dominated by a few very large TNCs that have been instrumental in diffusing Westernized forms of pop and rock across the planet. Some of the results – MTV, the Beatles, Michael Jackson, Elvis Presley and Madonna – in many ways epitomize the globalization of culture thesis.

In the past, the diffusion of music was limited by the absence of technology required to record and broadcast it. Thus the site of production was the live performance. Musical forms diffused through early globalized networks including world religions, empires and associated migratory patterns. Some flows, such as the slave-trade from Africa to the Americas, left an especially lasting legacy in terms of musical hybrids, leading to blues, gospel, and indirectly to rock and roll, soul music, RnB, and hip-hop/rap. The conquest of the Americas by the Spanish and Portuguese led to the formation of new musical forms combining flamenco, Western classical music and Andean music which can be heard clearly today in the music of Latin America. The globalization of music over the past 100 years illustrates the important facilitating role of technology as distinct and increasingly intense waves of globalization were linked to new infrastructures. Storage and reproduction of music has been revolutionized moving chronologically through vinyl record, cassette, compact disc, mini disc, and, most recently, the mp3 (digital storage), becoming increasingly 'footloose' with each successive invention. The rise of infrastructures of transmission associated with broader cultural globalization such as the radio, television and most recently the internet have lubricated this increased flow. Large cultural-production TNCs have been instrumental in the development of this technology and have taken advantage of the opportunities afforded by it. As a consequence of these trends music has globalized in three major ways: (1) music TNCs have evolved to foster global flows; (2) national markets have been increasingly penetrated by foreign performers; (3) Western styles and images have been diffused through the music industry.

Today's industry is dominated by a handful of very large TNCs that control an industry worth US$32 billion in 2003 with unit sales of 2.9 billion (IFPI, 2004). In 2002, five companies accounted for 75 per cent of total market share: Universal (27 per cent); Sony (14 per cent), EMI

(12 per cent); Warner (11 per cent) and BMG (11 per cent). These TNCs are often vertically and horizontally integrated and have been increasingly aggressive in terms of buying smaller, formerly independent, labels. These giant firms, which operate a host of linked labels beneath them, are based mostly in the USA and to a lesser extent the UK, and are generally anglophone in orientation. As noted above, the globalization of pop and rock has largely involved the diffusion of cultural forms from the USA and the UK to the rest of the world. In the aftermath of the Second World War, rock and roll, blues and soul were the major export styles. Important forms in approximate chronological order that have subsequently globalized include country and western, urban folk, heavy metal, punk, new romantic, gothic, acid jazz, house/rave, grunge, Britpop, rap, hip-hop, boy/girl bands, and new RnB.

Few forms have captured the global imagination as has rap/hip-hop however. Emanating from the ghettos of major US cities in the early 1980s, this form has risen to global predominance in the singles charts of virtually every country and is emulated, both as a musical form and lifestyle, by youth on all continents. Rap culture is diverse and complex, but it often goes hand in hand with other markers of neoliberal cultural globalization such as a global brand culture and ostentatious

Plate 6.9 *Hip-hop culture – New Zealand Polynesian rapper, King Kapisi*
(Photo: April Henderson)

materialism. The co-opting of rap/hip-hop by capitalist circuits was not necessarily intended or even predictable at its inception. The pioneering work of groups such as the Sugarhill Gang, and Grandmaster Flash and the Furious Five formed part of a political protest movement against 'mainstream' US culture. It could be argued that many eventually globalized musical forms such as blues, punk and grunge started life as counter-cultural movements only to be appropriated by cultural TNCs.

No countries have evolved globalized popular music to the same extent as the USA and the UK, and these markets are relatively impenetrable to artists from other markets. The Guinness British Hit Singles (Roberts, 2001) top 40 groups of 1952 to 2000, based on total weeks on the sales chart, features twenty UK acts, nineteen US acts and one Australian act (the Bee Gees, who were actually born in the UK). In other anglophone and part-anglophone countries such as Australia, Canada and New Zealand, dominance by US and UK artists remains pronounced despite government-led efforts in each case to promote national content on radio. The first two countries have produced a number of global 'stars' including INXS, AC/DC, the Bee Gees and Kylie Minogue in the case of the former, and Celine Dion, Bryan Adams, Neil Young and Rush in the case of the latter. New Zealand has found it especially difficult to penetrate global markets with only Crowded House achieving anything approaching global status. The New Zealand national market is particularly saturated by overseas acts. Of the Top 40 acts 1966 to 1996, based on weeks on the singles charts, only one (Dance Exponents) was from New Zealand, while nineteen were from the UK, fifteen from the USA, three from Australia (Bee Gees, Olivia Newton John, Jimmy Barnes), and one each from Ireland (U2) and Sweden (Abba).

Within Europe, France and Germany have strong domestic industries, but aside from a small number of 'europop' hit groups, such as Kraftwerk, Trio and Nena, they have not captured global markets. Scandinavian countries have enjoyed a little more success, producing Abba and, more recently, A-ha and Roxette. Spain has produced few global names in popular music, with notable exceptions being father and son Julio and Enrique Iglesias. Spanish artists are very popular in Latin America however. The diffusion of non-UK and US music through established networks is hindered by the general trend towards English global pop, and by the paucity of distributing companies located outside of these two countries. Consequently, European charts are heavily populated by US and UK acts. In Latin America, Spanish and US acts are overwhelmingly popular and national acts that make inroads to commercial success – such as Maná (Mexico), Los Tres (Chile) and Fito Paez (Argentina) – are relatively rare. In Asia, similar patterns are observed, although the rise of Bollywood as a cultural force in South and Southeast Asia is partially offsetting this trend.

One country that apparently bucks the above trend is Jamaica, which has been hugely influential in global markets through the reggae (Bob Marley) and, more recently ragga (Shaggy) forms. These forms are highly commercially successful in Western countries and have produced a number of emulative groups such as UB40. Reggae has become hugely popular across Africa and the Pacific Islands, and is often seen as a form of resistance to globalized music, encapsulated in the emancipatory messages of Bob Marley. Pacific Reggae has evolved a unique form, marrying the ska and off-beat of the Caribbean style with the melody, chord progressions and instrumentations of the Pacific. Another non-Western form which is achieving considerable global impact is the style broadly known as 'tropical', which incorporates salsa, merengue, cumbia and son. More traditional tropical rhythms and melodies were adapted and evolved by Puerto Ricans and Cubans in New York in the 1970s, including breakthrough artists such as Ruben Blades. These were popularized into the mainstream through the 'pop tropical' sounds of Gloria Estefan who sings in both Spanish and English. Carlos Santana, a member of the early 1970s psychedelic rock movement, has found a revitalized market for his particular brand of 'tropical rock' in the mainstream West and among the Hispanic diaspora in the USA. Across Latin America, 'tropical' artists such as Juan Luis Guerra, Willie Colon, Joe Arroyo and Celia Cruz challenge Western rock and pop styles to a certain extent.

Homogenization is not necessarily the outcome of the globalization of Western forms of music. One of the most enduring impacts of the diffusion of US and UK forms of music is the creation of localized hybrids. A few examples only can be provided here. Latin American rock, beyond being sung principally in Spanish, does have a quality which sets it apart from Western rock often involving the incorporation of 'latino' rhythms and flamenco-like guitar breaks in terms of both scales and timing (as in Carlos Santana); New Zealand hip-hop fuses US and Polynesian rhythms, tones and subject matter in a way which is unique (as in Che Fu); African reggae which mixes Caribbean sounds with the soukous style is unique; British Bhangra-Beat epitomized by Apache Indian from Birmingham in the early 1990s welded hip-hop, reggae, ska and bhangra. What is undeniable however is that these forms rarely flow back to where they came from and recondition the forms from which they were derived. That was, arguably, until the recent rise of so-called 'World music'.

Of course, outside of commercial circuits of capital infinitely rich and diverse musical forms exist that are reproduced daily. In the early 1980s, frustrated by the seeming impossibility of classifying such diverse styles, a small record store in London coined the phrase 'World music' (see Plate 6.10). Despite the neocolonial overtones of conflating the diverse musical heritage of over 90 per cent of the world's population, this new descriptor created a space for otherwise 'invisible' music and provided a model for larger record stores. Rising demand for World music led to increased airplay and the development of dedicated festivals such as WOMAD (World Music and Dance). A new industry, populated by relatively small-scale firms, has been stimulated, and there are now over 3000 World music labels. There has also been a flowering of devoted

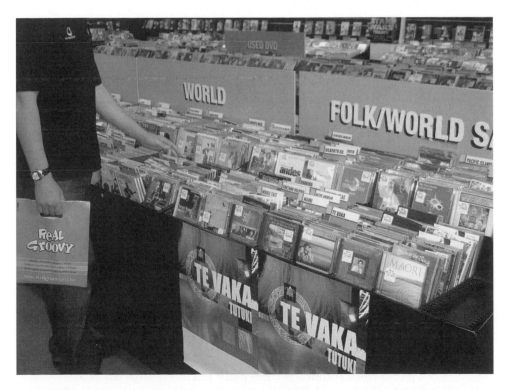

Plate 6.10 *World Music section and shopper at 'Real Groovy' record store, Wellington, New Zealand*

internet sites (somewhere in the region of 10,000) and specialist magazines such as *Global Rhythm*.

Despite the predominantly one-way traffic, then, it is now possible to hear alternative music in Western countries more than ever before. This music has been marketed in its 'authentic' form, and has also become acculturated into Western pop music through the work of artists such as Paul Simon (Southern African), and more recently Kula Shaker (Indian), Damon Albarn (African) and Ry Cooder (Tropical Caribbean) to mention but a few. This trend builds on earlier experiments with explicitly non-Western form by groups such as The Beatles, Led Zeppelin, The Jimi Hendrix Experience and The Doors in the late 1960s and early 1970s. There can be little doubt that this shift is correlated with increased travel and the facilitating role of communications technology. Indeed, critics would argue that the absorption of non-Western forms and its morphing into more marketable forms for commercial purposes has been central in the history of the global music industry – the foundation of contemporary music blues and jazz, after all, was rooted in Africa. The celebration of the rise of resistant musical voices may be premature however. World music accounts for a tiny proportion of total sales, standing at less than 1 per cent in the UK.

At the time of writing, the increasing role of the internet as a form of distributing music was undermining the hegemonic role of the global TNCs leading to industry-estimated losses of US$4.2 billion per annum (RIAA, 2004). Large corporations are reacting strongly to the increased accessibility of music downloads, as recent lawsuits against 'pirate' download sites such as *Napster* and *Kazaa* exemplify. Furthermore, they are moving rapidly into digital download technologies in an attempt to regain control of the market. New communications infrastructure has already created space for alternative artists and labels. For example, there are 50,000 independent bands currently listed at the website GarageBand.com. However, as long as access to this technology is stratified it may serve only to perpetuate the geographical asymmetries in power that have always characterized the global music industry.

Conclusion – towards progressive cultural spaces?

A global cultural shift has taken place and this is reflected in the academy by the evolution of perspectives that give greater precedence to cultural interpretations and explanations. The 'cultural turn' seemed to suggest that culture underpins change in all spheres. A more convincing argument is that the determining powers of culture, politics and economics are intertwined inseparably. In this vein Allen (2000, p. 466) argues:

> [W]e must be cautious about the current fashion for cultural explanations. If culture is understood to be the general context in which social action occurs, then all human behaviour which is not biological can be viewed as being cultural. . . . Culture therefore explains everything, and nothing in particular.

Culture gives people a sense of community and belonging – and is therefore one of the principal means through which identity is constructed and sustained. Until recently cultures have been thought of as relatively stable and spatially bounded, although historical evidence does not necessarily support this. Contemporary globalization has led to the growing recognition of ties between distant and disparate places, symbols and ideas. The emphasis in cultural geography has thus shifted from consideration of bounded spaces to flows of commodities, people, ideas, images and beliefs. This is an unsettling time for cultural identities as globalization has proceeded at a rate unmatched in any other sphere. Does this imply cultural homogenization and the deterritorialization of identity?

Homogenization arguments are based on dated views of cultural interaction. Although the Westernization and Americanization of global culture are powerful processes that really exist, the arguments that support these theses are usually anecdotal and often focus on the most readily visible aspects of cultural change. Homogenization is wittingly and unwittingly resisted by individuals and the cultures they inhabit on a daily basis. Thus the thesis of cultural imperialism is greatly exaggerated and eurocentric. Rather, contemporary globalization has led to the hybridization of culture to a greater degree than ever before. At the same time, as a reaction to homogenization, the reassertion of national culture is a trend that should not be underestimated. What makes this process all the more complicated is that 'national' identity and culture in general is increasingly 'located' in dispersed networks. The increasing incidence of diaspora is challenging the very concepts around which globalization has been constructed. Diaspora may plant the seeds of a *progressive* globalization of culture – so sorely required in today's society – which celebrates and builds from difference and diversity.

Further reading

Connell and Gibson (2003): This book explores the links between popular music, identity and space.

Crang (1998): This book is a thorough and lively introduction to the major debates in cultural geography.

Held *et al.*, (1999) ch. 7: A particular strength of the book *Global Transformations* is its attention to historical forces. This chapter places current cultural globalization in a solid historical framework.

Jackson (1989, 2002): The first reference here is a classic in cultural geography which should be read by those who are interested in understanding the roots of the cultural turn in geography. The second reference is an excellent chapter, which provides an updated overview of cultural geographies, consumption and globalization.

Johnston and Sidaway (2004) ch. 7: This chapter traces the rise of the 'cultural turn' in geography and is the best sourced entrance point into the broader literature.

Lash and Urry (1994): This work deals with issues of cultural commodification and its role in contemporary cultural globalization.

McEwan (2001): This chapter provides a succinct overview of definitions of culture, cultural interaction and the impacts of global change.

Pieterse and Parekh (1995): This work outlines the thesis of cultural imperialism in a compelling way. It represents relatively advanced reading however.

Part III
Global challenges

7 Inequality, development and globalization

- Unequal worlds – geographies of development
- Development and globalization – some links
- Discourses and measurements of development
- Patterns of global inequality
- History of global development – from the Second World War to the debt crisis
- Neoliberal development and globalization
- Responses from the 'periphery'
- Regional case studies of globalization and development
  - The Pacific Islands – off the global map?
  - Latin America – perpetual resource periphery?
- Conclusion – new development geographies?

Unequal worlds – geographies of development

As people wake up in the world this morning they do so in very different conditions. Some are healthy, well clothed, employed and secure. Three-quarters of the world's population are not so lucky. The majority find themselves in unhealthy, poor conditions with little economic security. Many of the world's poor live in Third World rural areas, but it is the ranks of the urban poor that are swelling most rapidly. Clearly, living standards vary widely between countries; it is less obvious that they vary widely within countries (see Plate 7.1). Latin American countries, for example, have among the highest income inequality in the world. There are also pockets of deprivation in 'rich' countries. The principal argument of this chapter is that accelerated globalization has exacerbated unevenness as it creates new networks of inclusion/exclusion and perpetuates old ones. Development geography studies patterns of inequality at various geographic scales – globally, within and between nation-states and localities and, more recently, on and off the networks. A relatively young subdiscipline, it has a number of overlapping aims:

Plate 7.1 *Inequality on Tongatapu Island, Kingdom of Tonga. These starkly contrasting peasant and royal homes are located within a kilometre of each other. Feudalism persists in this highly hierarchical Pacific Island society*

- To describe and map inequalities, especially as they relate to poorer regions;
- To explain and interpret the factors giving rise to such inequalities;
- To uncover historical trends of inequality across space;
- To study competing discourses of development and their spatial manifestations at various levels;
- To propose and assess frameworks for the reduction of geographic inequalities.

Development geography is an integrative subdiscipline in that it deals with multiple spheres of human activity – economic, cultural, political and environmental – in the context of understanding, interpreting and analysing differentiated patterns of well-being. It has changed considerably since its transformation from 'tropical geography' in the 1960s through the work of pioneers such as Keith Buchanan (1963, 1964) and Harold Brookfield (1975) (Power and Sidaway, 2004). Broadly speaking, the subject has evolved a greater concern for the political and cultural as opposed to the purely economic; it has shifted from positivism to a broader range of epistemologies and methods, including radical approaches and discourse analysis; and, the focus on the nation-state has moved in part to networks. Notwithstanding this transformation, elements of the old and the new permeate the way development geography is practised across the world. Two excellent, though partly contesting, overviews are offered in Potter *et al.*, (2004) and Power (2003). As this chapter argues, development geographers are yet to grapple convincingly with the implications of globalization for their subdiscipline and the Third World.

Development and globalization – some links

The proposition that globalization leads to uneven development has focused the attention of human geographers. To investigate this claim we need to carefully establish how globalization and development might be related and to assemble empirical evidence hung within coherent theoretical frameworks. There have been few attempts to link theses of globalization and theories of development explicitly, and Table 7.1 offers some first thoughts. Broadly speaking, three views with respect to the developmental implications of globalization may be envisaged – *neoliberal, neostructuralist* and *dependency/post-development*, all of which build upon established theoretical traditions in development studies. The former, which draws on *hyperglobalist* views, sees globalization as a positive force for development. The second, which bears much in common with the *transformationalist* thesis, argues that the impacts of globalization on development depend on the way it is regulated. The final perspective, which links with *sceptical* and radical *hyperglobalist* views, posits that globalization perpetuates underdevelopment, arguing that discourses of both 'globalization' and 'development' represent extensions of imperial strategies of dominance and are thus much the same thing. As is clear from Table 7.1, these perspectives turn largely on the definition of development – something to which we turn subsequently.

As discussed in Chapter 4, the rhetoric of *neoliberal globalization* discourses is that world income differentials are conflating and that prosperity will eventually be equalized. Evidence elsewhere in this book and beyond suggests that this is far from the case, and this is particularly so in terms of patterns of well-being and 'development'. Despite great absolute advances in poorer regions since the Second World War, the relative distribution of welfare, measured both between nation-states and individuals, is more uneven than ever before (Allen and Thomas, 2000; Potter *et al.*, 2004). These 'gaps' have a long heritage, rooted in particular in the period of colonial globalization. Arguably, however, it is during the postcolonial wave, and especially neoliberal phase, that

Table 7.1 *Theses of globalization and development theories – a schema*

Development theory	Perspective on globalization	Definition of development	Explanation for lack of development	Outcome of development	Main strategy and policy
Neoliberal	Pro-globalization hyperglobalist	Market-based economic growth modernization	State intervention corruption isolation	Convergence in incomes. Liberal democracy	Liberalization, deregulation, marketization
Structuralist/ neostructuralist	Alter-globalization transformation-alist	Holistic income growth that is sustainable	Nature of insertion into global system	Depends on how practised and regulated	Selective intervention for equity and sustainability
Dependency/ post development	Anti-globalization sceptical	Discourse to perpetuate capitalism	Exploitation by imperial and neo-imperial elites	Perpetuation of underdevelop-ment and marginality	Withdraw from capitalism, alternative lifestyles

welfare inequalities have crystallized as never before. The opportunities for those plugged into new networks of prosperity are not inconsiderable, but the vast majority of the planet's population are not linked in this way. This has the potential to fuel global conflict. Indeed, post-9/11 a number of commentators argued that the root cause of so-called religious fundamentalism was in fact inequality (Chomsky, 2004).

As this book has constantly argued, the nation-state is not necessarily the most appropriate unit for comparative purposes. Although national boundaries still play an important function in terms of regulating and containing politics, culture and economy, spaces of flows have become increasingly important. These networks of power and privilege increasingly transcend traditional boundaries. In general, such networks are concentrated most obviously in richer countries. Conversely, 'black holes' of marginalization and deprivation – that is, gaps in the net – are more common in the poorer world. Notwithstanding this, elements of both exist in both 'worlds'; there are pockets of privilege in the Third World and pockets of poverty in the West. As such, it could be argued that 'worlds' exist within 'worlds'. However, in the face of the processes of globalization it is increasingly meaningless to talk of separate 'worlds' in a territorial sense (although this was never *really* feasible, as *comprador* dependency theory elaborated). Networks of inclusion/exclusion are part of the same global system and the spatial implications of this are complex. It is these networks that geographers need to get inside in order to understand the relationships between inequality, development and globalization. In the remainder of this chapter we explore definitions, meanings and histories of development, and assess these issues in the context of two Third World regional case studies that consider the impacts of globalized networks.

Discourses and measurements of development

Not surprisingly, perspectives on development are wide-ranging and often conflicting. Box 7.1 offers three contesting options flowing from the highly orthodox to the very radical. Given the many meanings and discourses of development, no universal measure exists. When 'development' was first conceptualized following the Second World War it was assumed that social progress would trickle down through economic gains. The most common measurements were thus economic: GNP per capita; real GNP per capita; economic growth; the level of

---

Box 7.1

Three contesting definitions of development

The process by which a traditional society employing primitive techniques, and therefore capable of sustaining only a modest level of per capita income is transformed into a modern, high technology, high income economy.

(Williamson and Milner, 1991)

The process of improving the quality of all human's lives. Three important aspects of development are (1) raising people's living levels, through relevant economic growth processes; (2) creating conditions conducive to the growth of people's self esteem through the promotion of institutions which promote human dignity and respect;

and (3) Increasing people's freedom by enlarging their choice variables.

(Todaro, 1997)

Development has evaporated. The metaphor opened up a field of knowledge and for a while gave scientists something to believe in. After some decades, it is clear that this field of knowledge is a mined, unexplorable land. Neither in nature nor in society does there exist an evolution that imposes transformations towards 'ever more perfect forms' as a law. Reality is open to surprise. Modern man has failed in his effort to be god.

(Esteva, 1992)

---

industrialization and/or urbanization; and export structure. Clearly there are problems with these measures; they are not always comparable due to data inadequacies and, more importantly, they do not take social factors and inequality into account. By the end of the 1970s there was wide agreement that such yardsticks were not broad enough to capture what development *should* mean. For example, Edgar Owens said:

> Development has been treated by the economists as if it were nothing more than an exercise in applied economics, unrelated to political ideas, forms of government, and the role of people in society. It is high time we combine political and economic theory to consider not just ways in which society can become more productive but the quality of the societies which are supposed to become more productive – the development of people rather than the development of things.
>
> Edgar Owens, in Todaro, 1997, p. 15)

Given these criticisms there was an attempt to incorporate broader factors, and a number of *indices of development* were produced. Often, it was found that little correlation existed between GNP per capita and 'social' development. Although there are many more complex measures, the most broadly used is the UN's *Human Development Index* which combines the following three factors weighted equally to produce a score out of one (see also Map 7.1):

1 *Longevity* (life expectancy at birth);
2 *Knowledge* (adult literacy (1/3) and years of schooling (2/3));
3 *Standard of living* (real GNP per capita converted to purchasing power parity).

In academia, the debate over what development is and should be has broadened substantially over the past twenty years – in particular with the influence of the 'cultural turn' (see Chapter 6). Grand narratives and theories such as those espoused in the modernization and neoliberal periods have been roundly criticized (see discussion below). Influenced by postmodernism, local development issues and solutions based on participatory perspectives have been privileged in some quarters. Some have come to question the concept of development itself, arguing that it forms part of the neo-imperial discourse of the West (Power, 2003). The anti-development school, for example, sees development as part of the effort by the economically dominant West to expand its interests through the diffusion of market-based policy. In this sense globalization, if defined as an agenda of this nature, and development are much the same thing. Postcolonial thinkers argue the case further, claiming that development is merely an extension of discourses of development instrumental during the formal period of imperialism, and as such represents a further wave of globalization. Today's colonialism is more insipid and not always 'economic' in nature. It is increasingly cultural – thus the calls by postcolonial thinkers for the 'decolonization of the mind'. It is important to distinguish between development as an ideal construct and development *as currently practised*. In this sense, measuring development is increasingly meaningless to a new wave of development geographers.

Despite the richness of the academic debate, development in practice is largely still a top-down, modernist project. The bodies that are instrumental in the diffusion of orthodox development – principally the Bretton Woods institutions (IMF, WB and WTO) – are also the main bodies that espouse the virtues of economic globalization. The currently dominant discourse is that development may be achieved only by opening up borders to the processes of globalization – and this involves neoliberal reform designed towards that end. Fundamentally, the goals of orthodox development are much the same as they were after the Second World War. It is true that a number of ideas from the radical schools, such as sustainable development, participation and empowerment, have been co-opted by mainstream development trustees including transnational financial institutions and aid agencies, although policies often pay lip-service only to such concepts. The current 'Poverty Alleviation' consensus among aid agencies, for example, represents an extension of neoliberal discourse under another name (see Storey *et al.*, 2005). The hegemonic discourse at present then is 'globalize or perish', measuring development in terms of economic prosperity. It must be stressed again that this is only the mainstream view of what development means. The development literature is enormously rich and complex (see Brohman, 1996; Crush, 1995; Potter *et al.*, 2004; Power, 2003; Preston, 1996).

Patterns of global inequality

This section presents some empirical evidence to back up the claims of considerable and growing spatial inequalities of well-being at the global level. In the absence of comparable smaller scale data the nation-state and regions are focused upon. This information gives us a broad idea of the contours of inequality, but it must be viewed as a broad-brush outline only. Whether such inequality is a direct result of the processes of globalization per se is a proposition with which geographers and allied social scientists are currently engaged. There is a growing consensus in human geography that this is in fact so. Potter *et al.*, in their book *Geographies of Development* (1999, p. 103), argue that:

> [W]e can conclude that uneven and unequal development are still characteristic of the global capitalist system. Globalization is not all-encompassing and there is much that remains uneven about global relationships and global processes. All of these aspects of dynamic change are strongly skewed towards the developed North. The world may effectively be getting smaller, but the majority of its population does not yet have access to a telephone. For example . . . over half the world's population have never made a telephone call, and there are more telephones in the New York-New Jersey area alone than in all of Africa combined.

The use of statistics to describe patterns of development is problematic; they can hide difference, lead to labelling, may be unreliable and, crucially, miss very important *perceptions* of well-being. However, selective use can be instructive. Approximate absolute figures speak of vast global differences. At the start of the new millennium 1.3 billion people lived on less than US$1 a day; over 800 million were malnourished; 840 million illiterate; and 1.2 billion lacked access to safe water (UNDP, 2001). Children and women often suffer the highest incidence of deprivation. Differences in per capita incomes between individual nation-states are enormous, the ratio – measured in terms of purchasing power – between the richest (Norway, US$41,974 per capita GDP) and poorest (Sierra Leone, US$150 per capita GDP) standing at around 280:1 in 2003 (UNDP, 2004). The latest available figures at the time of writing on global inequality, measured in terms of income, life expectancy and education, are shown in Table 7.2.

Table 7.2 reveals a number of important spatial and empirical patterns. Incomes in the OECD nations are around six times that of developing countries as a whole. On average the least developed countries have GDP per capita incomes approximately six times smaller than the world average. There is significant variation between 'developing' regions, with Latin America, the Arab States and East Asia being relatively wealthy although if China were removed from this measure the latter region would appear much wealthier. Far and away the most deprived region is sub-Saharan Africa, the rise of AIDS having a major impact on life expectancy and livelihoods there. The United Nations divides the world into a three-part 'human development' classification, which illustrates the spatial inequalities that exist (see Map 7.1). Based purely on income, the World Bank's classification system reveals even deeper differentiation (see Map 7.2). Maps 7.1 and 7.2 reveal a clear spatial concentration of deprivation and affluence measured at the scale of the state. North America, Western Europe, parts of East Asia and Australasia are relatively prosperous, while Latin America, the Middle East, Eastern Europe, South and Southeast Asia, Africa and the Pacific are relatively deprived. The majority of the world's people (close to four-fifths) live in these latter regions. Overall, the UNDP (2001) estimates that the richest 1 per cent of the population receives as much income as the poorest 57 per cent.

Global poverty, in an absolute sense, is concentrated in South Asia, where half of the world's poor live; but as a proportion of the population, there are more people in poverty in Africa than anywhere else. Poverty has declined dramatically in East Asia (falling by 125 million between 1987 and 1998), particularly in China. However, over the past decade, poverty has been rising in South Asia, Eastern Europe during the

Table 7.2 *Global inequality in material terms: income, life expectancy and education, 2002*

Country grouping	GDP per capita[1]	Life Expectancy[2]	Education Index[3]
All developing countries	**4,054**	**64.6**	**0.71**
Arab States	5,069	66.3	0.61
East Asia and the Pacific	4,768	69.8	0.83
Latin America and the Caribbean	7,223	70.5	0.86
South Asia	2,658	63.2	0.57
Sub-Saharan Africa	1,790	46.3	0.56
Least developed countries	**1,307**	**50.6**	**0.49**
Central and Eastern Europe and the CIS	**7,192**	**69.5**	**0.93**
OECD	**24,904**	**77.1**	**0.94**
UN classification			
High human development	24,806	77.4	0.95
Medium human development	4,269	67.2	0.75
Low human development	1,184	49.1	0.50
World Bank classification			
High income	28,741	78.3	0.97
Middle income	5,908	70.0	0.84
Low income	2,149	59.1	0.59
World	**7,804**	**66.9**	**0.76**

Notes

[1] Gross Domestic Product per capita, purchasing power parity, $US.

[2] The number of years a newborn infant would live if prevailing patterns of age-specific mortality rates at the time of birth were to stay the same throughout the child's life.

[3] The education index is a composite indicator consisting of the adult literacy rate and the combined gross enrolment ratio for primary, secondary and tertiary schools.

Source: *Human Development Report 2004*, United Nations Development Programme
http://hdr.undp.org/statistics/data/index_alpha_indicators.cfm (21/09/04)

transition from communism, and especially sub-Saharan Africa where over 140 million people fell into poverty between 1965 and 2000 (UNTAD, 2002). There is a huge gap between rich and poor countries across a range of health measures. The number of children who die before reaching one year of age is around seven times higher in sub-Saharan Africa and South Asia than in the West. In many South Asian and African states, only half of the children of secondary school age are enrolled in education. Sub-Saharan Africa has a functional

Map 7.1 *Global disparities in human development, 2004*

Source: Based on measures of human development from UNDP (2004)

Map 7.2 *Global disparities in income per capita, 2004*

Source: Based on World Bank (2004) classifications

Income grouping

∴ High

Upper-middle

Lower-middle

Low

N/A

0   2,050   4,100   3,200 kilometers

illiteracy rate of over 63 per cent (UNDP, 2004). In many Third World regions gender inequalities remain stubbornly persistent, most tangibly in terms of access to education and political systems. The *technological gap* is especially troubling. Internet access is 'paralleling class systems of stratification, and threatens to splinter the globe into haves and have nots based on access to information/communications technologies' (Potter *et al.*, 2004, p. 103) (see also Map 7.3). For example, in 2002, 445 out of 1000 people used the internet in high-income countries, compared to only thirteen in low-income countries (UNDP, 2004).

But these static measures, however troubling, do not necessarily imply a worsening in the distribution of well-being over time. Figure 7.1 illustrates that in terms of real GDP per capita some Third World regions have converged with the OECD economics. This is particularly the case in East Asia (excluding China) where, in 1980, GDP per capita stood at 26.9 per cent of the OECD level; by 2000 this figure had risen to 51.6 per cent. Over the same two decades there was also a marginal income convergence between the OECD and Southeast Asia/Pacific, South Asia and East Asia (including China). However, in Latin America/Caribbean, the Arab states and sub-Saharan Africa incomes have declined relative to the OECD average. For developing countries as a whole (not shown in Figure 7.1) incomes first diverged, falling from 5.7 per cent to 5 per cent of the OECD average between 1980 and 1990. By 2000, however, this measure had risen to 5.7 per cent. When we compare the poorest countries with the richest countries, however, the evidence for global income divergence becomes clearer. Figure 7.2 shows income per capita (at constant prices) for the twenty richest countries and the least developed countries – a UNDP classification that includes forty-nine very poor economies. In the richer states, real income more than doubled between 1960 and 2000, while in the least developed countries it remained stagnant. Non oil-exporting poor countries performed particularly badly during this period. Potter *et al.*, (2004) estimate that between 1820 and 2000 the income ratio between the richest 20 per cent and poorest 20 per cent of nation-states has risen from 3:1 to approximately 70:1. Given negative economic growth in sub-Saharan Africa of –0.8 between 1975 and 2002, the extremes of inequality are set to increase.

When we consider trends in poverty in the poorest countries of the world the crisis of inequality becomes apparent. The proportion of people living on less than a dollar a day (the standard measure of poverty) increased from 48 per cent to just over 50 per cent between the periods 1965 to

Map 7.3 *Digital divide: national rates of internet use*

Source: World Bank (2004)

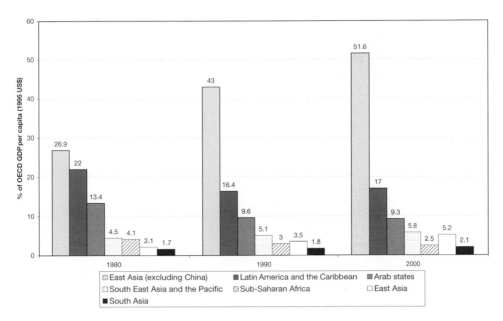

Figure 7.1 *Real GDP in Third World regions compared to the OECD (per capita at 1987 prices), 1980–2000*

Source: Calculated from UNDP (2002) and UNDP (2000)

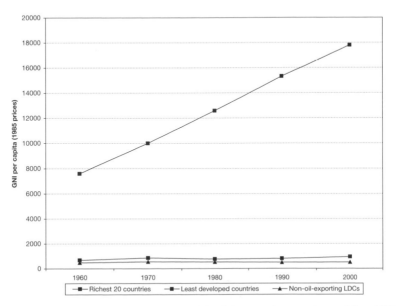

Figure 7.2 *Real per capita income in the world's richest and poorest countries, 1960–2000 (1985 US$)*

Source: Calculated from UNCTAD (2002)

1969 and 1995 to 1999. In absolute terms this represents a doubling in the numbers in poverty from 123 million to 279 million people. The situation is particularly poor in African least developed countries where rates have risen from 55.8 per cent to 64.9 per cent over the same period (see Figure 7.3). In the other twenty-two developing countries, however, the ratio of those in poverty has fallen consistently and stood at less than 8 per cent at the turn of the century. This corresponded to an absolute fall in poverty from around 760 million to 290 million.

There have therefore been some gains in certain development measures over the past thirty years. Potter *et al.*, (2004, p. 32) argue:

> One of the reactions to anti-developmentalist thinking is the argument that, in overall terms, impressive gains have been made to conditions in developing nations over the past thirty years . . . a child born today can on average expect to live eight years longer . . . the level of adult literacy has increased. . . . Average incomes in developing countries have almost doubled in real terms between 1975 and 1998.

However, the main point in the context of the discussion here is that absolute deprivation levels have risen to crisis point and that the gap

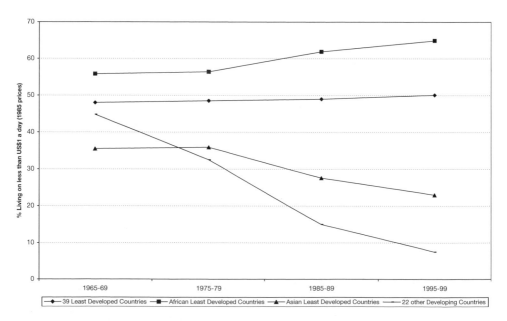

Figure 7.3   *Relative levels of poverty in developing and least developed countries, 1965 to 1999 (1985 prices)*
Source: Calculated from UNCTAD (2002)

between the upper and lower end has increased markedly. The United Nations recognized this when it announced its ambitious eight-point Millennium Development Goals, which aim to drastically reduce global deprivation and inequality (see UNDP, 2003), claiming that this was the 'world's last chance'. Despite the effort on the part of the UNDP, and its attempt to draw in other global institutions (the IMF and World Bank have also adopted the Millennium Development Goals, for example), development measures continue to deteriorate in some areas of the Third World. For example, between 1998 and 2002, the index of human development fell in twenty-one countries; economic growth declined in fifty-four countries and poverty increased in thirty-seven countries (UNDP, 2004). Within the developing world there is increased heterogeneity in terms of such measures, however. In this sense, relative levels of welfare are more unequal than ever across space.

History of global development – from the Second World War to the debt crisis

Before going on to look at particular regional cases we need to understand how the current development orthodoxy was arrived at and how this relates to discourses and concepts of globalization. The contention here is that development as conceived by mainstream thinkers in the post-Second World War era and the concept of globalization are wrapped up in the same desire to expand and speed up capitalism. In this sense the concept and goal of development is a defining feature of the second wave of globalization – although specific strategies have shifted across time. The agents instrumental in setting up the development agenda were the World Bank and IMF created after the Second World War in their effort to restructure the global economy and re-form the capitalist mode of regulation (see Chapter 5). We focus first on the modernization phase of this endeavour, and neoliberalism is considered subsequently. Note that in this context we are exploring the development concepts that have explicitly informed orthodox policy.

Modernization in the post-war development era

Modernization theory evolved from two real world events: the Great Depression (1930s) and the end of the Second World War. The latter provided resolve for liberal-capitalist powers, while the former supplied an example of how to lift nation-states out of recession through state intervention (based on the US New Deal). These factors converged in the form of the post-war reconstruction Marshall Plan in Europe. In general, the approach was a 'grand' paradigm (collection of linked theories) which argued that development equalled 'modernization' and that the 'level' of any given nation-state could be measured by comparison to 'developed' Western countries. Poorer societies had to throw off 'tradition-bound' ways to progress. In order to advance the vital processes of economic growth, urbanization and industrialization, 'modern' attitudes such as competition and individualism were seen as essential. Rostow's Stages of Growth model is the epitome of such a theory, involving five steps through which countries should progress in order to reach the end-goal of 'high mass consumption' (Rostow, 1960). It was in the context of the elaboration of the modernization argument that President Truman first used the terminology 'development' and 'underdevelopment' to refer to the challenges facing what became known as the 'Third World'.

What did the modernization paradigm imply in terms of policy for the Third World? Economically, the crux of Rostow's 'take-off' stage is stimulating investment. To do this, it was argued, urbanization and industrialization were required as these processes would raise savings through increased incomes and profits, feeding back into investment. These social changes would yield the necessary conditions for this, since backward and forward linkages, and thus multiplier effects, were higher in urban areas and industry than in rural zones and agriculture respectively. Industrialization, in particular, could be achieved through state intervention from national governments supported by external credit and aid (based on the Marshall Plan model). In this sense modernization promoted a state-centred 'developmentalist' approach, representing a form of state-assisted capitalism. Foreign capital played a central role in investment and flows increased markedly after the Second World War, especially to East Asia and Latin America, underpinning subsequent industrialization in the former at least. There was an important ideological component to modernization, forming part of the West's attempt to prevent the spread of communism. In this sense, it was the economic backbone of the West's Cold War geopolitics.

By the 1970s modernization approaches were largely discredited. In academia, the major criticism was that the paradigm represented an ahistoric and non-geographical interpretation of the world that was eurocentric. The approach was overly economic and universalizing

– failing to take into account cultural diversity and contesting perspectives on development. In the policy arena the approach was refuted because it did not work – economies did not grow as expected when pursuing the policies that were espoused. There was a transition sometime in the late 1970s from the state developmentalist/aid modernization model to the neoliberal model. It is important to stress, however, that the ultimate objective of these two approaches – in terms of what development *should* be – are the same. It is the route taken to stimulate economic growth and modernity that varies.

The debt crisis – a transition point

From the end of the Second World War until the early 1970s the global economy as a whole boomed. This was interrupted in the early 1970s by the *oil crisis* which saw a large hike in the price per barrel of oil charged by the OPEC nations. Inflated oil prices left the global economy awash with 'petro-dollars' (windfall profits of the OPEC group) and vast amounts were channelled through the World Bank to the Third World in the form of loans for modernization. When the second oil crisis hit in 1979 to 1980, it had a devastating effect on the whole global financial system. Cost-push inflation rose sharply in the OECD, and to fight it the monetarist solution of raising interest rates was pursued. This had an impact on Third World countries' debt to both transnational institutions and richer countries, exemplifying the increased interconnectedness of the global economy. In many cases, given extensive borrowing in the 1970s, it was not possible for countries to pay back the interest on their loans (never mind the capital). In some Latin American countries, for example, debt rose to levels higher than total GDP per annum. Led by Mexico in 1982 there was widespread default on debt to the World Bank especially, resulting in what became known as the Debt Crisis (see Box 7.2). In response, the World Bank and IMF developed and applied 'structural adjustment programmes' (SAPs) based on neoliberal principles. Ostensibly, these were designed to foster economic self-sufficiency and to make the most of globalization. In reality, they were designed to stabilize the global financial system, which underpinned the economic prosperity of the West.

Box 7.2

The debt crisis

Virtually all Third World countries that were not oil-exporting nations, which is the vast majority, were hit hard by the oil crises and the ensuing interest rate hikes and general recession of the early 1980s. In an absolute sense, Latin America shouldered the highest burden; by 1982 countries in the region had accumulated debt of over US$315 billion (Gwynne, 1985). In sub-Saharan Africa by the same year, total debt equalled US$200 billion. However, absolute figures are not the most revealing; it is the ratio of GDP and debt that is the most relevant. In 1981, in Latin America as a whole, debt stood at around 30 per cent of GDP, and this figure was substantially higher in sub-Saharan Africa. In individual countries, such as Bolivia, debt rose to levels that outstripped GDP. The ratio of debt to export earnings is also important as the latter provide the foreign exchange to service the former. In the early 1980s in Latin America, debt represented approximately 220 per cent of export earnings, whereas in East Asia it was closer to 50 per cent. The lower ratio in East Asia, given far higher per capita export earnings there, formed part of the explanation for the region's subsequent economic success. In the Third World in general a 'lost decade' of development resulted from the debt crisis, which, it could be argued, is continuing into the present (see Corbridge, 2002). The 'lost decade' was characterized by the following processes:

- Direct foreign investment declined markedly and an outflow of profits to the OECD took place;
- Continued high interest rates in the 1980s meant that countries were paying out more to cover interest on debts than they were actually repaying in terms of original capital;
- Non-oil commodity prices continued to decline, further lowering poor countries' terms of trade;
- Aid and lending declined as the recession cut into the global financial system;
- Protectionism grew in the OECD – despite the fact that this organization recommended that the Third World open its doors and reduce regulation;
- Austere structural adjustment policies were introduced, raising poverty and unemployment.

This combination of factors had major economic, social and political ramifications:

- *Economic crisis* – Many Third World countries were plunged into negative growth and declining exports. This was particularly the case in countries extremely reliant on primary product exports (apart from oil), i.e most of Africa, Latin America, South and Southeast Asia. Latin American per capita incomes fell by 10 per cent in the 1980s, for example (Gwynne and Kay, 2004).

Box 7.2
continued

• *Social crisis* – Income inequality increased as government spending was cut and other social programmes were pulled back through the application of SAPs. Poverty rose dramatically, with women and children being the hardest hit (Corbridge, 1993).

• *Political crisis* – The ensuing socio-economic chaos provided the rationale for military interventions in many countries. Latin America, for example, was entirely governed by dictatorships by the mid-1980s (Silva, 2004).

The debt crisis is only over in the sense that debts have been rescheduled. In absolute terms, the foreign debt of developing countries has continued to rise. From US$658 billion in 1980, it had soared to US$2560 billion by 1999 (World Bank, 2002). It currently stands at close to US$3000 billion, a figure which many have argued is simply not payable. Between 1980 and 1999 sub-Saharan Africa's debt tripled to around US$220 billion, and throughout the decade Africa returned US$81.6 billion to its creditors. Latin America's debt had grown to around US$820 billion by the start of the twenty-first century. Individual country figures point to the overwhelming challenges which many poorer countries still face (UNDP, 1999). The World Bank considers that there are close to forty-five heavily indebted countries including Angola, Bolivia, Ethiopia, Laos, Nicaragua, Sierra Leone and Vietnam. In close to twenty of these, debt is equal to or above GDP per annum. The Mexican crisis of 1994, the Asian crisis of 1997 and the ongoing Argentinean crisis

have all had the effect of increasing debt in these respective areas. More than any other single factor, it is debt burden that prevents meaningful progress in poorer countries.

Calls for the cancellation of Third World debt have grown from across the NGO sector, and some European governments are now actively involved in debt-relief initiatives. Oxfam, a transnational NGO, has fought for two decades on this front, for example. In the 1990s a significant global movement pulling together anti-debt groups and powerful figures from politics and entertainment was launched. Jubilee 2000 aimed to 'cancel the unpayable debts of the poorest countries by the year 2000 under a fair and transparent process' (Jubilee 2000, 2002, p. 2). A global petition signed by twenty-four million people was assembled to support this objective, giving birth, in 2001, to the Jubilee Movement International. Partly in response to this general groundswell the World Bank and IMF established an enhanced 'heavily indebted poor country' initiative and the issue of debt was placed on the agenda at G7 summits. This has seen debt relief extended to around twenty-five countries to date, with decisions pending in other cases. In the case of Bolivia total external debt, for example, has been reduced by 50 per cent. More recently the UK government, led by the efforts of Chancellor Gordon Brown, lobbied the G7 and international financial institutions for 100 per cent debt relief for thirty-seven of the worst cases, as part of Labour's *Marshall Plan for Africa* promoted under the slogan of 'making poverty history'. In February 2005, G7 ministers and Russia

Box 7.2
continued

committed to these objectives and confirmed immediate debt relief for the countries affected by the 2004 Boxing Day Tsunami in the Indian Ocean. They remained divided on how such initiatives should be funded, however. The USA, in particular, was sceptical of Gordon Brown's proposal for an international financial facility that would channel US$10 billion per annum into debt relief over the subsequent decade, and rejected plans for increased overseas aid.

Significant and symbolic as these trends might be, they represent unfulfilled pledges that have not been fully ratified by the appropriate institutions and nation-states. Even if they are instituted, the amounts involved represent a drop in the ocean of total debt. Poorer countries will remain locked in the vicious cycle of debt for the foreseeable future, as they are obliged to pursue neoliberal solutions for their relative deprivation.

Neoliberal development and globalization

Neoliberalism is the term used to refer to the economic paradigm and cultural ideology that is now virtually hegemonic across the world (see Chapter 4). Neoliberalism recommends the elimination of government intervention in the economy, arguing that 'governments fail' – causing inefficiency, crowding out private investment, and ultimately lowering competitiveness on the global stage. Some neoliberal arguments have linked state developmentalism and corruption. According to this argument, 'rolling back the state' will foster efficient and sustainable economic growth. A deregulated, privatized, economy is argued to be the best way to maximize welfare at global level, and global free trade based on the exploitation of comparative advantage is given precedence (see Chapter 4). The most important point from a globalization perspective is that neoliberalism opens the doors to flows in investment and trade, facilitating the penetration of TNCs, which allows the greater articulation of global networks into localities. Neoliberalism is argued by some to be an emerging culture, which forms part of the wider mode of regulation of capitalism (Peck and Tickell, 2002). It is ideologically grounded in Western values of modernization, 'civilization', individualism, materialism, accumulation and rationality. In this sense it forms part of what Cowen and Shenton term 'doctrines of development' which are rooted in the Enlightenment period (Cowen and Shenton, 1996). The application of neoliberal policies has important cultural impacts in terms of the rhythms and spatiality of everyday life, leading to the increased flow of Western cultural symbols especially.

Neoliberalism emanates from the University of Chicago and is associated with Milton Friedman's monetarist economics. It was first applied in Chile, after the military government which seized power in 1973 contracted Chicago economists to develop a model that would totally reverse the socialist path taken hitherto (Barton and Murray, 2002). The 'Chicago boys', as they became known, effectively used the country as a laboratory for their theories, aggressively privatizing and opening up the economy, and reducing the size of the state to a bare minimum. Because neoliberalism requires austerity, which hits the poor the hardest, neoliberalism could only be applied in Chile, and later Latin America in general, under military rule. Neoliberalism is not just part of the socio-economic history of the poorer world; from the early 1980s it was also applied across the OECD as the post-Second World War 'welfare consensus' was eroded following the oil crises of the 1970s. Neoliberal principles were applied in the UK under Thatcher from 1979, and this was followed in the USA under Reagan from 1980. New Zealand adopted such reform with particular purity from 1984 onwards, and, despite rising income inequality, marginalization and deprivation since that time (Kelsey, 1995) is seen as something of a model in neoliberal and hyperglobalist circles (Ohmae, 1995).

Academic supporters of the neoliberal paradigm often inhabit schools of macroeconomics or management. Analysis in these areas tends to avoid engagement with social and cultural issues, partly explaining their enthusiasm for the model. The rapid economic development of East Asia over the past thirty years is sometimes claimed as a victory for neoliberalism and outward orientation, although many have argued that what sets East Asia apart is its special combination of neoliberal and structuralist/developmentalist policies (Coe *et al.*, 2003). In human geography there is a general consensus that neoliberal policy is in fact tantamount to jungle law (McKenna and Murray, 2002; Peck and Tickell, 1994), which increases inequality both within and between countries. Thinkers are divided on the extent to which 'pure' neoliberalism has diffused across the world; there is no such thing as a pure free-market economy and all are 'mixed' to a certain extent (LeHeron, 1993). However, there can be no doubt that there has been a paradigmatic shift towards the market end of the political spectrum and that this has facilitated heightened globalization and increasingly intense restructuring impacts on the ground.

Structural adjustment policies

As noted previously, neoliberalism dominates development policy and
has eclipsed a range of alternative development theories from academia
and beyond (see Box 7.3). SAPs are the major policy platform for
neoliberalism, and IMF and World Bank loans, as well as some aid
flows, became conditional upon their adoption from the early 1980s
onwards (see Box 7.2). As Map 7.4 shows, they have diffused widely
across the Third World as a consequence. By 1990, for example, nearly
every country in Latin America had been subject to such a programme.
Through conditionality, these policies confer great power on the
institutions, industries and state elites that are involved in their design
and implementation. These bodies can become the effective regulators
and governors of the affected economies.

Structural adjustment policies have been criticized for a one-size-fits-all
approach although they do vary over time and space to a certain extent
(see Stiglitz, 2002). They generally include the following sets of
measures:

- *Downsize* – reduce public expenditure (e.g. education, health and
  public infrastructure) and taxation to raise incentives for private
  investment and enterprise, and to reduce budget deficits.
- *Privatize* – sell off the state-owned enterprises and privatize
  government functions where possible to release an efficiency-inducing
  profit motive and reduce government expenditure of the market.
- *Deregulate* – reduce the intervention of the state in the economy, such
  as regional policy or sector subsidies, to allow the market to find its
  natural optimum or equilibrium. Reduce the red tape required to
  establish private enterprise (see Cartoon 7.1).
- *Globalize* – reduce tariffs and any other protectionist measures in
  order to open borders to inward TNC investment; to facilitate the
  transfer of technology, to stimulate competition with world producers;
  and to attain economies of scale from larger market size in order to
  enhance efficiency and exports. Devaluation of the domestic currency
  in order to stimulate the latter is often a further characteristic.

The impacts of SAPs are hotly contested. In general it is recognized by
proponents and detractors alike that the austerity required of structural
adjustment will bring regressive impacts in the short term. All countries
adopting SAPs experience increases in poverty, income inequality,
unemployment and informal sector activity (see Plate 7.2 and 7.3).

13,800 kilometers

6,900

3,450

0

Map 7.4 Countries undergoing structural adjustment, 1980s

Cartoon 7.1 *Deregulation*

Source: Kirk Anderson

Plate 7.2 *Urban poverty in Bangkok, Thailand*

Source: Donovan Storey

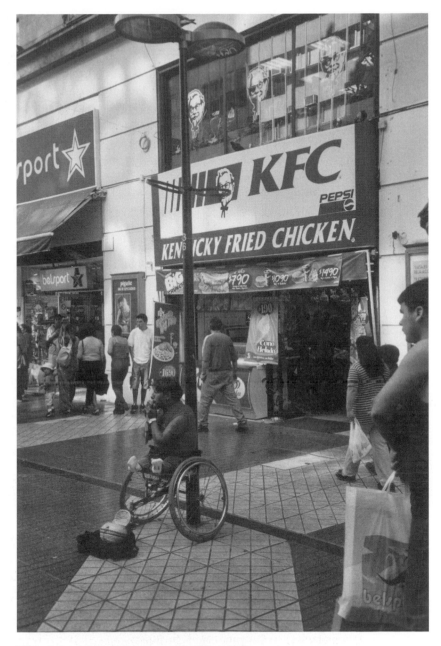

Plate 7.3 *Informal economy, Santiago, Chile*

Neoliberals are not troubled by this, seeing it is as part of the 'shake-down' required to advance efficiency in the medium and long term. Critics point to the fact that many countries adopting SAPs have experienced persistent problems with income distribution, poverty and other social measures of development. Empirical studies of SAPs have revealed that it is women, children and the poor who bear the heaviest burden, particularly as state support mechanisms are withdrawn (Radcliffe, 2004). In the case of women in Latin America, for example, double or triple days became increasingly common in the reform period of the 1980s as women took on both formal and informal employment in order to make ends meet. This was combined with the fact that female-headed households were becoming increasingly common, due in part to the poverty unleashed
by SAPs (Chant, 1999). Neoliberalism in general privileges capital over all else and creates networks that are difficult for the poor and already marginalized to access.

Responses from the 'periphery'

In spite of the hegemonic position attained by modernization and later neoliberalism, there is a rich tradition of alternative development theory. Some of this has emanated from the periphery itself; especially Latin America but also Africa and Asia. Structuralist and dependency theories, two important Latin American contributions, are implicitly tied up with globalization (see Box 7.3). A unifying characteristic of these linked approaches is the anti-modernization idea that all development is 'contingent'. That is to say, development turns in part on the historical, social, cultural and economic specificity of place and nation-state. Central in both accounts is how globalization has unfolded according to these specificities. Although these ideas are considered outdated in Western policy circles there has been a revival of academic interest recently, as thinkers search for realistic ways of interpreting the developmental failure of neoliberalism. In the places that they were evolved, such perspectives are becoming increasingly relevant. The current economic model in Chile may be termed 'neo-structuralist', for example (See Gwynne and Kay, 2004).

Despite innovative theorizations from the periphery such as those outlined in Box 7.3 – and to these could be added many more examples such as marginalization and anti-development – they have been largely superseded in orthodox circles. Structuralism and dependency influenced policy for a few short decades between the Second World War and the debt crisis; economic policy in Latin America and East Asia built on structuralist ISI models, for instance, and a number of revolutions in Latin America and Africa in the 1970s were informed by dependency analysis. Generally, however, doctrines of development have diffused from the core to the periphery. There has been widespread resistance to, and calls for reform of, the dominant globalist model (see Power, 2003; Routledge, 2002). Collective and individual resistance to Western development in Latin America has a long and violent history, for example (Kay, 2001). It could be argued that ongoing conflicts across the Middle East form part of this collective resistance to the penetration of Westernization. There has been significant work undertaken to attempt to explode the myth that Third World populations are passive victims of 'development' – practising everyday resistance in the form of livelihood strategies and life-cycle shifts (Bebbington, 2004). There have been a number of institutionalized attempts to counter the transnational institutions of the OECD and Bretton Woods. Significant examples include OPEC, the non-aligned movement and the G7 (see McGrew, 2000). It is possible to interpret increased regionalization efforts in poorer regions as a form of resistance (see Chapter 5), with examples including MERCOSUR, SADC, COMESA and ACS. In general however, the dominant ideology of development has been articulated so powerfully and imposed so forcefully that it is difficult to see how it can be reversed. This proposition is illustrated through two regional case studies in the following section.

Box 7.3

From the margins? Globalization and dependency/structuralism

Two theoretical innovations from Latin America conceptualize the relationship between the global, the national and the local in much more explicit ways than modernization or neoliberalism and have informed policy there, and beyond, at points in recent history.

Dependency

Dependency analysis is a collection of theories that emphasize the 'dependence' of the Third World on international capitalism. Based on Marxist analysis – which emphasizes class struggle – exploitation is theorized at different geographical levels. This approach became

Box 7.3
continued

popular during the 1970s, as Cold War struggles continued and the post-war boom faded (Kay, 1989). At the global scale Frank (1969) argued that connections between 'metropolitan' and 'satellite' states (trade flows, aid flows, technology transfer, capital flows, cultural exchange) were channels through which rich countries extracted 'surplus' (profit) from the colonized and formerly colonized periphery. This was referred to as 'unequal exchange', as occurred in Marxist theory between capitalists and the working classes. As such, dependency is actually an early example of network theory. TNCs were argued to be the major agents in this process and the neocolonial relations that resulted would persist indefinitely, thus perpetuating the 'development of underdevelopment'. Similar relationships could evolve within nation-states also. The suggested action to break this cycle was, in radical forms of dependency analysis at least, political revolution.

Structuralism

Dependency theory is a more radical take on an earlier structuralist paradigm. This collection of associated theories traces the impact of historical evolution on the structure of the global economy and the resultant patterns of well-being. Raul Prebisch of the Economic Commission for Latin America (ECLA) was the leading structuralist (Prebisch, 1950, 1964). His major contribution was the 'secular decline' hypothesis, which related to the problems of specialization in natural resource exports created by the historical insertion of peripheral economies into the colonial system as suppliers of resources.

It was argued that primary products face income-inelastic demand, while manufactured products face income-elastic demand. Therefore, as global income increases, the demand for, and therefore the price of, primary products will decline relative to that of manufactured products. In countries that specialize in primary product exports this has serious implications. Over time they will have to export more primary products to pay for the same amount of manufactured products. This vicious circle is referred to as a decline in the terms of trade. In the long run, secular decline will tend to increase the gap between the 'core' and 'periphery', as the centre captures the benefits of boom, and growth in general. On the basis of this, Prebisch argued for rapid state-sponsored industrialization in order to escape the 'primary product trap'. Import Substitution Industrialization (ISI) was the policy prescription which would involve the replacement of imported items, moving through progressive stages. In many ways, structuralism represented an early elaboration of ideas concerning resource peripheries and state developmentalism (Hayter et al., 2003; Murray, 2002a).

Regional case studies of globalization and development

There have been relatively few human geographies which seek to explicitly trace out the impacts of globalization on poorer regions. It is suggested here that an ideal framework for the exploration of these issues is provided by new regional geography (see Bradshaw, 1990). This approach, which builds on the criticisms of 'traditional' regional geography and takes on some of the points made by both radical and cultural geographers, provides a multi-scaled analysis of the outcomes of globalization. New regional geography searches for links, rather than breaks, within and between regions. Recently, geographers in this field have elaborated the concept of 'resource peripheries' in the global economy, partly building on structuralist ideas (Hayter *et al.*, 2003).

The two case studies offered below reflect the research interests of the author and focus on economic globalization in particular. In both cases, regional trends and outcomes relating to globalization – both historical and contemporary – are identified and neoliberalism is placed in a long-run context. These transformations are then illustrated in more detail with reference to examples of change at national and local levels. These studies illustrate that the unfolding of globalization is path-dependent; that is to say, it evolves in distinctive ways across space according to historical, environmental and geographical contingencies. Those who are interested in the impacts of globalization on development in East and Southeast Asia should see Rigg (2003) or Lin Sien (2003). Geographies of globalization and development in Africa have been understudied; an entrance point is Aryeetey-Attoh (2003) or Cline-Cole and Robson (2003).

The Pacific Islands – off the global map?

The Pacific region is the largest region on Earth in terms of surface area (see Map 7.5). In terms of land mass, however, it is one of the smallest. Some of its nation-states and territories are among the most diminutive on Earth in demographic and economic terms. Despite this, it is a complex region in terms of history, economy, culture and politics, which is divided into three subregions that exhibit some, albeit limited, internal homogeneity: Polynesia, Melanesia and Micronesia.
The processes which have unfolded in the region tell us much about the relationship between globalization as an agenda and a set of processes, and development as both a discourse and a process. It is a region little

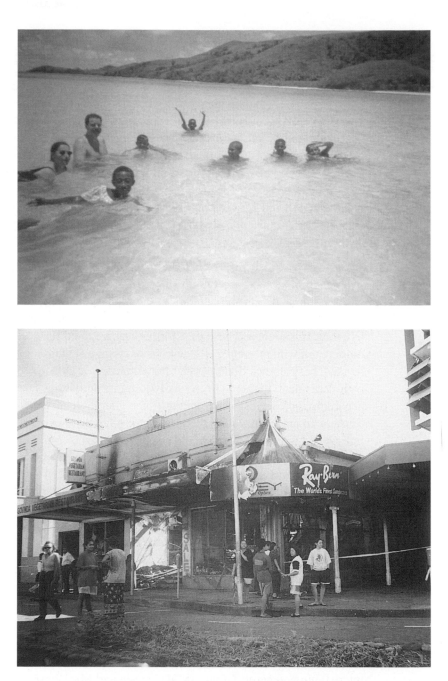

Plate 7.4 *Contrasting images of the Pacific Islands. The first picture epitomizes the stereotypes of tropical island paradise; the second shows the stark realities of political conflict taking place, in this case following the Fijian coup of 2000*

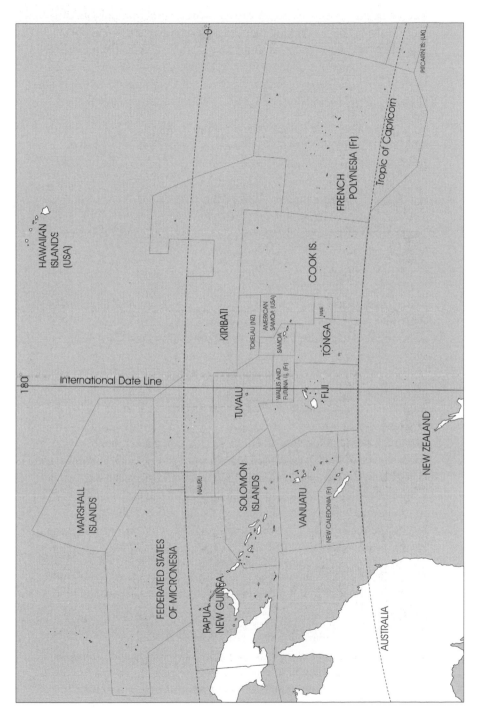

Map 7.5   The Pacific Island region

known or understood by the rest of the world (it is often invisible in world atlases, for example), including those countries nearest to it in a territorial sense (Murray and Storey, 2003; Overton and Scheyvens, 1999). Constructions of an exotic and friendly paradise, perpetuated through the tourist sector, led to a pervasive 'orientalism' of the Pacific (Nicole, 2000) (see also Plate 7.1).

At a global level, the Pacific Island region is marginalized in economic and political terms. Material standards are especially low in parts of Melanesia and Micronesia. They are highest in French colonies, especially French Polynesia, although income distribution in these territories is highly regressive. Since the start of decolonization in the 1960s, Pacific Island Countries (PICs) have found it difficult to prosper in the competitive, globalizing world – many of the problems of its special geography plus its subordinate position in the global economic hierarchy weigh against it. Modernization is causing massive social, economic and environmental tensions as new systems of resource management, and governance systems, interact with traditional ones. Absolute poverty is high in some countries (e.g. Fiji 25 per cent), although it is not as high in relative terms as it is in Latin America or Pacific Asia. The region is sometimes argued to be characterized by 'subsistence affluence', a concept which suggests that abundant land and sea resources confer a level of prosperity far greater than economic indicators might suggest. To a large extent this is an invention of the Western imagination, which idealizes daily life in the region. Economic marginalization and environmental degradation make livelihoods as precarious as they are anywhere else in the Third World. Notwithstanding this, it is true to say that traditional systems of land ownership have led to a relatively equal income distribution, in the independent states at least. Some human development indicators for the independent countries of the region are set out in Table 7.3, illustrating the relative impoverishment of conflict-torn Melanesia.

The region's unique geography has created five important challenges in terms of the participation of its constituent nation-states in the global economy.

- *Territorial* – isolation in terms of absolute distance;
- *Environmental* – vulnerability to extreme weather and global environmental change;
- *Political* – low political bargaining power due to diminutive nation-state size;

Table 7.3 *Human development indicators for Oceania, 1999*

	Population (000s)	Annual population growth (1990–1998)	GDP per capita (US$)	Life expectancy	Urban population (%)	HDI
Fiji (Mel)	797.8	0.9	2,684	66.5	46	0.667
PNG (Mel)	4,412.4	2.5	1,196	54	15	0.314
Solomon Islands (Mel)	417.8	3.3	926	64.7	13	0.371
Vanuatu (Mel)	182.5	2.7	1,231	65.8	18	0.425
FSM (Mic)	114.1	2.1	2,070	65.7	27	0.569
Kiribati (Mic)	85.1	2.1	702	61.6	37	0.515
Marshall Islands (Mic)	61.1	3.6	1,182	65	65	0.563
Nauru (Mic)	11.5	2.4	3,450	58.2	100	0.663
Cook Islands (P)	16.5	–1.8	4,947	72	59	0.822
Niue (P)	2.1	–1.0	3,714	74	32	0.774
Samoa (P)	174.8	1.2	1,060	66.6	21	0.590
Tonga (P)	98	0.3	1,868	68	32	0.647
Tuvalu (P)	11	3.0	1,157	67	42	0.583

Notes: P = Polynesia; Mel – Melanesia; Mic – Micronesia
Source: Murray and Storey (2003)

- *Economic* – limited economies of scale and low diversity due to small economic size;
- *Factor endowment* – limited land and land-based natural resources in some cases, especially on atoll islands.

Orthodox analysis has often blamed these 'inherent' problems for the economic deprivation of the region. Along with these constraints to development, some also add 'traditionalism' and 'communalism' – although this approach has been widely criticized. Although the above five factors are of influence, the underlying explanation for the region's relative economic deprivation is far more complex. Since colonial penetration the region has served as a resource periphery and geopolitical playground for global powers, and this plays a role in its continuing political-economic marginalization. A structuralist/transformationalist approach would argue that the current position of the Pacific is best understood through an analysis of the way the Island region has fitted into the wider world over time.

Waves of globalization in Oceania

The history of globalization in the Pacific Islands has been divided into two waves (Firth, 2000; Murray, 2001), punctuated by a period referred to as the MIRAB era (Bertram and Watters, 1985). The following periodization is being suggested here:

1 The first wave – colonial globalization (1830s–1960s)
2 MIRAB era – 'indigenous' globalization (1960s–1980s)
3 The second wave – postcolonial neoliberal globalization (mid 1980s–)

Colonial globalization

The first wave began with the arrival of the 'buccaneers of global capitalism' (Firth, 2000, p. 180) who were part of the evolving global division of labour made possible through improved maritime technology. Thus initial contact was made through whalers, missionaries and traders who were largely British and French, beginning the process of the diffusion of Western socio-cultural values, such as monetization, Christianity and eurocentric constructs of progress and civilization. Political rivalries meant that the South Pacific became an important strategic location (Rapaport, 1999). There were also economic benefits to be extracted. Based on calls from settlers and traders for the establishment of law and order in their new peripheral economies, formal colonization began in the mid-1800s in the French colonies first, followed by British and German territories thirty to fifty years later. The German territories were lost following the First World War, ran as mandates and then handed over to the UK, Australia or New Zealand (see Table 7.4). The first wave of globalization drew the PICs into the mercantilist circuit of production, placed firmly at the periphery of the global economy. As Firth (2000, p. 182) argues:

> [G]lobalisation *when combined with colonial rule*, meant incorporation into the global economy on terms that suited the interests of the colonial powers. . . . The place of the tropical world in the first globalisation was to be subordinate to the temperate and developed world.

Table 7.4 *Colonization and independence in the Pacific Islands*

Colonial power	Incorporated into colonial power*	Dependent colony/territory	Self-governing in free association	Independent
UK and France				Vanuatu (1980)
United Kingdom		Pitcairn (c1838)		Fiji (c1874, i1970) Tonga (i1970)* Tuvalu (c1916, i1978)** Solomon Islands (c1920, i1978)** Kiribati (c1916, i1979)**
France		French Polynesia (c1904) New Caledonia (c1853) Wallis and Futuna (c1842)		
New Zealand		Tokelau (c1916)	Cook Islands (c1900, sg1965) Niue (c1901, sg1974)	Samoa (c1900, i1962)
Australia		Norfolk Island (c1788)		Nauru (c1920, i1968) Papua New Guinea (c1884, i1975)
USA	Hawai'i (c1898, in 1959)	American Samoa (c1899) Guam (c1898)	Marshall Is (c1885, sg1986) FSM (c1885, sg1986) Palau (c1885, sg1994)	
Chile	Rapa Nui (Easter Island) (in 1888)			

(i1970) = date of independence. (c1874) = date of colonisation. (sg1986) = date of self-government. (in 1959) = date of incorporation

Notes: * Tonga was a protectorate of the United Kingdom from 1901 not a fully fledged colony. ** These countries were protectorates for approximately twenty years before formal colonialism. In many cases the colonial power changed hands once or twice (especially after the First World War when Germany lost its possessions) the latest colonial power is reported here.

Source: Adapted from Murray and Storey (2003)

The classic colonial era in the region persisted until the 1960s, although a number of territories remain colonies into the present. Independence came late to the PICs relative to the rest of the Third World. This late decolonization can be explained in terms of two factors – economic non-viability and the geopolitical advantage to US-allied powers of maintaining Pacific territories in the Cold War context. In those countries that became independent in the 1960s and 1970s, the influence of Keynesian and structuralist policies was strong and many turned to inward oriented development strategies, including ISI (Chandra, 1992). These were often funded by aid flows from former colonial powers in attempts to modernize the Pacific economies. This, combined with increased flows of remittances from ex-patriot islanders who migrated in large numbers to the Pacific Rim in the 1960s and 1970s, led to the development of networked MIRAB economies (see Box 7.4). According to Bertram and Watters, such economies represented a form of 'dependent development' that offered long-term sustainability. This was rapidly eclipsed however.

Box 7.4

MIRAB – indigenous globalization?

MIRAB is an acronym used to describe countries where migration, aid, remittances and bureaucracy play a central role. The hypothesis, developed by Bertram and Watters (1985) to refer to postcolonial Pacific Island economies, seeks to explain why Pacific Islanders have higher average standards of living than per capita GDP measures predict. Following independence a number of linked processes were observed. Migration, from the islands to former colonial and other powers on the Pacific Rim (predominantly NZ, Australia and the USA) established 'transnational corporations of kin', through which individuals, acting in a utilitarian manner, channel remittances both financially and in kind. Aid formed a significant proportion of national income,

driven originally by the strategic imperatives of the Cold War and also by postcolonial 'obligations', sometimes constitutionally enshrined, of former powers. The combined magnitude of aid and remittances allowed consumption at levels well above GDP. Having inherited relatively large colonial bureaucracies, operating in small labour markets, the state sector became the major employer.

The model is applicable to varying degrees in the different PICs, and has been applied less successfully to other island regions. It remains relatively salient in Polynesia where remittances and aid play significant roles. In Niue, for example, aid accounts for over 50 per cent of GDP, while in Samoa remittances can account for up to 40 per

Box 7.4
continued

cent of GDP (Bertram, 1999). In 1985, Bertram and Watters argued that MIRAB constituted a form of dependent development that was both rational and sustainable. This has been called into question, however, as third and fourth generation migrants lose physical and psychological ties with their 'homelands' (although there is, as yet, no recorded decline in remittances); aid from traditional sources declines as the Cold War geopolitical comparative advantage of the region fades (although aid from East Asia – attracted by fish supplies – is possibly replacing this); and public sectors have been radically downsized on the advice of regional and global development institutions. Within the region itself MIRAB remains a highly controversial topic that still frames much development and globalization debate at academic and policy levels (see Bertram, 2006).

Postcolonial globalization

Aid and finance donors do not share the optimism of Bertram and Watters with regard to the sustainability of MIRAB economies. Across the Pacific Island region there is a notable trend towards neoliberal economic restructuring. This change mirrors patterns elsewhere in the Third World – although it has come relatively late to the Pacific Island region. Following the end of the Cold War, economic policies of donor and creditor countries shifted from the traditional aid/modernization approach towards that which was intended to stimulate autonomous economic development. Assistance, both in aid and loans, is now usually accompanied by a range of restructuring conditions such as downsizing the public sector, privatization, tariff reduction, subsidy elimination, and other 'competitiveness-inducing' policies. Among the main objectives of the neoliberal outward orientation is the stimulation of productive efficiency and export competitiveness.

The first country to adopt this strategy was Fiji, following the coup of 1987. The inward-oriented strategy of the state, which represented the most advanced attempt at industrialization in the region, was reversed and reforms intended to stimulate export-oriented growth enacted (Murray, 2000). Policies included a number of devaluations, the establishment of export-processing zones, widespread privatizations, and a reduction in the government sector and social expenditure (Chandra, 1992). It is only recently however that neoliberalism has begun to approach hegemony in the region, opening the doors for a renewed wave of capitalist expansion. The PICs have little choice but

Plate 7.5 *The changing role of women in the Pacific Islands: female seasonal workers in an export tuna cannery, Levuka, Ovalau, Fiji*

to adopt such restructuring, and in this context Firth (2000, p. 186) laments:

> Just as the place of the Pacific Islands in the first globalization was to be sub-ordinate to the temperate, developed world, so their place in the second globalization is also to be subordinate, this time to a set of international institutions that have set the rules of the global economy. The new globalization, *now combined with independent sovereign rule*, means incorporation into the global economy on terms that suit the interests of the financial markets, the aid donors, and those relatively few Pacific Islanders who are in a position to benefit from the new situation.

What are the impacts of the diffusion of neoliberal globalization for the region? A number of points can be made:

- Studies have shown that neoliberalism has led to the development of inequitable and environmentally and socially unsustainable export systems, especially in the agricultural sector (Murray, 2000; Storey and Murray, 2001).
- In urban areas, unemployment and marginality have risen as state support mechanisms and traditional networks are eroded (Connell, 2003a).
- In general the economic policy of the region is determined by outside powers with a vested interest in exploiting the region, and island governments themselves have very little role in decision-making (Henderson, 2003).

- Poverty has risen sharply in urban and rural areas affecting children and women disproportionately (Sriskandarajah, 2003).
- Privatization has led to a number of scandals in the pubic sector, and buy-outs by government.
- The development of enclave economies, such as those which characterize tourism sectors, have proliferated.
- Pressures to reform land structures from communal systems to private property regimes have grown, despite the role these have played in maintaining relative welfare equality (Overton, 2000).
- Linked to this latter point, and in general, some have argued that Pacific Island cultures, which are millennia-old, are increasingly unsustainable as Western constructs become increasingly pervasive (see Plate 7.6).
- Finally, some have argued that neoliberalism and the inequalities it has produced have led to overt political conflict (Murray and Storey, 2003; Overton, 2003).

Neoliberals argue that it is because the region has not globalized enough that the problems outlined above are being encountered, and

Plate 7.6 *This food outlet, in Nadi, Viti Levu, Fiji, reflects the influence of 'local' and 'global' forces in Fijian fast food diets*
Source: Author

that internal problems such as corruption explain the poor performance of Pacific economies (Hughes, 2003). Such explanations, though theoretically redundant and morally questionable, remain influential in policy-determining circles. The move towards neoliberalism represents nothing short of a paradigmatic shift in the region. It is important that questions are raised as to the validity of pursuing such a model in the context of the economically and environmentally fragile microstates of the Pacific Islands. The Pacific Islands are at an important crossroads; globalization cannot be turned back but the region requires a model which will engender sustainable and equitable development.

Latin America – perpetual resource periphery?

Despite improvement in some development indicators in the 1990s, Latin America remains one of the poorest regions in the world. The average GDP per capita stood at approximately US$6500 per annum in 2004, and very high levels of inequality, absolute and relative poverty, and debt plague the continent (see Plate 7.7). Globalization has affected the region profoundly (Gwynne and Kay, 2004). Within the global economy Latin America's role has long been that of a 'resource periphery' and this largely remains the case. It is a fascinating place to study development models and globalization, as policy reversals have been common and have often moved between extremes. Nowhere is the search for new and exotic resources to fuel European development both now and in the past better exemplified than in Latin America; to early explorers the continent was the 'land of gold'. Latin America has proceeded through various periods with respect to its insertion into the global capitalist economy, and these are explored briefly in the discussion that follows (Murray, 2004).

1 Colonialism and mercantilism (1500–1820s)
2 Independence and neoclassical theory (1820s–1930)
3 The Great Depression and structuralism (1930s)
4 Revolutions and dependency (1960s)
5 The debt crisis and neoliberalism (1980–?)

Plate 7.7 *Poverty and inequality, Chile: this slum on the banks of the River Mapocho is located in one of Santiago's richest comunas (districts), Lo Barnechea. A high-cost apartment complex is pictured in the background*

Colonialism to the debt crisis

As colonialism was established and consolidated through the sixteenth and seventeenth centuries the Spanish and Portuguese focused on the extraction of a wide range of agricultural and mineral resources, paying no attention to the local environmental impacts of this or its long-run implications. Colonialism established a relationship of dependency with the imperial core, seeing nearly all surpluses expropriated outside of the region. Independence arrived beginning in the 1810s, but did little to alter social conditions or the continent's role as a resource periphery. As long as the world economy grew, so would exports and progress, argued neoclassical thinkers. However, many of the activities were 'enclave' in nature, presenting few multiplier linkages into the socio-economy and concentrating surplus among a small landholding elite and foreign-oriented class.

In general then, Latin America was a broadly outward-oriented region during colonial globalization and in the immediate postcolonial period.

This changed somewhat following the First World War and the Great Depression however, as the continent hit severe recession. Increasingly, doubt was cast upon the idea that simply exporting raw materials and unprocessed foodstuffs could sustain economic progress. The structuralist school of thought emerged out of this crisis (see discussion above). The recommended policy of ISI formed part of a wider move to 'inward-oriented' policies combined with efforts towards regional integration. ISI dominated most economies until the 1980s, with some success in Brazil and Mexico. However, by as early as the late 1960s it had become obvious that structuralist solutions were not delivering broad economic development. Explosive rates of urbanization led to a crisis in the region's cities and the rural sector was left out of policy initiatives (Gilbert, 1996). Land reform progressed only slowly despite the insistence of structuralist academics (Kay, 2002). Dependency theory grew out of dissatisfaction with the performance of structuralist theories, suggesting that participation in the global economy was perpetuating underdevelopment in the continent. Based in part on Castro's 1959 revolution in Cuba, left-wing revolutionary movements were formed in many countries. In the context of the Cold War, socialist revolution was hotly contested by the USA through a mixture of economic, political and military interventions and revolutionary dependency was soon eclipsed. Non-revolutionary dependency has remained influential in regional academia, if not policy circles, until today however (Kay, 1989).

The debt crisis and neoliberalism

The debt crisis was the major factor causing the shift to free-market neoliberal policies and a return to outward orientation in Latin America. The combination of debt and SAPs in the 1980s (see discussion above) led to the worst decade ever in terms of development indictors; income growth fell from 5 per cent between 1965 and 1980 to 1.4 per cent between 1980 and 1989, and was negative for Argentina and Bolivia over the latter period; unemployment and poverty rose – open unemployment stood at over 10 per cent in six countries in 1985; inflation sky-rocketed, reaching an average of 1502 per cent per annum in Bolivia between 1984 and 1993; and, Western-backed dictatorships seized the reigns of power across the continent (Silva, 2004). Once again the continent became articulated into the global economy as a resource periphery as SAPs insisted on export orientation based on comparative advantage. This saw

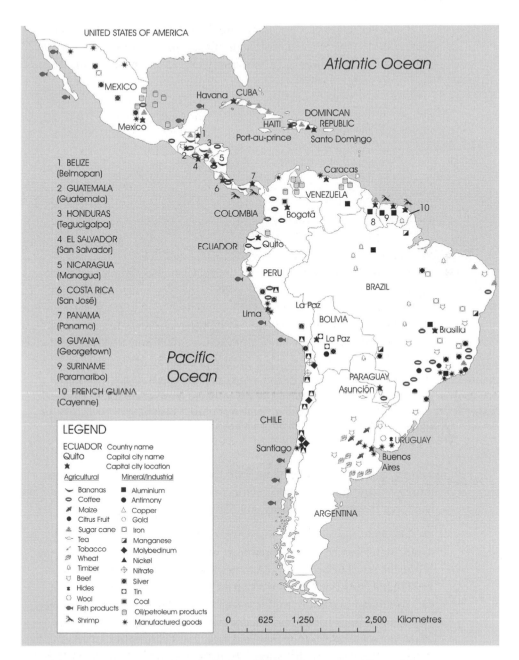

UNITED STATES OF AMERICA

Atlantic Ocean

MEXICO

Havana CUBA

DOMINCAN
REPUBLIC

HAITI
Port-au-prince   Santo Domingo

Mexico

1 BELIZE
(Belmopan)

2 GUATEMALA
(Guatemala)

3 HONDURAS
(Tegucigalpa)

4 EL SALVADOR
(San Salvador)

5 NICARAGUA
(Managua)

6 COSTA RICA
(San José)

7 PANAMA
(Panama)

8 GUYANA
(Georgetown)

9 SURINAME
(Paramaribo)

10 FRENCH GUIANA
(Cayenne)

Caracas

VENEZUELA

10

COLOMBIA   Bogotá

8   9

ECUADOR   Quito

PERU

BRAZIL

Pacific
Ocean

Lima   La Paz

BOLIVIA

La Paz   Brasilia

PARAGUAY

Asunción

CHILE

URUGUAY

Santiago

Buenos
Aires

ARGENTINA

LEGEND

ECUADOR   Country name
Quito   Capital city name
★   Capital city location

Agricultural   Mineral/Industrial

⌣ Bananas   ■ Aluminium
⊘ Coffee   ● Antimony
✦ Maize   △ Copper
● Citrus Fruit ○ Gold
⬟ Sugar cane   ▢ Iron
⌢ Tea   ◪ Manganese
✎ Tobacco   ◆ Molybedinum
✿ Wheat   ▲ Nickel
♤ Timber   ⊕ Nitrate
♨ Beef   ◉ Silver
✖ Hides   ▫ Tin
○ Wool   ◼ Coal
⬱ Fish products ▩ Oil/petroleum products
➤ Shrimp   ✱ Manufactured goods

0   625   1,250   2,500   Kilometres

**Map 7.6   Resource exports from Latin America. Popular representtions of the resource
export sectors of Latin America such as the above reveal a continent endowed with a
startling array of 'gifts of nature'**

Note: Symbols are indicative of the spatial distribution of export activities in a general sense and do not reflect
quantities or values

Plate 7.8 *Grapes and the global economy: the town of El Palqui (middle distance) nestled among table grape vineyards in Chile's semi-arid Norte Chico region. This zone has become one of the major supply sources for the Northern Hemisphere winter market and as a consequence has been entirely transformed socially and environmentally from an inward-oriented traditional economy to an outward-oriented export monoculture*

a partial return to reliance on mineral exports and the creation of non-traditional agricultural export sectors serving affluent markets in the Northern Hemisphere (Murray and Silva, 2004).

By the end of the 1990s, democracy had returned to Latin American nation-states as Cold War geopolitics, which had led to the implicit and explicit support of neoliberal dictatorships by the USA, faded (Barton, 1997). The transition has seen a partial return to policies informed in part by structuralism, referred to as neostructuralism, involving greater concern for social equity and the environment as well as a revival of economic regionalism (Gwynne and Kay, 2000). In reality however, these policies represent an extension of the neoliberal orthodoxy. The legacy of neoliberalism in terms of poor socio-economic conditions is apparent. For example, in Chile income distribution was worse in the early 2000s than it was in 1966 (Barton and Murray, 2002). Despite some diversification at the continental level, most countries remain dangerously reliant on highly specialized primary product export sectors. In 2002, in sixteen of the twenty largest countries, primary products accounted for over 50 per cent of total export earnings. In Nicaragua and Venezuela this measure stood at 90 per cent. The export in eighteen of the twenty major Latin American countries is a primary product, with only the largest countries, Brazil and Mexico, bucking this trend (see Table 7.5 and Map 7.6).

The central role which resource exports play suggests that the original concerns of the structuralists remain relevant today, and are likely to remain so as neoliberalism induces further specialization based on comparative advantage. Latin America's role as a resource periphery has arisen precisely because of its natural endowment in a wide range

Table 7.5 *Export specialization in Latin America, 2001 proportional values*

	Major export	%	Next two major exports	%	Top 3%
Venezuela	crude petroleum	58.9	petroleum products,aluminium	28.6	87.5
Belize	raw sugar	25.7	fruit juice,shell fish	43.8	69.5
Ecuador	crude petroleum	43.5	bananas,shell fish	22.5	66.0
T.and Tobago	petroleum products	38.9	crude petroleum,natural gas	26.4	65.3
Nicaragua	coffee	27.1	shell fish,meat	27.1	54.2
Paraguay	soya beans	32.8	raw cotton,vegetable oil	17.8	50.6
Panama	bananas	19.3	fresh fish,shell fish	28.2	47.5
Chile	refined copper	25.6	copper ores,fresh fish	19.6	45.2
Colombia	crude petroleum	30.6	coal,coffee	14.6	45.2
Costa Rica	office machinery	29.7	bananas,coffee	15.4	45.1
Peru	gold	16.7	fish meal fodder,refined copper	23.5	40.2
Guatemala	coffee	21.3	raw sugar,bananas	15.7	37.0
Honduras	coffee	23.8	soaps,lumber	11.1	34.9
Bolivia	zinc ore	11.7	vegetable oil,aircraft	20.0	31.7
El Salvador	coffee	22.2	petroleum products,medicines	8.4	30.6
Uruguay	meat	15.5	leather,rice	14.3	29.8
Barbados	raw sugar	14.0	petroleum products,food preps.	15.4	29.4
Argentina	crude petroleum	10.8	vegetable oils,petroleum products	13.5	24.3
Mexico	motor vehicles	9.9	crude petroleum,statistical mach.	13.8	23.7
Brazil	aircraft	6.3	iron ore,soya beans	9.5	15.8
Latin America	**crude petroleum**	**12.5**	**motor vehicles,petroleum products**	**9.2**	**21.7**

Source: Murray (2004, p. 121)

of items. In this sense resources can act as a curse and lead to what Karl calls a 'paradox of plenty' (see Box 7.5). With the rolling out of the Free Trade Area of the Americas it is possible that this will become entrenched further, although neoliberal thinkers see the FTAA as a potential saviour – illustrating the contested discourses which shape the geographies of the region. It is time to reconsider the principles of structuralism and dependency analysis in order to build policy which reforms the nature of globalization in the continent and makes development from within possible (Sunkel, 1993). As Karl argues with

---

Box 7.5

Resource peripheries and the paradox of plenty

Development history over the recent past challenges the orthodox concept of the relationship between resources and development, and calls into question policies which encourage poorer countries to specialize in primary product exports. This is based on the 'paradox of plenty' observed by Karl (1997). Two points are of special relevance in this context:

1 East Asian economies are relatively 'resource deficient' but they have achieved high rates of economic growth from the 1980s onwards.
2 Latin American economies are relatively 'resource rich' but have performed poorly from the 1980s onwards.

In resource peripheries such as Latin America 'favourable' resource endowment can prevent progressive economic development – Auty (1995) refers to this as *resource curse* (see also Hayter *et al.*, 2003). 'Resource cure thesis' argues that there is a range of problems associated with placing heavy reliance on the primary product export sector (see also Gwynne, 2004). Among others the following trends may evolve:

● Decline in the terms of trade
● Increased TNC control and expropriation of surplus
● Negative environmental impacts
● The perpetuation of low value-added activity
● Increased vulnerability to fluctuating global commodity markets
● Problems associated with management of windfall profits in mineral sectors
● The evolution of enclave economies
● Focus on the needs of external markets leading to problems of domestic food security

Although the paradox of plenty is not necessarily inevitable, solving these issues requires regulation and intervention. Neoliberalism is not capable of facilitating such policies, and serves only to fuel the paradox.

reference to the paradox of plenty, 'it is not inevitable. Paradoxes can be resolved and development trajectories altered' (Karl, 1997, p. 242).

Conclusion – new development geographies?

Historically, globalization has created and perpetuated inequalities in levels of well-being and development. During the first wave this was reflected in terms of the colonial division of labour that evolved to subordinate the periphery. In the postcolonial wave, when the concept of development was invented, patterns of marginalization and deprivation have become more complex. Neoliberalism in particular is leading to new networks and flows that threaten to raise inequality to greater heights within, as well as between, nation-states. The old North/South classifications and core–periphery models are less relevant today as new networks of inclusion/exclusion are forged. The poorer world as a whole, however, remains saddled with poverty and debt and, despite economic advances in East Asia, there is little consensus on what might solve such problems.

There are three ways of conceptualizing the relationship between globalization and development – neoliberal (hyperglobalist), structuralist/neostructuralist (transformationalist), and dependency/ post-development (sceptical). The evidence presented in this chapter lends support to the second view. That is to say, globalization – when conceived of as stretched flows across space – is not *inherently* negative for the poorer world if it is regulated and managed effectively. However, history has taught us that this has rarely been the case, and globalization as currently practised is exacerbating global inequalities, failing to raise people out of relative deprivation, and locking whole regions into an exploitative capitalist global economy.

Is there a way that globalization can be turned to the advantage of poorer nation-states and those which inhabit networks of deprivation? We are seeing the evolution of global movements that are resistant to the regressive impacts of globalization and development as currently practised. Whether such movements effectively capture the multiple voices that are vying to be heard is a matter for further geographical action-based research. Discourses of globalization and development are closely linked. The pressing task for development geographers is to map the concrete outcomes of these discourses and get inside new networks of power in order to understand how they restructure the multiple spaces of development. Only then can we hope to inform eclectic and holistic policy designed to face the immense global challenge presented by inequality.

Further reading

Cowen and Shenton (1996): In this classic book the authors explore the links between the Enlightenment, imperialism and the current concept of development.

Gwynne and Kay (2004): This collection of work from an international panel of academics presents a systematic and broad-ranging assessment of the impacts of globalization and modernity in Latin America.

Kay (1989): In this book the Latin American contributions to development theory are discussed in detail.

McGrew (2000): This chapter takes a first look at the relationship between development and globalization and presents a useful framework for viewing these connections.

Overton and Scheyvens (1999): This edited collection pulls together a range of detailed chapters on various development issues in the Pacific Islands organized around the concept of sustainable development.

Potter *et al.*, (2004): This is the most comprehensive book on development geography currently available and deals with a broad range of debates in a balanced way. It includes a chapter on globalization.

Power (2003): This book urges us to rethink geographies of development, taking a radical approach that seeks to unpack powerful imperialistic discourses of development.

Environment, sustainability and globalization

- The environment, globalization and geography
- Environmental degradation – definitions
- A brief history of environmental degradation
- Global environmental problems?
- Population growth, globalization and resource consumption
- Globalization of environmental awareness
- Global environmental regulation from above and below
- Conclusion – transformed environments

The environment, globalization and geography

The concepts of 'one world', 'interconnectedness' and 'globalism' have the environmental movement to thank for their broad diffusion around the globe and into the popular imagination. Since the first flight to the Moon in 1969 and the beaming back to Earth of pictures of the world as one fragile and floating globe, the movement has made increasing inroads into conventional politics, academia and the media. The idea that the global environment is at risk due to anthropogenic activity, and that something must be done quickly to abate this, is accepted by a growing majority of people across the world – especially in the relatively affluent West. Environmentalist 'worldviews' are now more or less mainstream in the academic literature. Left-wing ecologists talk of the fragility of the Earth and the imbalance that humans have created, and relate it to concepts such as 'spaceship earth' and the *gaia* thesis (O'Riordan, 1981). Almost all are agreed that processes which are global in scope and yet local in terms of their source threaten the Earth's environment.

In the manifestos of virtually all major political parties across the West the environment figures centrally. In some countries, such as Germany and New Zealand, Green parties have themselves carved out significant representation in Parliament. We have entered a period where those who deny that environmental problems exist, or that technology and human innovation in response to scarcity will solve all issues, seem like extremists, and the isolated George W. Bush US administration should be seen in this light. Just twenty-five years ago it was the other way around, and environmentalists were broadly seen as idealistic and esoteric. This in itself is great testimony to the globalization of concern for the environment and the building of unprecedented global networks that articulate and circulate these issues.

Many definitions of geography cite 'interactions between the environment and humans' as a core focus of the discipline. Human geography has moved through a number of phases in terms of its conceptualization of the human–environment relationship. In the early part of the twentieth century the concept of *environmental determinism* was seen as *the* organizing principle of the discipline. In this argument, the environment set strict boundaries to human activity in any given place and conditioned resultant patterns. Although this concept has been widely discredited, some environmentalists still appeal to its logic when talking of 'limits to growth' at the planetary level. Human geography has moved through phases of *environmental possibilism* and *human determinism*. The former sees the environment presenting a range of opportunities for humans, and the latter argues that human ingenuity allows complete domination over the environment. Over the past twenty years there has been a steady rise in human geography's engagement with the environmental issues, from a possibilistic perspective in particular. This has involved both applied studies of human impacts, work on the politics of environmental change, and – more recently – critical geographies concerning the boundaries between and social construction of 'natural' and 'human' activity (Whatmore, 2002). Ideas from political ecology have had an important influence on environmental geographers over the past fifteen years. This field traces how patterns of power shape the interaction of humans and the environment, and how the outcomes of change are mediated and distributed (i.e. in whose interests is the environment used, and who bears the costs of change). Human geography, then, is a natural home for the study of interaction between ecological and social systems. What makes the discipline particularly useful in the current era is its attention to global–local relations, allowing an analysis of the implications of globalization for the environment at different scales.

What is the relationship between globalization and the environment? This question is explored in this chapter through three interlinked observations. First, it is widely argued that the *processes of globalization are part of the explanation for environmental degradation*. In the first part of this chapter we discuss the concept of degradation, then seek to relate its history to the unfolding of the different waves of globalization. Second, some *environmental problems, including population growth, have become global-scale concerns*. In a subsequent section we look at what constitutes a 'global' environmental problem and provide a number of case studies to back up the arguments. We then move on to look at the relationship between demographic change and environmental issues. Third, *concern for the environment has become globalized and the search for sustainability requires global-scale action*. We conclude by considering the evolution of the global environmental movement and attempts to regulate the system from above and below.

Environmental degradation – definitions

In order to analyse the implications of the processes of globalization on the environment it is necessary to have a working definition of environmental degradation. A practical definition is offered by Tyler-Millar (2002, G5):

> Depletion or destruction of a potentially renewable resource such as soil, grassland, forest or wildlife that is used faster than it is naturally replenished. If such use continues the resource can become non-renewable (on a human time scale) or non-existent (extinct).

Examples of such degradation could refer to the following processes: urbanization of productive land; waterlogging and salinization of soil; soil erosion; deforestation; groundwater depletion; overgrazing and desertification; biodiversity loss; pollution; ozone depletion; and anthropogenic climate change. Humans need not cause such degradation, although most environmentalists agree that human systems are depleting resources and degrading the environment at unprecedented rates. In reality, the definition of environmental degradation is complex and contested and turns on one's worldview. A deep ecologist, for example, who views 'nature' as having equal existence rights as humans, would have a very different viewpoint from a cornucopian who believes that the environment's purpose is to serve humans. We distinguish here between two very broad models in order to establish a perspective for the remainder of the chapter:

1 *Anthropocentric* (human system approach) – Under this view degradation occurs when ecosystems, or parts of them, are transformed in such a way that the consequences have net negative impacts upon the lives and/or health of humans. This may also include

aesthetic degradation, which will not necessarily have negative ecosystem effects. This definition is deceptively simple. For example, what do we mean by negative? To what extent are aesthetic values to be included? To what extent should the distribution of environmental degradation be considered?

2 *Ecocentric* (whole system approach) – In this definition environmental degradation takes place where ecosystems, or parts of them, are transformed in such a way that the consequences have net negative impacts on the ecosystem as a whole. Again, this definition is deceptively simple. For example, are all parts of the ecosystem to be valued equally? Science is not always clear on what the impacts of a given change will be in the short, medium and long term – how do we judge therefore? Finally, by measuring and possibly prioritizing certain parts of the ecosystem we are implying that human judgement can and should be applied which leads towards a more anthropocentric perspective.

In this chapter we will be referring to debates and works that often adopt an anthropocentric perspective, as this is the approach that pervades most writing and thinking on the topic. Deciding on what broadly constitutes degradation is one thing; actually measuring it is another, and scientists are deeply divided, as Box 8.1 on global warming illustrates. The issue in the context of this chapter, then, is whether we can reasonably argue that processes or agendas of globalization are causing such degradation.

A brief history of global environmental degradation

While scientific opinion is fairly unified on the general proposition that many aspects of the global environment are being degraded, and that humans are playing the major role in this process, the links between this and the processes of globalization are not immediately obvious. How has globalization related to environmental degradation across the various waves?

Environmental degradation of global significance during the pre-modern globalization period (see Chapter 3) included a range of global extinctions of mammals and birds from over-hunting (for example, the Moa in New Zealand hunted by Mäori). The diffusion of microbes due to large-scale migratory movements led to epidemics and population collapse – such as the Black Death in Europe during the Middle Ages. Localized pollution began to accumulate during this period. The main driving forces of environmental degradation during this time included urbanization, large-scale migrations, overpopulation in local areas, warfare and poor agricultural practices. In general the extensity, velocity and impact of such processes were relatively limited as both nomadic and settled peoples were incapable of transforming their environments to any great degree (Held *et al.*, 1999).

The colonial wave of globalization set in motion a range of significant changes. The diffusion of European populations, methods and principles to Latin America and parts of Asia led to unprecedented change in the former especially. The isolation between Europe and the Americas was broken during this period, which set a number of important linked changes in motion. The introduction of non-native species, intensive hunting and disease impacted on both animal and human populations greatly, leading to the near extinction of many indigenous groups, for example. The exploitation of both mineral and non-renewable resources for use in the colonies and beyond transformed landscapes. Within Europe, the agricultural revolution and its particular constellation of innovation, technological change, market incentives and higher investment led to a rapid expansion in cultivated areas and land degradation as populations grew. Forests were cut down, marshland drained and a number of species, including wolves and bears, were greatly diminished by the growing 'organic economies' of the time (Held *et al.*, 1999).

The industrial revolution in Northern Europe and its subsequent diffusion across that continent and beyond combined with the way it pulled 'peripheral' economies into the new globalizing industrial complex is *the* watershed in terms of the human impact on the environment. Within Europe new production techniques combined with intense urbanization raised the potential for local degradation and initiated systemic global environmental degradation through industrial pollution, principally via the use of coal. In the colonies, possibly the major impact of the industrialization phase was deforestation, as land was cleared for plantation agriculture to supply imperial heartlands and the colonies. Southeast Asian tropical rainforest was greatly depleted beginning in this period. This phase of globalization involved a steep acceleration in the extensity of European impacts and an increase in the velocity of environmental change.

The postcolonial wave of globalization from the end of the Second World War led to increasingly profound environmental change. The rise of modernization as the organizing principle for the new concept of 'development', which resulted in the diffusion of industrialization and urbanization to the periphery, had major ramifications for environments in the newly independent 'developing' countries (see Chapter 7). In these places, new pressures on resources required for economic growth, combined with unprecedented population growth, placed extra strain on local environments. In countries that remained locked in as resource peripheries, such as many of those in Latin America and Africa, these effects were amplified. The rise of socialist industrialization in the USSR and China also increased both local and global degradation significantly, often in ways that ravaged ecosystems more than in the West. This was due to the fact that Marxist theory was partly explicit that the environment existed to be exploited by humans. In the West the post-war boom fuelled rapid increases in per capita resource consumption, especially in the USA, where a new consumer culture grew and the motorcar was widely adopted. All combined, these factors led to historically unprecedented increases in the velocity, extensity and intensity of human impact on the environment. Many of the problems which remain salient today, including ozone depletion, the anthropogenic greenhouse effect, deforestation, desertification, marine and air pollution, acid rain, biodiversity decline and nuclear waste, were initiated during this period.

The neoliberal phase of the postcolonial wave has had the effect of magnifying the above problems, through the shift to free market economics, reduced state regulation, and the accelerated diffusion of Western industrial/urban/consumerist models of development to poorer countries. Free-trade reform under the WTO, for example, has led to the dismantling of environmental regulations under 'unfair' trading rules in some cases. The continuing debt crisis in poorer countries and the increasing global wealth gap divert resources away from environmental conservation and towards short-term solutions to poverty. Deprivation also fuels conflict, both overt and otherwise, which often has deleterious environmental impacts. Overall the diffusion of capitalism and its associated production and consumption patterns is greatly perpetuating global environmental degradation. A number of examples including the boxed case studies of global environmental change illustrate this point throughout the chapter.

Global environmental problems?

In this section we look at the idea that environmental problems, or at least an increasing amount of them, have become global in their scope. There is nothing inherently good or bad about global environmental change. 'Problems' and 'issues' are generally defined only in terms of their relationship to humans – that is to say, through an anthropogenic model of degradation. In this sense environmental 'problems' are socially constructed. So what do we mean by global environmental problems? This depends very much on the definition of 'global', and this echoes earlier debates concerning how we define different scales (see Chapter 2). On the one hand, no environmental change is global; nothing impacts on the Earth's surface and ecosystems in a uniform way. On the other hand, all environmental change is global in the sense that all transformations are inextricably linked. It is now widely accepted that the global ecosystem is the sum total of intricately interlocking parts and that there are very few, if any, closed ecological systems.

Given the above, how can we conceptualize global environmental degradation in a practical way, and how can this be linked to globalization? It is useful to divide global degradation into global systemic degradation and global cumulative degradation. Examples of the former include climate change, sea-level rise and ozone depletion. Examples of global cumulative degradation include desertification, flooding, deforestation, air and water pollution, soil erosion, acid rain and biodiversity loss.

Global systemic degradation refers to change which affects systems that are inherently global in scope. These systems are sometimes also referred to as global 'commons', the main examples being the atmosphere and the oceans. Within such systems borders are relatively meaningless (although the Law of the Sea has achieved this for the oceans to a certain extent). The essential point is that degradation may come from many sources, which may be quite specific in location, but the impacts affect the system as a whole, as illustrated through the case study of global warming in Box 8.1. This does not imply however that the impacts across the system are necessarily even, as the example of sea-level rise in Box 8.2 exemplifies.

Cumulative global environmental degradation occurs when local incidents of degradation become multiplied to such an extent that they begin to constitute a global-scale concern. As ever, what exactly is meant

Box 8.1

Systemic global degradation? Global warming and globalization

There is considerable scientific and political controversy over the issue of climate change, and in particular 'global warming'. The process of global warming is based on a naturally occurring phenomenon known as the 'greenhouse effect'. Short-wave solar radiation passes through the Earth's atmosphere, is absorbed by the Earth's surface, and reradiated as longer wave radiation back into the atmosphere. Here, it is absorbed by atmospheric elements, principally water vapour and carbon dioxide.

Some of this radiation is then reradiated, heating the Earth's surface and creating a positive environmental feedback loop. Without this effect global temperature would average –19C, compared to the current average of 15C, making life on Earth impossible. There is evidence to suggest that the CO_2 content of the atmosphere is rising markedly due to anthropogenic activity and that through the feedback loop referred to above this is increasing the average surface temperature (see Figures 8.1 and

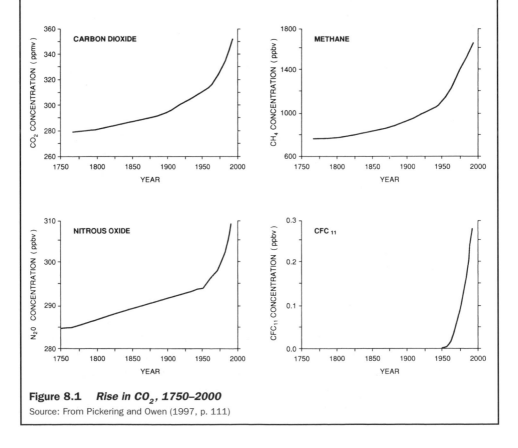

Figure 8.1 *Rise in CO_2, 1750–2000*
Source: From Pickering and Owen (1997, p. 111)

Box 8.1
continued

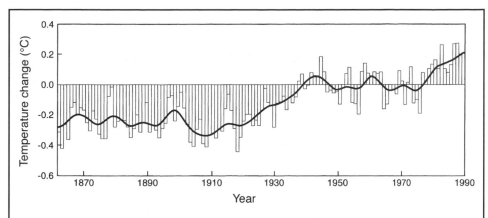

Figure 8.2 *Global mean air and land temperature rises, 1861–1989*
Source: Pickering and Owen (1997, p. 135, fig. 3.14)

8.2). The International Panel on Climate Change (IPCC), a body formed by members of the World Meteorological Organization and the United Nations Environment Program, estimates that the temperature has risen by 0.6C over the past century (IPCC, 2001) and that the magnitude of atmospheric greenhouse gases has increased by around 35 per cent since 1750. Whether this is due to human activity or part of a longer term climatic cycle is contested, although the IPCC claims the former. According to the IPCC the 1990s was the hottest decade on record.

General circulation computer models used by the IPCC have predicted temperature rises of between 1.4C and 5.8C by 2100. This could have devastating impacts, including the melting of the polar ice-caps, sea-level rise and related flooding, alterations of the gulf stream, and major agro-system transformation (see Figure 8.3). The distributional impacts of climate change are uncertain, but it will cause substantial shifts in the climatic regimes across

latitudes, leading to drought and potential famine in many mid-latitude locations, for example (IPCC, 2001). The IPCC estimates greater heatwaves, a more intense hydrological cycle and unprecedented sea-level rise. It is argued that the greatest proportion of negative impacts will fall on Third World countries, and poor persons within other countries. Other scientists from industrial lobby groups have argued, however, that circulation models are not sophisticated or reliable enough and that rises of 0.5C over the twenty-first century are more likely. Notwithstanding these arguments, there is a growing scientific consensus that global warming represents a serious and significant problem that must be addressed.

How does this relate to globalization? The rise in greenhouse gases is traceable to the industrial revolution. As industrialization has diffused across the world the effect has been magnified. Given discourses of modernization, many poorer countries aggressively pursue industrialization

Box 8.1
continued

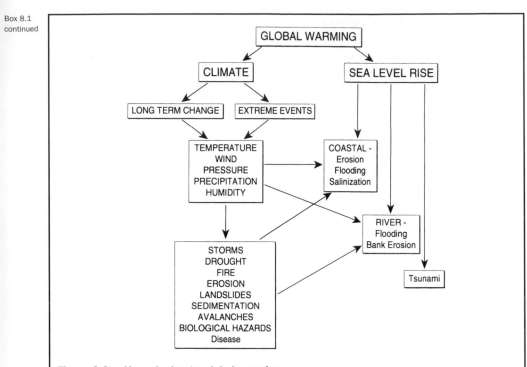

Figure 8.3   *Hazards due to global warming*
Source: Pickering and Owen (1997, p. 149)

strategies and are adopting global consumer culture – including demand for petrol-driven cars and fossil fuel-powered appliances. Industrial expansion in heavily populated countries such as China and India has the potential to add greatly to greenhouse emissions. However, as we discuss elsewhere in this chapter, it is per capita consumption in Western countries that is most responsible for the growth in greenhouse pollutants (see section below on consumption overpopulation). Some poorer countries have argued that the industrialized countries of today paid scant attention to the environment as they modernized, so why should they not be

afforded the same luxury? A distributional dimension is thus inherent in this global problem.

As such the anthropogenic greenhouse effect has become the cause of major political controversy, and initiated globalized political reactions in the form of international cooperation, protocols and treaties which attempt to stabilise/reduce/eradicate polluting emissions (see discussion below on the Kyoto Protocol, for example). Some national governments have erred on the optimistic side and have either failed to ratify or have not followed the guidelines. Across the industrialized world, and in the USA, there

Box 8.1
continued

are a number of pro-industrial organizations that have sponsored media and research activities which play down the proposed seriousness of the global warming issue. The USA, under the administration of George W. Bush (2000–), was especially sceptical of global warming predictions. Many commentators pointed to the administration's links to the oil industry to explain the zeal with which the environmentalists have been fought. Overall, from a total ecosystem perspective, it is unclear whether global climate change of this nature can rightly be considered 'degradation'. From an anthropogenic perspective, however, it clearly is.

Box 8.2

Globalization, sea-level rise and the Pacific Islands

Significant scientific evidence exists which suggests that average sea-level is rising across the planet. The IPCC claims that sea-level has risen by 10–20cm over the past 100 years, and estimates that by 2100 it will have risen between 9cm and 88cm. The melting of the ice-caps and the expansion of the surface of the ocean given raised temperatures across it bring this about. Island regions in general are especially vulnerable to the impacts of climate change and sea-level rise in particular (IPCC, 2001). In the Pacific Ocean there are a number of linked effects that arise from the combination of sea-level rise and warming (Nunn, 2003). Most crucially for the Pacific Island region such conditions are likely to increase the frequency, intensity and impact of cyclones. Such weather systems are devastating when they hit islands, as recent destruction in Niue (2003) and the Cook Islands (2005) exemplify. In the case of the former, the country's infrastructure and agricultural production was almost completely destroyed by Cyclone Heta. Raised sea-levels exacerbate the problems cyclones bring, increasing the intensity of storm surges and thus wave damage along coasts. A further problem is that protective coral reefs, which help dissipate the effects of wave damage, are under pressure, especially in tourist localities. This is a particular concern on the southern coast of Viti Levu, Fiji, for example, where resort tourism is concentrated.

Other effects of sea-level rise are troubling. Across the region, the vast majority of the population lives on the coast. As sea-level rises it can salinate freshwater tables and lead to crop failure. These problems are magnified on atoll islands – ringed ancient reefs surrounding a submerged volcanic island – where limestone rock forms allow easier percolation. Processes such as this have been observed in Kiribati, Tokelau, the Federated States of Micronesia and the Marshall Islands. Although the press has tended to exaggerate claims that islands are likely to disappear and some politicians

Box 8.2
continued

in the countries themselves have latched onto such claims for their own benefit (Connell, 2003b), contingency plans for the removal of people are being made in some cases. Ultimately the sovereignty of a number of smaller island nations such as Niue and Tuvalu could be under threat from processes whose source lies entirely outside of the Pacific Islands' territory and control.

by 'global' here is subject to debate; however, where degradation accumulates to the extent that it impacts upon a large proportion of the Earth's population or has the ability to threaten global atmospheric, biological or hydrological systems it can be deemed as such. An example of this type of problem is deforestation (see Box 8.3). Other examples include the increased incidence of flooding, desertification, air and water pollution, soil erosion and biodiversity loss. The nature of these problems is such that it is hard to regulate them within the context of existing state borders and there are thus overspill effects (*negative externalities*). Air pollution, for example, is subject to atmospheric system movements and may be carried across borders. This was the case in the 1980s during the height of the acid rain controversy in Europe, where pollution from the east caused tree loss in Scandinavia and other parts in the northwest of the continent. In another example, deforestation in the Indian Himalayas has caused extensive flooding of the Ganges downriver in Bangladesh. Although the negative externalities caused by cumulative degradation are generally not of the same magnitude as those caused by systemic degradation it is clear that environmental governance to solve cumulative problems *requires* moving beyond nation-states. There are many overlaps between cumulative and systemic global environmental problems, and facing them requires interpreting them as part of one system.

Overall, there are a number of uncertainties in attempting to define and analyse global environmental degradation. Even if we were to arrive at a point of consensus with regard to what constitutes degradation, this would not imply agreement on its magnitude or what should be done about it. Science is highly imperfect, and our measurement and predication of environmental change in general is laced with ambiguity. This uncertainty can lead to inaction and lack of resolve. Given this, many environmentalists favour a *precautionary principle*, arguing that we should restrict our consumption and restructure activities on the basis of worst-case scenarios. Other groups have argued that forgoing current consumption on the basis of unequivocal scientific evidence is foolhardy

Box 8.3

Cumulative global degradation: deforestation

Deforestation refers to 'the removal of trees from a locality. This removal may be either temporary or permanent, leading to partial or complete eradication of the tree cover. It can be gradual or rapid process, and may occur by means of natural or human agencies, or a combination of both' (Jones, 2000, p. 123). The major reason for deforestation at the global scale, however, is human activity. Among other things deforestation may take place to create new land for urban or agricultural use; to provide wood for fuel or construction; to allow the exploitation of mineral deposits; or to create reservoirs or build highways. In adjacent areas the practice degrades the environment in a range of linked ways. Among other things it can lead to: a loss of nutrients; increased run-off and flooding; heightened erosion; increased landslides; and coral reef depletion due to sediment buildup. Deforestation has occurred since human populations roamed the Earth and was especially common in pre-modern slash-and-burn societies. In England it is believed that deforestation goes back at least 8900 years. However, change at this time was relatively localized and minimal in terms of its scope. In Europe, from the eleventh century, extensive deforestation took place as land was cleared for use as fuel and settled agriculture. In 200 years the forests of Central Europe were almost completely depleted (Pickering and Owen, 1997).

With the rise of colonial globalization, deforestation in the periphery began both to serve local needs and supply raw materials to the imperial heartlands (see Chapter 7). In North America, before the arrival of colonial peoples, the forests occupied 170 million hectares of land; currently only approximately 10 million hectares remain (see Map 8.1). European settler colonies established in the eighteenth and nineteenth centuries, including New Zealand, the USA, Canada and Australia, have exhibited very high rates of tree loss, clearing the equivalent of that felled in Europe in hundreds rather than thousands of years.

More recently, in Third World postcolonial states, exploitation for export has formed an important part of national development strategies. Deforestation rates are highest in the humid tropics, where great swathes of hardwood rainforest are felled daily in regions such as Southeast Asia, Central Africa and the Amazon (see Map 8.2 and Plate 8.1). Commercial pressures are such that this forest is unlikely to ever be regenerated, causing unquantifiable biodiversity losses and unknown climatic impacts. Rainforest covers approximately 7 per cent of the Earth's surface today; a few thousand years ago it covered 14 per cent, and the majority of this has been felled over the past two centuries. The rate of loss is accelerating rapidly according to the FAO (Pickering and Owen, 1997), and rainforests

Box 8.3
continued

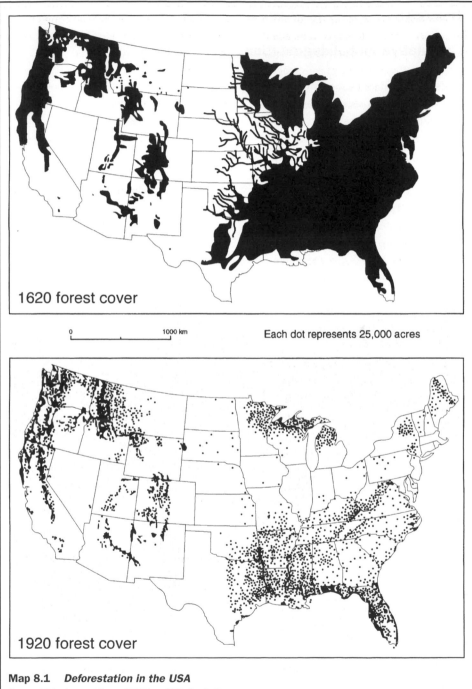

1620 forest cover

0   1000 km

Each dot represents 25,000 acres

1920 forest cover

Map 8.1 *Deforestation in the USA*

Source: Pickering and Owen (1997, p. 351, fig. 9.1)

Box 8.3
continued

are now disappearing at the rate of one acre per second – that is to say an area equal per annum to the combined size of England and Wales. Among the most threatened regions is the Amazon where between 1978 and 1988 annual rainforest clearance had risen from 78,000km^2 to 230,000km^2. The case of the Amazon clearly illustrates the dilemmas facing poorer countries in this position. Farmers expanding the agricultural frontier undertake some of the clear felling. The building of highways to support this endeavour has an especially devastating effect, as does the building of hydroelectric power dams. In general, for poorer governments, a trade-off between environmental conservation and economic development often exists. International bodies have only just begun to understand these linkages and to appreciate that in order to break out of the cycle of environmental degradation alternative livelihood strategies need to be generated in affected areas.

Deforestation has become a global problem in three ways. First, it is affecting an increasing proportion of the Earth's surface, constituting an *extensive* problem. Second, it has an important impact on the global

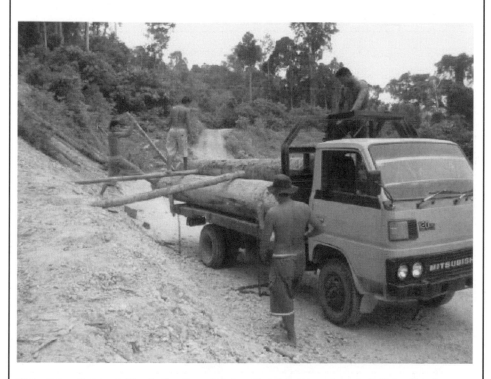

Plate 8.1   *Deforestation in Southeast Asia – illegal loggers at work in Palau Pandang, Malaysia*
Source: Potter and Cooke (2004, p. 340)

Box 8.3
continued

Map 8.2   *Present and past extent of rainforest*

Box 8.3
continued

climatic system, reducing the carbon-absorbing capacity of the biosphere and adding to pollution through burning. Third, the driving force in deforestation is often not 'local'; pressure to deforest has come from processes associated with globalization as practised such as colonialism, modernization and neoliberalism (see Chapter 7). The demand for hardwoods in the West, for example, has a direct impact on poorer countries where forests exist. In many richer countries reforestation is taking place, but does not offset increasing rates of deforestation in the Third World. In sum then, deforestation is a global environmental problem because both the consequences and the causes of degradation are felt beyond national borders and at the scale of the whole world.

and possibly dangerous. In the case of the Third World current needs are more pressing than elsewhere; hence the tension between resources and development is greatly amplified. However, irrespective of worldview and debates over the details of environmental degradation, there can be little doubt that the Earth's environment is being radically altered and that economic and demographic change will place extra pressure on finite resources, as we discuss in the subsequent section. Meyer and Turner (1995, p. 317) confirm this point, arguing that:

> Though the patterns, sources and impacts of environmental change are not uniform across the globe, humankind has so altered the physical conditions of the earth that we must recognise overall an 'earth transformed' and an earth certain to be transformed further. These changes will have different expression and consequences in different regions, but they will be significant almost everywhere. The question for humankind is whether it will be able to . . . adapt to the foreseeable and unforeseeable consequences without destroying the only home it has.

Population growth, globalization and resource consumption

One of the most important factors in the environmental degradation debate is demographic change. Is systemic and cumulative global degradation due to population growth as is commonly heard? Or, are other processes that have unleashed unprecedented levels of consumption, especially in the West, at work? The answer is a mixture of the two. Consumption in the West is far greater in magnitude than in the Third World, despite the fact that close to four-fifths of the planet live in the latter. However, as poorer regions' economic bases change and per-capita consumption rises, continued high population growth has the potential to shift the relative contributions to global environmental degradation.

Population grows geometrically; that is to say, it increases by a proportion over a given time period, leading to exponential growth. In 10,000 BC the population of the Earth was approximately ten million. By the time of Christ it was around 200 million. By 1650 it reached 500 million and, by 1750, one billion. Eighty years later it equalled two billion, rising to three billion by 1960, four billion by 1976 and five billion by 1987. It now stands at substantially over six billion (see Figure 8.4). Current projections estimate a global population of nearly nine billion by 2050 (UNDP, 2003). Based on this explosive growth, concern has been raised regarding the prospect of 'overpopulation' and its links to environmental degradation. This echoes the Malthusian view of the relationship between population and resources, which was accepted wisdom for much of the nineteenth and early twentieth centuries. It is wrong to think that the rate of population growth is increasing; it is actually falling at the scale of the globe as a whole. Between 1965 and 1970 total population grew on average at 2.04 per cent per annum, and between 1995 and 2000 this fell to 1.33 per cent per annum. There is the real prospect that by the mid-twenty-first century global population

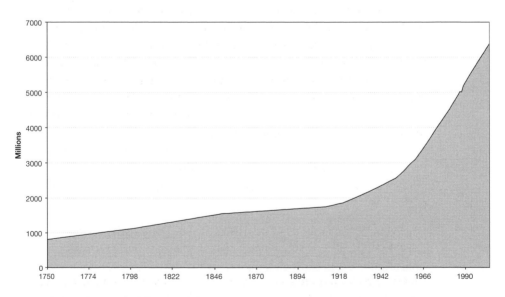

Figure 8.4   *Global population growth*

Source: UNDP (various years)

will actually begin to fall. Sixty-one countries now have fertility rates below the replacement level.

Geographies of demographic change

Patterns of change and growth have varied widely across history and have been associated with important historical changes. After the industrial revolution in Europe there was a major population expansion as death rates declined. In the latter half of the last century, the major process was the decline in death rates in poorer countries brought about by the diffusion of Western medicine, which formed part of the modernization paradigm. It is this technological diffusion that explains the population 'explosion' in poorer countries. Despite these very broad generalizations there are important spatial variations, and as we progress down the geographic scale generalizations become more difficult to make, which is one reason why abstracted models such as the 'demographic transition' are no longer useful. Today, rates of growth are generally lowest in the more affluent regions of the Earth – although there are significant variations, and a strong negative correlation between women's participation in the labour force and population growth has been observed (Buckingham-Hatfield, 2000). In the Third World, rates of growth are still relatively high, particularly in sub-Saharan Africa, and the Middle East and North Africa (see Table 8.1). There is undoubtedly a correlation between economic development and population growth. Based on differential growth rates at the regional level the distribution of the global population is shifting markedly. The greatest challenge we face is that growth is projected to be greatest in region and localities that are poorest. Migratory flows from rich to poorer regions will offset this trend only partially, not least as significant restrictions to migration have been erected in the West in the face of growing global disparities. People in high-growth regions will increasingly demand resource-intensive living standards as they emulate rich and 'globalized' lifestyles of the West.

For many years it was assumed, based on Malthusian ideas, that population growth was *causing* environmental degradation, and that this was leading to underdevelopment and poverty (see Ehrlich, 1971; Hardin, 1968). In more recent times the causality between these factors has been questioned however. Today, *poverty is seen as a cause of population growth rather than a symptom*. High levels of poverty cause

Table 8.1 *Current population, population growth rates and projected population by country groupings (millions unless otherwise indicated)*

Country grouping	Population 2002	Average annual growth rate (%), 1980–2002	Projected: 2015
East Asia and Pacific	1838.3	1.4	2036.9
Europe and Central Asia	472.9	0.5	478.2
Latin America and Caribbean	524.9	1.8	619.4
Middle East and North Africa	305.8	2.6	382.7
South Asia	1,401.5	2.0	1683.7
Sub-Saharan Africa	688.9	2.7	882.1
Low Income	2494.6	2.1	3044
Middle Income	2737.8	1.3	3039
High Income	966.2	0.7	1007.7

Source: Development and Globalization: Facts and Figures, United Nations (2003, p.11)

people to want more children in order to break the vicious circle of poverty. Children make economic sense, and form a system of social security in many poorer countries. This view carried through at the 1994 Cairo UN Population Summit, for example. This implies that solving population growth is contingent upon reducing poverty, which in turn is likely to reduce local environmental degradation. In general then there has been a broadening in the debate on the relationship between population, development and the environment, which is now far more complex than was once thought (see e.g. Batterbury and Forsyth (1997) for an alternative take on the relationship between population density and environmental change in Africa).

Resource overpopulation and the spread of modernization

At the same time as the demographic debate has widened, we have seen the evolution of neo-Malthusian concepts. These have more to do with concerns over the nature of consumption than population growth per se. There has been an increasing focus on the *global* environmental impacts of population growth *combined* with increased industrialization and consumption and the use of finite resources that this entails. Tyler-Miller (2002) refers to the current problem as one of 'consumption overpopulation' as opposed to 'people overpopulation'. In describing the environmental impact of human activity a simple identity may be used:

$I = P * A * T$ *(where I = Impact; P = Population; A = Affluence; T = Technology)*

The above implies that impact is equal to population multiplied by the amount of resources each person uses, multiplied by the environmental effects of the technologies used to provide and consume each unit of that resource. In poorer countries population size and the resulting degradation of resources is the major issue. Overall, however, resource use per capita in such societies is relatively low. In richer countries high per-capita resource consumption and the resulting pollution and degradation are the major factors causing environmental problems. As Tyler-Millar notes (2002, p. 14);

> it is estimated that the average US citizen consumes 35 times as much as the average citizen of India and 100 times as much as the average person in the world's poorest countries. Thus poor parents in a developing country would need 70–100 children to have the same lifetime resource consumption as two children in a typical US family.

Such discrepancies give rise to major differences in terms of the relative contribution of different regions to global environmental problems. Table 8.2 shows the relative levels of resource consumption (per 100 people) in the 'industrialized' and 'developing' worlds as they are referred to there. Figure 8.5 shows contributions of different regions to annual greenhouse gas emissions. It is clear that it is the industrialized world which is adding significantly more to global environmental problems.

Table 8.2 *Relative levels of resource consumption in the industrialized and developing world*

Country grouping	Vehicles per 1000 people (1996)	Paper and paperboard, kg/person/per year (2002)	Energy consumption, thousand metric tons oil equivalent (2001)
Developed countries	448	171.93	1,088.8
Developing countries	25.5	19.78	827.9
World	**120.9**	**52.08**	**1,631.3**

Source: World Resources Institute, http://www.earthtrends.wri.org/

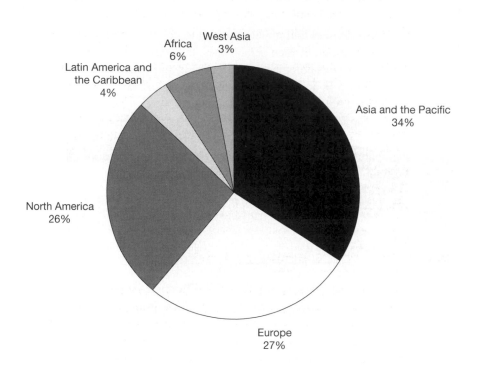

Figure 8.5 *Regional contributions to annual net greenhouse gas emissions, 1998*
Source: UNCTAD (2000)

Global development, population and environment – links and challenges

It is too simplistic to blame population growth for environmental degradation and global environmental problems. Population growth per se is not the cause of resource depletion – resource consumption is. At the present time this consumption is concentrated principally in the West – but this will not be the case for ever. We simply do not know if the world can accommodate a 'catch-up' by poorer nations if they pursue globalized Western economic models. Some argue that it has been in the interest of the West to curtail Third World populations rather than shoulder the burden of reducing the global impact of the industrialization that sustains their economies. Findlay (1995), for example, argues that the West has turned its attention to the 'overpopulation' explanation at the various summits and conferences which have taken place in order to avoid more costly solutions to the problem – such as emissions cutting. However, this is not to say that population growth is not a problem; accompanied by certain conditions in certain places it clearly is. In general, the approach required lies somewhere between the extreme views that exist. That is to say, we should accept that there are limits to growth but that appropriate technology can, in theory, reduce environmental pressures while at the same time increasing prosperity. However, alternative technologies require sustained investment and the world remains dangerously reliant on finite resource-fuelled development at present. We need ways of increasing local well-being that do not degrade the local and global environment if we are to face off the 'population-consumption bomb'.

Globalization of environmental awareness

The modern environmental movement is often argued to have begun in the 1960s (Castree, 2000). Defined as the 'organised political expression of environmentalism' (ibid., p. 220) it was influenced by works including *Silent Spring* by Carson *et al.*, (1962), and *The Limits to Growth* by Meadows (1972). It is associated with the post- Second World War generation and the 1960s liberal social revolution in the USA, the UK, and the West more generally. As recently as the late 1970s, members of the environmental movement were seen as radical, but there has been a significant mainstreaming since then. Whether in the West this general concern has evolved out of genuine unease regarding the long-term evolution of the planet or out of individualistic and utilitarian notions of the loss of aesthetic and use value is a debatable point, however.

It is from within this movement that the maxim 'think global, act local' evolved, representing an early articulation of a globalization concept that has entered the popular imagination. According to Castree (2000) the environmental movement comprises six categories:

1 Environmental NGOs (e.g. Friends of the Earth)
2 Environmental new social movements (e.g. Namada Dam Movement)
3 Green parties
4 Governments with Green sensibilities
5 Businesses with Green sensibilities
6 'Green' consumers

To this list a seventh group may be added, namely *intergovernmental multilateral agencies* (such as the United Nations Environment Program or the IPCC) that seek to coordinate intergovernmental regulation as well as to accommodate the views of pressure groups and other NGOs.

Environmentalism, defined as 'a concern that environment should be protected, particularly from the harmful effects of human activity' (Castree, 2000, p. 220) pre-dates the rise of this movement however. The early Greeks expressed concern over human–environment relations and religions such as Buddhism have profound environmentalist elements. In the USA, early environmentalism is associated particularly with the works of John Muir who in the early twentieth century helped establish the US National Parks system. Environmental concern and management is not just a concept pertaining to Western cultures. Many precolonial cultures in the South Pacific, Latin America and Africa had elaborate systems for the conservation of natural resources and many evolved concepts which viewed humans as an inextricable part of the environment (e.g. Mäori and aboriginal culture in Oceania, *Vanua* in Fiji). We must be careful not to commit the error of 'eurocentrism' when considering the evolution of environmentalism.

Notwithstanding the above, the genesis of the *global* environmental movement took place in Western society during the late 1960s, and this should be seen as a watershed period. It may be argued that the movement's rise at this time had three roots: (1) the evolution of the 'post-materialist' society which grew out of post-Second World War affluence. This led to a changed value system which involved concern for the environment among other things (see Chapter 5); (2) space flight which introduced the concept of *spaceship earth* and brought the idea of limits to growth into explicit view; (3) advances in science and linked technology which allowed an understanding of the complexity of the environment that was hitherto impossible. The evolution of a globalized movement does not imply that there is agreement on the nature of the relationship between humans and the environment. On the contrary, the movement has become a very heterogeneous entity characterized by diverse groups of many political and philosophical shades. There are, according to O'Riordan (1981), many different 'environmentalisms' flowing from different ideologies and worldviews, which result in very different regulation strategies (see Box 8.4).

Since the early 1970s then, there has been an explosion of concern across the globe as well as resistance to, and regulation of, environmental

Box 8.4

Worldviews and environmental policy

Increasing ecological awareness has seen the development of a number of worldviews concerning the relationship between humans and the environment. Many of these have their roots in older philosophies and development theories. At the extreme of a continuum there are two perspectives, which often clash in debates over appropriate policy:

● *Ecocentrism* – Science and technology is the problem. Humans must live *with* the Earth. There are inherent limits to growth. Ecocentrics are likely to be sceptics or radical transformationalists with respect to their views on globalization.
● *Technocentrism* – Science and technology is the solution. We will

devise ways of using the Earth more efficiently. There are no inherent limits to growth. Technocentrics are likely to be conservative transformationalists or hyperglobalists with respect to their views on globalization.

These worldviews have clearly different implications in terms of environmental policy. Essentially, the technocentric worldview dominates at the present time in the actions of governments and transnational agencies. The ecocentric view is institutionalized by some NGOs, new social movements and, significantly, Green parties, although these bodies occupy the fringes of policy-making in most democracies.

degradation in a variety of forms (see Figure 8.6). Environmental concern may be said to have globalized in at least three, interlinked, ways, and we discuss points two and three in detail subsequently.

1   *A rapid increase in the physical and intellectual resources committed to researching the environment*, and in particular the human impact upon it. In universities, environmental science and environmental studies have become mainstream undergraduate and postgraduate degrees. At the research level, dozens of new international journals have evolved that are committed to environmental issues. Results are widely disseminated through such publications, international academic conferences and other network flows of academic activity. Unifying many of these approaches is the desire to define and implement *sustainability* (see Box 8.5).

2   *A flourishing of intergovernmental agencies which seek to manage*

Box 8.5

Sustainable development – organizing concept or smokescreen?

Sustainability was *the* buzzword of the 1990s and replaced 'environmentalism' as shorthand for concerns over the environment and society. It has a wide range of meanings, and all have to do with continuity and are broadly positive. Longman's dictionary gives the following definitions of 'to sustain': to give support or relief to; to supply with sustenance; to cause to continue; to support the weight of; and to endure. In debate around environment and society the word 'sustainable' usually means to allow change while maintaining the best of what already exists. The most widely accepted definition of sustainable development, for example, comes from *Our Common Future* (Brundtland, 1987, p. 2):

> Development that meets the needs of the present without compromising the ability of future generations to meet their own needs.

Sustainability has come to include a holistic concern for the nature of environmental change and incorporates the critical interlinkages between environment, economy, politics, society and culture. The concept has been derived from scientific theories (such as 'carrying capacity') and social scientific principles (such as 'maximum sustainable yield') – even if it

does not always say much about them in action. It has become a term around which much work in both the physical and the natural sciences is now organized. The concept first emerged at the United Nations Conference on the Human Environment in Stockholm (1972) as a compromise intended to allow economic growth and environmental conservation at the same time. The meanings and usages of the term have proliferated since then and it has been hijacked as a concept to rationalize environmental exploitation taking place under politics of all shades. Some have argued that the use of the term by some, such as oil TNCs in their advertising and right-wing conservative parties in their manifestos, amounts to the 'Green-washing' of society. Is the term *more* than just a buzzword? In many ways the term just relabels the debate over human–environment interaction. As such, it hides many of the important intricacies and details that one must appreciate if meaningful policy is to be designed. To some, sustainable development is a 'theoretical black hole' (Adams, 1999). Overall, the concept per se is useful as an ideal construct to frame debates over the environmental consequences of the processes of globalization, even if it has been widely misappropriated.

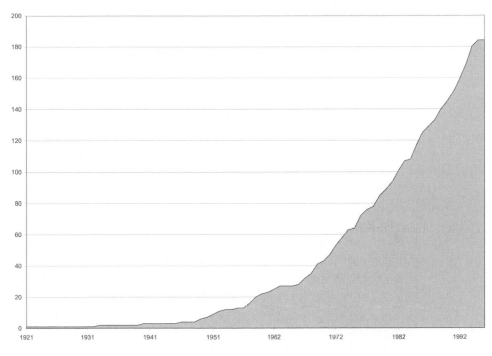

Figure 8.6 *International environmental organizations registered by the UN*

and regulate the global environment from above, giving rise to new international laws, protocols and conventions (see Figure 8.6).

3 *A flourishing of political responses to environmental degradation from below*, including NGOs, new social movements, Green parties, environmentalist businesses and Green consumers.

Global environmental regulation from above and below

The environmental movement is characterized by the evolution of political engagement at all geographic levels. In this context Held *et al.*, claim that '[t]here has been an enormous growth in the number and scope of international institutions, laws and treaties that regulate the environment alongside the development of complex international alliances of environmental movements and organisations' (Held *et al.*, 1999, p. 376). In other words there has been an evolution of attempts at regulation from above and below and activities at these different scales interact, as we explore below.

Regulation from above

Over a century ago there was only nascent regulation from above. International agreements on issues such as the trade in exotic and rare species existed, but the magnitude of agreements and their scope pales in comparison to today. The proliferation of international agreements/ conventions/treaties outstrips globalizing trends in any other sphere (Calvert and Rengger, 2002). The first international convention arose after the Second World War and sought to regulate international whaling (1946). In the 1950s there were various agreements on habitat protection schemes, the nuclear waste cycle and the passage of toxic waste.

It was the 1972 Stockholm Conference hosted by the UN which gave birth to the international regulatory bodies we see today. The agenda that was established there set the scene for subsequent decades, leading directly to the World Conservation Strategy of the 1980s. In the 1970s and 1980s there were a number of important developments. These included the regulation of international waters and marine pollution (e.g. the London Dumping Convention of 1972 and the UN Law of the Sea of 1982); and wildlife protection (e.g. agreements on Antarctica, wetlands, migratory birds, polar bears and seals). Through the 1980s, wide-ranging conventions were established on the passage of hazardous wastes (Basel Convention 1989); control of CFC emissions (Vienna and Montreal protocols of 1985 and 1987) and acid rain deposition in Europe. The EU has been especially active in terms of environmental regulation since the 1970s, passing directives on issues such as air pollution, water pollution, waste disposal and vehicle emissions, and lobbying consistently for global agreements in these and other areas.

The *Bruntland Report* of 1987 theorized and discussed the possibility of 'Green' growth and opened the door to a global-scale debate about sustainability (see Box 8.5) that eventually resulted in the 1992 Rio Summit on Environment and Development. Organized under the auspices of the United Nations Environment Program (UNEP), the summit attempted to evolve far-reaching policies to deal with a comprehensive set of environmental problems and was attended by nearly every national government and hundreds of NGOs. Agreements were forged on biodiversity, climate change, deforestation and desertification. The North/South division was highly visible in the agonizing preparatory and conference negotiations. Some poorer countries maintained that they should be allowed to use the natural resources within their boundaries as they saw fit, and industrialized countries should pay the price for environmental degradation in the past. Agenda 21 was the detailed policy-guidance document which resulted from Rio. This 600-page document was divided into four sections (environment, development, actors and policy) of ten chapters each and was intended to provide a blueprint for local sustainability within a global framework. Some critics have argued that the Agenda, while sound in its intentions, has allowed agencies and governments who are practising clearly non-sustainable policy to hide behind a whirlwind of buzz words and spin. Many of the agreements put in place at Rio have not been implemented and progress on others remains slow.

A significant development born out of the Rio Summit was the Kyoto Protocol (1997) on greenhouse emissions and global warming which forms the most important example of an international agreement/ convention on climate among the many which have been put in place since the Second World War. The Protocol evolved out of the UN Framework Convention on Climate Change (UNFCCC) signed by 167 nations in Rio and enacted in 1994. The UNFCCC required parties to voluntarily commit to reducing greenhouse gas emissions to 1990 levels by the year 2000. In 1997 the signatories met at Kyoto, Japan, and developed a document that would be legally binding if ratified. This flexible Protocol allows countries to set varied targets, undertake to reach them on different dates (but before 2012), and trade their emissions-reduction goals with other countries. The Protocol requires an average reduction of 5.2 per cent in greenhouse gas emissions compared to 1990 levels. Required reductions are higher in some places than in others according to their proportional contribution to emissions. For example, Europe is required to reduce its emissions by 8 per cent, while a number of low carbon-emitting countries in the Third World may actually increase their reductions. Full ratification required signatures from countries that together accounted for 55 per cent of total emissions in 1990. When Russia ratified the document in November 2004 this paved the way for the Protocol to come into force by February 2005, committing signatories to the appropriate reductions by 2012. The USA and Australia are the two largest nations to not ratify. The former accounted for 23 per cent of all emissions in 1990 and its output has since grown by 18.1 per cent. In 2001, the USA pulled out of the negotiations entirely and was only coaxed back into talking through the watering-down of the agreement. At the Buenos Aires UN climate convention of late 2004 the US administration stated that it would never sign the Protocol, given what it saw as unfair terms for developing countries. The USA also sought actively to block discussion of what should happen after the Protocol was completed. The targets of the agreement have not been met to date. Between 1990 and 2000 emissions from industrialized countries increased by 8 per cent (although this was offset significantly by declining industrial output in the former USSR and Eastern Europe). Per-capita emissions in high-income OECD countries rose from 12.2 to 12.5 metric tonnes between 1980 and 2000, and from 1.3 to 1.9 metric tones in the developing world (UNDP, 2004). Globally, the IPCC estimates that emissions will rise a further 10 per cent by the 2012 deadline.

The UN *World Summit on Sustainable Development* (alternatively called Rio+10 or the Earth Summit) took place in September 2002 in Johannesburg, South Africa, and was intended to build on the Rio Summit of ten years earlier. Hopes were high for the meeting as it brought together over 150 countries, a wide range of NGOs and representatives from new social movements. The Zimbabwe issue, and the impending War on Iraq, overshadowed the proceedings however. US President Bush did not attend and many saw this as a significant snub in terms of the international consensus which it was hoped was evolving. Despite agreements in water and sanitation, global warming, energy, biodiversity, trade, human rights and health – these were significantly toned down from original goals. In many ways Rio+10 was a disappointment. A clear gulf has developed between the USA and EU in terms of environmental regulation and this was clearly evident at the Summit, leaving the former isolated on many issues, not least its failure to ratify the Kyoto Protocol. In general, the USA has entered a phase of extreme aggression *vis-à-vis* the environmental agenda. In the long term the current administration's environmental stance is likely to be judged as the most backward-looking and self-interested in the history of the USA.

Although progress has sometimes been disappointing there can be little doubt that the evolution of global regulation in this area represents a significant and positive step. Environmental issues have now been placed on the agendas of orthodox institutions including the World Bank (see World Bank, 1992), the IMF, the G7 and the World Trade Organization. Much of the impetus for this evolution has come from below as the grassroots movement has consistently lobbied national governments to act in a cooperative way with other states.

Regulation from below

There has been a widening in the range of political environmental actors that seek to regulate global degradation from below the scale of the state. The combined growth in the number of transnational environmental NGOs and environmental new social movements has outstripped the rise in intergovernmental agreements. McCormick (1989) argues that the grassroots movement has built upon a commonality of environmental interests providing fertile ground for a struggle against those in power and the technocentric approaches they employ. As already discussed, the 'grassroots' or 'bottom-up' environmental movement began largely in richer countries and was institutionalized through the formation of NGOs such as Greenpeace and Friends of the Earth (both established in 1971). Often built by civil society, and usually on a voluntary basis, environmental NGOs played a major role in raising public and state awareness. New social-environmental movements have also played a crucial role in evolving grassroots environmentalism. Such organizations, including, for example, the Namada Dam Movement and the Chipko Movement, are often ephemeral, local issues-based, relatively informal and flexible. They have provided the NGO sector with salient and topical issues that are employed to illustrate their broader goals. Environmental NGOs and NSMs are increasingly intertwined at the global scale and a range of global networks representing these interests exists.

It has been argued that in the West, environmentalism has been concerned primarily with issues of lifestyle and amenity (Escobar, 1995). There are some environmentalist NGOs based in poorer countries which share these concerns – often supported by the elite and urban middle classes. Many environmental NGOs in the poor world are branches of Northern NGOs or are set up by European expatriate or settler groups (as in Greenpeace Fiji). In general, however, these issues have not captured the general public's imagination as they have in Europe, North America or Australasia over the past thirty to forty years. However, many new social movements have emerged among lower socio-economic groups directly affected by environmental degradation in poor countries (Routledge, 2002). In this way Third World environmentalism is often concerned with livelihoods rather than with lifestyles.

Formal Green parties have played a major role in mobilizing grassroots action on the environment. The German Green Party was established in the 1970s and was the first to win significant formal representation in a national Parliament. A number of similar parties have evolved and have attempted to emulate this success in the UK, Australia, New Zealand and the USA among others. None has yet gained a controlling influence in any Parliament and is unlikely ever to do so given the appropriation of ecological policies by mainstream parties. In New Zealand, representation is relatively strong and the party has benefited from the existence of a proportional representation system, polling around 5 per cent in the 2000s. In the USA a similar electoral system would deliver a considerable presence in the Congress for the Greens. The current system, however, sees this party locked out of mainstream politics. As it stands, Green parties tend to influence policy by lobbying or through issue-based alliances or coalitions with larger parties. A global Greens network, which draws together parties from across the world, represents an important platform for the shared vision of these groups.

A final important aspect of the grassroots environmental movement is the rise of Green consumerism. A rising number of consumers has been voting with its wallet by actively seeking *eco-friendly products*. Industry has responded to this and perpetuated this increasingly profitable niche. Taking both local and global environmental issues into account is more or less an industry standard in many richer countries and in markets for elites in poorer ones (see Plate 8.2).

Conclusion – transformed environments

Environmental problems have become global in their scope. Systemic and cumulative degradation has occurred on an unprecedented scale, causing geographically differentiated outcomes that threaten the whole Earth system. This has been driven by the processes and agendas of globalization – particularly since the industrial revolution and colonialism, and more recently through neoliberal economic policy. This has led to a global consciousness of unprecedented scope and scale concerning environmental issues. Many international agreements as well as less formal globalized networks have evolved in order to attempt to regulate this both from above and below. In the face of issues that threaten collective welfare in the global commons and create trans-boundary issues, nation-states are too small by themselves to supply the required regulation. Furthermore, they are often too large to deal with localized environmental concerns, which can accumulate into global problems. This is not to say that the nation-state does not have a role to

Plate 8.2   *An eco-friendly product*

play in regulating the global environment; it does, but it will only achieve success in conjunction with a wide range of interests operating at various geographic scales.

Despite the rising concern that is evident almost universally, commitment to action varies greatly across the world. Some of the largest states, in particular the USA, have been far too willing to pull out of agreements which they perceive as damaging to their self-interest. The global environmental movement will apply increasing pressure on governments and international bodies over the coming years in debates on climate change, sea-level rise, air quality, water scarcity, food safety and genetic modification. The issue of degradation cannot be resolved, however, until global inequality is addressed in a coherent way. The global environmental movement is divided, not least because the interests of the rich and poor in the context of globalization processes are more divergent today than ever before. In this sense environment and development are inextricably linked and represent a single global challenge of enormous magnitude.

Future reading

Adams (2001): This book seeks to link discourses of environmentalism with discourses of development and engages critically with the concept of sustainable development.

Bradshaw (2004b): This chapter provides a useful overview of the energy crisis and some of the responses that have been put in place to address it.

Held *et al.*, (1999) ch. 8: This chapter traces out environmental degradation over history and seeks to link it explicitly to the evolution of globalization.

Johnston (1996): This book is particularly strong on the relevance of geography to the current environmental debates.

O'Riordan (1981): *Environmentalism* is the classic book on the rise of the environmentalist movement which traces its early nature and its philosophical roots.

Pickering and Owen (1997): This text provides a fact-packed and readable guide to environmental issues from a global environmental science perspective.

Long live geography? Progressive globalization

- Controversies and signposts
- Geographies of globalization – a schema
- An agenda for human geography
- Some final reflections

In this chapter the major questions posed in the introduction to this book are returned to, the contribution of geographers to the study of globalization summarized, and the geographical elements of globalization theses drawn out. An agenda for human geographical research on globalization is presented, calling for a decentred new regional geography that approaches big issues in a holistic way through the mapping of spatial differentiation on the ground. In general, it is argued that although globalization as currently practiced is increasing global inequality and degrading the environment, it need not necessarily be a regressive process. Globalization, if re-conceptualized, researched and regulated effectively, has the potential to improve global well-being, equity and sustainability. Cultivating a critical globalized imagination, which celebrates difference without idealizing it, is crucial in constructing an alternative and inclusive globalization process.

Controversies and signposts

By way of conclusion we return to the questions posed in Chapter 1. Summary answers are offered below on the basis of the evidence and discussion presented throughout the book. The highly contested nature of globalization implies that each response is one of a number that could be made. In general, the arguments below adopt a transformationalist worldview. It is this perspective that sits best with *most* theoretical and empirical work undertaken by geographers. The transformationalist 'theses', as argued in Chapter 2, sits on a continuum and the general tone of research on globalization in human geography might be more appropriately termed *radical transformationalist*. That is to say that while accepting that globalization is real, has concrete impacts and can be transformed by human action, geographical accounts have also emphasized the power of discourses of globalization and the driving role of capitalist/imperialist exploitation over space and time. The 'answers' presented here are expressed in relatively abstract form, and interpreting them requires revisiting the empirical evidence and theoretical arguments presented in relevant chapters.

How can globalization be defined?

A formal definition was offered in Chapter 1 and followed throughout the book. This represents only one of many possible definitions however. Most agree that globalization involves the *stretching of social relations across space* in all three spheres and that this results in *transnational flows which transcend borders*. This latter point gives rise to a *network society*, whose population is increasingly *conscious of the shrinking world*. Many definitions also refer to the *role of capitalism* as the driving force of the process. Of course, for sceptics who do not believe that globalization as a set of processes exists, it is seen as a discourse employed at various times in diverse ways to serve specific interests.

Is globalization new?

This book has argued consistently, and Chapter 3 has attempted to illustrate in detail, that globalization has *a long history*. When exactly it began is a matter of debate however. Some argue that it started with the first intercontinental flows of people and ideas over 10,000 years ago. Others date it from the end of the Second World War. The periodization of globalization turns on its definition. If you believe, as this book has argued, that globalization is intimately tied to the expansion of capitalist circuits and the stretching of flows of accumulation across the world, then it *began in about 1500 with the rise of the Hispanic empires*. It then proceeded through *two waves*, colonial and postcolonial, each of which was further subdivided into two phases: mercantilist and industrialist in the case of the former, and modernization and neoliberal in the case of the latter. Each of these waves and phases was brought to an end by restructuring crises in global capitalism.

What drives globalization, now and in the past?

This is a source of great contention in globalization studies. Some see technology determining its evolution; others place culture or economics at the centre. Many accounts see capitalism as intimately involved, as part of a wider human imperative to shrink space. This book has argued that *capitalism is the central driving force now and in the past* and that time–space compression has resulted from the inherent need of capitalism to reduce the turnover time of capital and speed up its circuits, but it has also been constantly stressed that this is not an economic determinist position. *Capitalism is conceptualized as a broader culture* born of ideologies rooted in the Enlightenment period resulting in specific traits that drive its desire and ability to expand.

Is contemporary globalization different from the globalization of the past?

The current wave, especially the neoliberal phase, is characterized by increased extensity, intensity and velocity of global flows and the rise of dense networks that transmit them. According to the argument throughout this book and which is not an uncommon one, *we are living in an era of vastly accelerated globalization*.

Is it homogenizing global society?

Permeating this book is the idea that *globalization exacerbates difference and fragments society*. There are powerful forces (e.g. cultural imperialism, superpower hegemony, development, modernization, Fordism) which, if unfolded on a homogeneous pre-existing landscape, have the potential to create a universalized society. However, these forces encounter locally specific histories and geographies and subsequent *interaction leads to hybridized or glocalized outcomes in culture, economy and politics*. Globalization works both ways at the same time, constantly reproducing scale and forging an ever-changing, heterogeneous, social landscape. Furthermore the pervasiveness of networks means that 'global' processes do not penetrate territory in a ubiquitous way and 'black holes' are left between the conduits of the net.

Is globalization a process or an agenda?

Globalization comprises both processes and agendas that interact in complex ways. It has been argued in this book that agendas of globalization are tied up with imperatives of capitalism. These agendas unleash processes and flows that reconstitute agendas, which then go on to create another cycle of processes. Ideas from the regulation school are important here, since they provide arguments that support the idea of mutually constitutive agendas and processes.

How does globalization alter our concepts of space, place and scale?

Each of these core geographical elements is profoundly transformed by globalization, as was argued at length in Chapter 2. *Space is relativized, place is hybridized and scale is collapsed*. In general, deterritorialization and territorialization exist simultaneously and spaces of flows are disrupted and punctuated by spaces of places. In other words, networked flows emanate from particular localities, and some of these localities – like world cities – play a commanding role in directing such flows. A good way of summing this up is to talk of glocalization.

Is globalization the same thing as internationalization?

The sceptical argument posits that what we refer to as globalization is in fact just more intense internationalization. This book has argued that *internationalization is a component part of the broader process of globalization*, and contemporary globalization is qualitatively different to that of the past. All globalizing processes must take place in localities within nation-states and networks link these processes to other localities in other nation-states. However, the networks that result are transnational, in that they exist beyond the sphere and, sometimes, influence of the state. New transnational networks do not necessarily use the nation-state system, which is a constructed entity, as a reference point. *The whole is greater than the sum of the parts* and the rise of a global consciousness exemplifies this.

Is globalization eroding the power of the nation-state?

Chapter 5 argued that *the state is being transformed by globalization as it seeks to regulate flows from above and below.* Some have argued that this implies a less important role for the state in a multi-layered governance system. Most evidence points, however, to the continued salience of the nation-state as a container of economic activity, a reproducer of culture and a body that represents political will. Nation-states will proliferate but they may become more closely tied to actual nations – that is, ethnic and racial groupings – as they exert their identities beyond the confines of the state.

Is globalization bad for the environment?

As Chapter 8 illustrates, *the impact of globalization as currently practised has almost certainly degraded local and global environments seriously*, and this is associated with the spread of industrialization and rising consumption inherent in capitalism. Yet at the same time *globalization has created space for the expression of concern with this situation.* The global environmental movement has the potential to resolve the tensions if it can capture political will at the nation-state and global levels, although the continued reticence of the USA to face global environmental issues does not bode well.

Is globalization bad for the 'poor world'?

As Chapter 7 argued, *globalization as currently practised has undoubtedly led to increased inequality within and between countries and regions*, and has spread discourses that result in the subordination of millions of people. But the *new networks that are evolving from the bottom up seek to challenge the hegemony of neoliberalism and provide potential space for the practice of progressive development* which could yield equitable and sustainable improvements in well-being. Having said that, it is only the *potential* which exists. The current power relations are such that achieving this kind of outcome will involve great struggle.

Does globalization create more winners than losers?

This is not the same question as the previous one because *winners and losers are created in both the Third World and the West*. Globalized society has become increasingly complex and, instead of seeing the existence of two or three 'worlds of development', or cores and peripheries, we are seeing the proliferation of networks of inclusion/exclusion. Some people fall through holes in the net in rich countries, and the majority do so in poorer ones. There can be little doubt that as currently practised, *globalization leads to greater concentration of power and affluence* and, relatively speaking, the majority of the world's population is left behind.

Can globalization be reformed?

Globalization is not an inexorable force that exists outside of human control; it is socially constructed and perpetuated. People can reform the process through the construction of global regulation building on grassroots concerns and actions. But this is obviously no easy task and will require a vigilant, well-informed and committed civil society.

If so, how should it be reformed?

This is a crucial question that this book does not pretend to go near to answering. Some have argued for deglobalization and a return to localist strategies. But delinking is probably not possible and arguably not desirable; it could lead to even greater relative deprivation and prevent the capturing of the opportunities presented by globalization. In approaching this question we need to move beyond simplistic concepts of globalization as good or bad; it is inherently neither. That is why the evolving *alter-globalization movement provides a more realistic option for meaningful reform* than the anti-globalization movement. Democratizing global institutions is a start, promoting local affirmation is important, and universal education is crucial to creating the kind of civil society that can face the challenges of globalization.

What role can geography and geographers play in forging an alternative progressive globalization?

This is the subject of a subsequent section in this conclusion where an agenda for more progressive geographies of globalization is suggested.

Geographies of globalization – a schema

In Table 9.1 the geographical implications of the three theses of globalization in terms of the spheres and challenges considered in this book are summarized. Note that the hyperglobalist column refers to neoliberal, not radical, hyperglobalist theory. This book has argued consistently that it is the transformationalist viewpoint that most accurately captures the evolution of historical and contemporary globalization. The frustrating thing about this perspective, however, is that it is broad and offers a range of possibilities. If we adopt this view we must therefore be careful to back up our conceptual claims with solid

Table 9.1 *Geographies of globalization: a schema*

	Hyperglobalist	*Sceptical*	*Transformationalist*
Economic geographies	*Market is omnipresent and TNCs powerful.* Networks replace core–periphery division Activity disembedded from nation-states. Deregulation and convergence.	*Trading blocs formed.* Core–periphery division increased. Activity embedded in nation-states. TNCs reflect national strategies. Regulation and divergence.	*Networks and structural patterns co-exist.* Nation-states and TNCs govern markets. Re-regulation and convergence and divergence at the same time.
Political geographies	*Market-led global governance.* Nation-state replaced by 'natural' region states. Boundaries dissolve and sovereignty surrendered to global market.	*Core-led regionalism.* Sovereignty surrendered to regional groupings designed by powerful nation-states. Boundaries retrenched.	*Multi-layered governance at three scales* – global, national and local. Nation-state remains central. Centralization and devolution at same time.

Table 9.1 *continued*

	Hyperglobalist	Sceptical	Transformationalist
Cultural geographies	*New global civilization.* Homogeneous consumerist culture and global brands dominate. Universalization of cultural identities.	*Clashing cultures.* Civilizational blocs entrenched and cultural identities differentiated and relativized.	*New global and local hybrid cultures.* Possibility of progressive cultural change. But Westernization currently dominant.
Environment/ sustainability	*Sustainability solved by the market.* Market will price the environment efficiently and technology will solve scarcity.	*Sustainability threatened by global capitalism.* Spread of modernization and consumerism pushes environment to limits.	*Sustainability attainable through political action.* Environmental problems and concerns globalize. Regulation from above and below required.
Development/ inequality	*Development attained by taking part in globalization.* Poverty solved by market. Integration unavoidable.	*Development threatened by global capitalist expansion.* Marginalization increased by spread of the market.	*Globalization offers threats and opportunities for development.* Progress contingent on careful management.
Moral message	*Pro-globalization.* Globalization is real, and is a moral good and force for progress.	*Anti-globalization in some readings.* Globalization is a discourse propagated by powerful interests.	*Alter-globalization.* Globalization is real and requires regulation to optimize it.

empirical work. We are compelled to make concrete suggestions in terms of how we can improve the world and, as addressed in the subsequent section, to do this we need to alter the way we approach geographies of globalization.

An agenda for human geography

In a recent article 'Geographers and "globalization" (yet) another missed boat?' Dicken (2004) argues that geographers have 'developed a disturbing, even dysfunctional, habit of missing out on important intellectually and politically significant debates, even those in which geographers would seem to have a role to play' (p. 5). Dicken is particularly concerned that, as a group of scholars, geographers have had a limited impact on the evolution of the wider debate concerning globalization. This is despite the fact that geography is usually considered to be the 'world discipline' (p. 6). In a survey of some of the most influential books on the issue he finds that less than one-third have references to geographers' work, and that of the total 14,000 references only 2 per cent are of geographers. Dicken (2004, p. 7) argues that:

> [t]he evidence shows unequivocally that geographers are, indeed, marginal – at best – to the wider globalization debates. Geography is rather like the small child in the playground who always gets missed out when the big children are picking teams.

The fact that nobody wants to play with us is troubling, first, because geographers cannot afford to miss out on the big issues of the day from a self-sustenance point of view, and second, their insights are valuable for broader work in this field. So, why are we left out? Geography is poorly understood by other academics and is misrepresented in society as a whole. In the case of the former there is concern from some quarters, even within geography itself, that the discipline has become divorced from the 'big issues'. Some have blamed the 'cultural turn' and the influence of postmodernism on this (see Johnston and Sidaway, 2004). Hamnett (2003, p. 1), for example, argues:

> [The] rise of a 'post-modern' human geography, with its stress on textuality and texts, deconstruction, critique, 'reading' and interpretation has led human geography into a theoretical playground where its practitioners stimulate or entertain themselves and a handful of readers, but have in the process become increasingly detached from contemporary social issues and concerns. The risk is that much of human geography will cease to be taken seriously in the world beyond the narrow confines of academe.

Under the influence of postmodernism geography has, according to some, become increasingly parochial in its outlook and less willing to engage with important shifts in actual places and regions across the world. Instead, it has moved towards increasingly intricate studies of minute events, processes and phenomena at a small scale. Arguably, geography has lost its 'global studies' status in the desire to get 'local'. Dicken notes this in his 2004 article, and this echoes the criticisms of a number of geographers over the past three decades including Stoddard (1987), Johnston (1986) and Potter (1993). Although the above concerns have some merit, they are probably exaggerated. The 'cultural turn', while leading to esoteric geographies in some cases, was founded on the belief that geography ought to be more critical of hegemonic paradigms.

Many of the propositions of this 'turn' have provided richer and subtler ways of interpreting the globalizing world (see Crang, 2002; Imry, 2004). In this area then, geographers need to stand their ground.

Outside of academia, however, the perception of geography is quite different and even less favourable. The discipline is seen largely as a descriptive compendium of facts – ironically one of the shortcomings the cultural turn sought to address – rather than a discipline that engages in the analysis, interpretation and explanation of differentiation in global society. The perception of geography as a glorified atlas may also be discerned in other disciplines on occasion. Ironically, a number of subjects, such as economics, have recently discovered a new spatiality. However, the employment of a 'geographical perspective' elsewhere often seems to misrepresent or under-acknowledge what has gone on in geography. Dicken (2004) and Lee (2002) worry that the type of geography which is being diffused across disciplinary boundaries is reductionist and simplified. Given these multiple misconceptions, the discipline is often sidelined in important debates where it should, by rights, be leading the pack. Therefore, addressing questions of what geography is and what it is relevant to focus upon remain important tasks. Articulating clear and compelling responses to academia and the broader world represents an even more important challenge.

The central task for geographers, according to Dicken, is to pay more attention to the *outcomes* of globalization. In the past (building on the argument of Bridge (2002)) there has been a tendency for economic geographers in particular – although the same could be said of all human geographers – to focus on processes. Bridge says, '[T]he commitment by many economic geographers to a processual view of the world has the effect of collapsing conventional process–outcome distinctions: processes themselves become the object of explanation and are in themselves regarded as outcomes' (Bridge, 2002, p. 362). To elucidate outcomes geographers need to conduct empirical work at fine-grained levels (i.e. below the nation-state level) building geographies of globalization from the ground up. Dicken (2004) notes a number of 'silences' in empirical research on the outcomes of globalization in economic geography. He argues that the subdiscipline has tended to be productionist rather than focusing on consumption; in manufacturing, certain areas have been concentrated upon (such as motor vehicles) at the expense of others; in the service sector less is known about distributional services than financial ones; agriculture is relatively understudied; work on resource peripheries has been scarce; less is known about trade flows than investment flows; environmental impacts of economic globalization are not understood well; gender issues require more attention; and resistance to globalization, however conceived, has not been seriously elaborated. A key task for geographers in other areas of the discipline is to identify similar lists of areas where knowledge gaps exist.

Overall, while geographers have talked to each other fairly extensively about the outcomes of globalization, they have been largely ignored by the world outside of the discipline. To face big challenges such as globalization we need to better define what we are and better represent this to the outside world. Massey, for example, argues that geography needs to be 'more confident of its own specificity' (2001, p. 5) and this, according to her, refers to the human–physical interface as well as our relatively sophisticated notions of space and place. Based on the above arguments, building on Dicken (2004), a seven-point agenda for improved geographies of globalization is discussed below.

1 *Conceptualization* – Geographers need to develop more sophisticated frameworks of globalization upon which empirical work may be hung. Of crucial importance in this endeavour is the use of geographical concepts of space and place, often naively expressed in non-geographical globalization studies where space is seen as a fixed/territorial entity and places as concrete/bounded entities (see Chapter 2). A second crucial insight which geography has to offer is its relatively sophisticated notion of scale (see Chapter 2). Much of the non-geographical debate assumes a simplistic global/local dichotomy with the former determining the latter. As this book clearly argues, globalization is leading to a multi-scalar world. It is important to develop these debates in relation to research on concrete outcomes however, as work of this nature can quickly become esoteric.

2 *Empirical 'mapping'* – Much of the globalization literature is characterized by grand statements based on little empirical evidence. Studies that have sought to outline discourses of globalization, and the ways these are practised and performed, are important. However, such work should be complemented by empirical studies that seek to elucidate both the processes and outcomes of globalization. Again, geography is ideally placed. The subject has a long tradition of studies that seek to link change at lower scales to processes at higher scales and how restructuring at different levels interacts in mutually determining and complex ways. It is crucial, as noted previously, that we get below the level of the nation-state in what is researched. What

geography needs are locality studies that have one eye firmly on the global.

3 *A new regional geography* – We require what Dicken refers to as 'a revitalisation of "regional geography" within a relational framework' (2004, p. 19). This call is not a new one; Bradshaw (1990) made the case for a 'new regional geography'. Different regions exhibit distinct responses to globalization and the knowledge that critical area studies bring to research is crucial. This is not a call for a return to the regional studies of the 1950s and 1960s, which were descriptive and overgeneralized. Rather, it is a call for an analysis of the relationship between unique localities, places and regions and the wider global system. In this endeavour geography needs to pay far more attention to research on regions outside of the Anglo-American zone. Non-Anglo-American geographers, located in the regions of concern, are undertaking a great deal of this research. However, their work is often marginalized.

4 *A decentred critical geography* – To understand globalization, and to provide suggestions for progressive reform, we need a truly global and democratic human geography. Contemporary geography is far from this; it is ironic that in a discipline where postcolonial perspectives have become so prominent the actual practice of research and publication is in many ways neocolonial. This is reflected by the fact that most geographical investigations of globalization study the West and are undertaken by Western academics. Furthermore, the geography journals where globalization research is published are located largely in the USA or the UK, dominated by a handful of individual departments. Work by non Anglo-American-based geographers on non-Anglo-American geographies should be encouraged more explicitly in the mainstream journals, and non-Western avenues of publication accorded greater status. We need a much less eurocentric geography before we begin to describe and explain globalization in all its diversity.

5 *Intra- and cross-disciplinary interaction* – Geography is ideally placed to understand globalization given its inherently cross-disciplinary nature. Of particular importance are studies that seek to span the human/physical divide. As Chapter 8 illustrated, it is in the area of environmental globalization that some of the major challenges exist. In general however, given the multi-dimensional and multi-thematic nature of globalization, we need greater interaction across the subdisciplines. There have been great strides taken towards this; new economic geography, for example, has built in many ideas from

cultural geography. Naturally, more interaction is required with other disciplines but this will be fruitful only if we have a strong discipline characterized by cooperation across its component parts.

6 *Big issues* – There is an inherent tendency in academia to want to push back barriers and to shift to new frontiers of research. Many academics have a vocation towards this end, but the pressure to publish and the desire to differentiate work from that of others also drive the process. Globalization has been on the agenda in the social sciences, in a serious way at least, for only around fifteen years. What must be avoided is the desire to add a 'post' prefix to globalization prematurely. We are just beginning to understand its importance and nature, and it is not yet time to move on. There is a habit in geography of shying away from 'big issues', focusing instead on smaller, less obvious concerns. This is partly why geography missed the environment and development boats. Focusing on big issues does not mean that geographers should concentrate on the global at the expense of the local however, since to do so would be to throw away one of the greatest strengths of the discipline.

7 *Moral geographies and globalized imaginations* – There has been something of an upsurge in moral geographies over recent years (see e.g. Lee and Smith, 2004) building on the relevant debates of the 1970s. Although we need strong empirical studies that flesh out critical discourse-based work, we need to be less reticent in saying what we feel the world *should* be like. Geography would do well to combine empirical studies with *globalized imaginations* that suggest alternatives to the status quo. This requires going beyond simplistic anti- or pro-globalization arguments towards more sophisticated analyses that acknowledge the potential opportunities and threats of the process.

By way of summary then, it is clear that globalization creates both opportunities and challenges for geography (Martin, 2004; Yeung, 2002). As Dicken (2004, p. 20; emphasis in original) states:

> The challenge of 'globalization' is for us to get beneath the simplistic hype, to recognise that serious things are happening, to understand what the processes and the outcomes are and to use our understanding to make a better world. The *opportunity* of 'globalization' is that it should enable us to use our potential disciplinary strengths to increase the visibility of geography within the academy, among policy makers and to a more popular audience.

Some final reflections

The rise of globalization adds a new urgency and dynamism to human geography. The recognition that space, place and scale are important and that a better appreciation of these, and other geographical concepts, might help inform progressive policy implies a responsibility however. We are thus duty-bound to continue to interpret, model, test, measure, read, gaze at, present and represent the unfolding of globalization and its geographies. In terms of this book, what is perhaps frustrating is that it has not provided *definitive* answers to the many questions it has posed. Answers to such questions turn as much on one's own political and philosophical make-up as anything else. What I hope the book has provided, however, is a framework for evaluating contesting perspectives on globalization. I also hope that the reader has been convinced that the geographical take on the debates is distinct and valuable.

My own 'thesis' is that globalization is real and we live in unprecedented times. However, the links to the past are clear; the current wave of globalization is reproducing and intensifying the unequal world. We live in a neo-imperial age (Gregory, 2004; Harvey, 2003). The imperialism of today is insidious, it is rarely territorially explicit, and the corporations, governments and elites that drive it are non-accountable. The globalization agenda, and the processes which have unfolded from it, are intimately linked to the expansion of monopoly capitalism – although in rather Orwellian terms this is often referred to as '*free*-market' economics. This imperative creates dependency and vulnerability and is distinctly 'unfree' in many ways; it is detrimental for the majority and it is built upon aggressive, sometimes military, penetration of marginalized economies and regions designed to maximize the wealth of the core. The only freedoms afforded through this process are fleeting, material benefits for those very few who are fortunately placed on networks of inclusion. But globalization is not like a lottery – some people will never win – there is too much at stake for it to be any other way. The capitalist elite has appropriated the process and made it its own – and is pursuing its own objectives at the expense of broader society, culture and the environment. As it is currently practised, globalization is perhaps the single greatest threat to human society.

However, there is another way, and globalization as currently practised is only one manifestation of a much broader potential. The cross-fertilization and hybridity which globalization creates could underpin a progressive society with the end-goals of welfare equality, global security and environmental sustainability. It is time to rewrite the agenda and to forge a new globalization from below that not only celebrates difference but also builds from it and perpetuates it. This will entail the construction and application of new and alternative 'globalized imaginations'. If we believe that the global is constituted of the local and vice versa, then we can challenge globalization as currently practised through our own actions and carve a 'glocalized' future. Those of us who live on global networks in 'spaces of places' that have been made affluent by exploitative 'spaces of flows' have a duty to contribute to the construction of a better world for those who are not so privileged. The articulation of a more progressive discourse of globalization is the place to start.

References

Adams, W.M. (1999), 'Sustainability', *in* Cloke, P.J., Crang, P. and Goodwin, M. (eds), *Introducing Human Geographies*, London, Arnold.

Adams, W.M. (2001), *Green Development: Environment and Sustainability in the Third World* (2nd edn), London, Routledge.

Aguiton, C., Petrella, R. and Udry, C. (2001), 'A very different globalization: the globalization of resistance to the world economic system', *in* Hourtart, F. and Polet, F. (eds), *The Other Davos*, London, Zed Books.

Albrow, M. (1996), *The Global Age: State and Society Beyond Modernity*, Cambridge, Polity Press.

Allen, J. and Hamnett, C. (1995), *A Shrinking World?: Global Unevenness and Inequality*, Oxford, Oxford University Press.

Allen, T. (2000), 'Taking culture seriously', *in* Allen, T. and Thomas, A. (eds), *Poverty and Development into the 21st Century*, Oxford, Oxford University Press.

Allen, T. and Thomas, A. (2000), *Poverty and Development into the 21st Century* (revised edn), Oxford, Open University in association with Oxford University Press.

Amin, A. (ed.) (1994), *Post-Fordism: A Reader*, Oxford, Blackwell.

Amin, A. (1997), 'Placing globalization', *Theory, Culture and Society*, 14, 123–137.

Amin, A. (2002), 'Spatialities of globalization', *Environment and Planning A*, 34, 385–399.

Amin, A. (2004), 'Regulating economic globalization', *Transaction of the Institute of British Geographers*, 29, 217–233.

Amin, A. and Thrift, N.J. (1994), *Globalization, Institutions, and Regional Development in Europe*, Oxford, Oxford University Press.

Amin, A. and Thrift, N.J. (eds) (2004), *Cultural Economy: A Reader*, London, Sage.

Amin, S. (1997), *Capitalism in the Age of Globalization*, London and New York, Zed Books.

Angel, D. (2002), 'Studying global economic change', *Economic Geography*, 78, 3, 253–256.

Armstrong, W. and McGee, T.G. (1985), *Theatres of Accumulation: Studies in Asian and Latin American Urbanization*, London, Methuen.

Aryeetey-Attoh, S. (2003), *Geography of sub-Saharan Africa* (2nd edn), Upper Saddle River, NJ, Prentice Hall.

Auty, R.M. (1995), *Patterns of Development: Resources, Policy and Economic Growth*, London, Edward Arnold.

Barlow, M. and Clarke, T. (2001), *Global Showdown: How the New Activists are Fighting Global Corporate Rule*, Toronto, Stoddart.

Barrientos, S., Bee, A., Matear, A. and Vogel, I. (1999), *Women and Agribusiness: Working Miracles in the Chilean Fruit Export Sector*, London, Macmillan.

Barton, J.R. (1997), *A Political Geography of Latin America*, London, Routledge.

Barton, J.R. and Murray, W.E. (eds) (2002), *Chile: A Decade in Transition*, Special edition of the *Bulletin of Latin American Research*, 21 (3).

Batterbury, S.P.J. and Forsyth, T.J. (eds) (1997), 'Environmental transformations in developing countries', *Geographical Journal*, 163, 2, 126–224.

Baudrillard, J. (1988), *Selected Writings*, Stanford, CA, Stanford University Press.

Bebbington, A. (2004), 'Livelihood transitions, place transformations: grounding globalization and modernity', *in* Gwynne, R.N. and Kay, C. (eds), *Latin America Transformed: Globalization and Modernity* (2nd edn), London, Edward Arnold.

Beck, U. (1992), *Risk Society: Towards a New Modernity*, London, Sage.

Beck, U. (2000), *What is Globalization?*, Cambridge, Polity Press.

Bell, D. (1974), *The Coming of Post-industrial Society: A Venture in Social Forecasting*, London, Heinemann Educational.

Bell, D. and Binnie, J. (2000), *The Sexual Citizen: Queer Politics and Beyond*, London, Polity Press.

Bell, D. and Valentine, G. (1994), *Consuming Geographies: We Are What we Eat*, London, Routledge.

Bell, D. and Valentine, G. (1995), *Mapping Desire: Geographies of Sexualities*, London, Routledge.

Bello, W.F. (2002), *Deglobalization: Ideas for a New World Economy*, London, Zed Books.

Bertram, I.G. (1999), 'Economy', in Rapaport, M. (ed.), *Pacific Islands: Environment and Society*, Honolulu, University of Hawaii Press.

Bertram, I.G. (2006), 'The political economy of small islands in the 21st century: beyond MIRAB?', Special issue, *Asia Pacific Viewpoint*, 47, 1, in press.

Bertram, I.G. and Watters, R.F. (1985), 'The MIRAB economy in South Pacific microstates', *Pacific Viewpoint*, 23, 3, 497–519.

Blunt, A. and McEwan, C. (eds) (2003), *Postcolonial geographies*, Continuum, London.

Boyer, R. and Drache, D. (1996), *States Against Markets: The Limits of Globalization*, London, Routledge.

Bradshaw, M.J. (1990), 'New regional geography, foreign area studies and perestroyka', *Area*, 22, 4, 315–322.

Bradshaw, M.J. (2004a), *A New Economic geography of Russia*, London, Routledge.

Bradshaw, M.J. (2004b), 'Resources and development', *in* Daniels, P.W., Bradshaw, M.J., Shaw, D.J.B. and Sidaway, J.D. (eds), *Human Geography: Issues for the 21st Century* (2nd edn), London, Longman.

Brecher, J. and Costello, T. (1994), *Global Village or Global Pillage: Economic Reconstruction from the Bottom Up*, Boston, MA, South End Press.

Brecher, J., Childs, J.B. and Cutler, J. (1993), *Global Visions: Beyond the New World Order* (1st edn), Boston, MA, South End Press.

Brecher, J., Costello, T. and Smith, B. (2000), 'Globalization from below', *Nation*, 271, 18, 19–22.

Brenner, N. and Theodore, N. (2003), *Spaces of Neoliberalism*, Oxford, Blackwell.

Bridge, G. (2002), 'Grounding globalization: the prospects and perils of linking economic processes of globalization to environmental outcomes', *Economic Geography*, 78, 3, 361–386.

Britton, S. and Clarke, W.C. (1987), *Ambiguous Alternative: Tourism in Small Developing Countries*, Suva, University of the South Pacific.

Brohman, J. (1996), *Popular Development: Rethinking the Theory and Practice of Development*, Oxford, Blackwell.

Brookfield, H.C. (1975), *Interdependent Development*, London, Methuen.

Brundtland, G.H. (1987), *Our Common Future*, Oxford, World Commission on Environment and Development, Oxford University Press.

Bryceson, D., Kay, C. and Mooij, J. (eds) (2000), *Disappearing Peasantries? Rural Labour in Africa, Asia and Latin America*, London, Intermediate Technology Publications.

Bryson, J. and Henry, N. (2001), 'The global production system: from Fordism to Post-Fordism', *in* Daniels, P.W., Bradshaw, M.J., Shaw, D.J.B. and Sidaway, J.D. (eds), *Human Geography: Issues for the 21st Century* (1st edn), London, Prentice Hall.

Buchanan, K.M. (1963), 'The Third World: its emergence and contours', *New Left Review*, 18, 5–23.

Buchanan, K.M. (1964), 'Profiles of the Third World', *Pacific Viewpoint*, 5, 2, 97–126.

Buckingham-Hatfield, S. (2000), *Gender and Environment*, London, Routledge.

Burton, J.W. (1972), *World Society*, Cambridge, Cambridge University Press.

Bygrave, M. (2002), 'Where have all the protestors gone', *Guardian Weekly*.

Callinicos, A. (2001), 'Where now', *in* Bircham, E. and Charlton, J. (eds), *Anti-capitalism – A Guide to the Movement*, London, Bookmarks Publications.

Calvert, P. and Rengger, N.J. (2002), *Treaties and Alliances of the World* (7th edn), London, John Harper.

Carson, R., Darling, L. and Darling, L. (1962), *Silent Spring*, Boston, MA, Riverside Press.

Castells, M. (1996), *The Rise of the Network Society*, Malden, MA, Oxford, Blackwell.

Castles, S. and Miller, M.J. (1993), *The Age of Migration: International Population Movements in the Modern World*, New York, Guilford Press.

Castree, N. (2000), 'Environmental movement', *in* Johnston, R.J., Gregory, D., Pratt, G. and Watts, M. (eds), *The Dictionary of Human Geography*, London, Blackwell.

Champion, T. (2001), 'Demographic transformations', *in* Daniels, P.W., Bradshaw, M.J., Shaw, D.J.B. and Sidaway, J.D. (eds), *Human Geography: Issues for the 21st Century*, Harlow, Prentice Hall.

Chandra, R. (1992), *Industrialization and Development in the Third World*, London, Routledge.

Chang, T.C. (ed.) (2005), 'Place, Memory and Identity in "New" Asia', special edition of *Asia Pacific Viewpoint*, 46, 3, London and Melbourne, Blackwell.

Chant, S. (1999), 'Urban livelihoods, employment and gender', *in* Gwynne, R.N. and Kay, C. (eds), *Latin America Transformed: Globalization and Modernity*, London, Arnold.

Cho, G. (1995), *Trade, Aid, and Global Interdependence*, London, Routledge.

Chomsky, N. (2001), *9–11*, New York, Seven Stories.

Chomsky, N. (2004), *Hegemony or Survival? America's Quest for Global Dominance*, London, Penguin.

Christopherson, S. (2002), 'Changing women's status in a global economy', *in* Johnston, R.J., Taylor, P.J. and Watts, M. (eds), *Geographies of Global Change: Remapping the World*, London, Blackwell.

Cline-Cole, R. and Robson, E. (eds) (2003), *West African Worlds*, London, Pearson Education.

Cloke, P.J., Crang, P. and Goodwin, M. (1999), *Introducing Human Geographies*, London, Arnold.

Cloke, P.J., Philo, C. and Sadler, D. (1991), *Approaching Human Geography: An Introduction to Contemporary Theoretical Debates*, London, Paul Chapman.

Coe, N., Kelly, P.F. and Olds, K. (2003), 'Globalization, transnationalism and the Asia Pacific', *in* Peck, J.A. and Yeung, H.W.C. (eds), *Remaking the Global Economy: Economic-geographical Perspectives*, London, Sage.

Connell, J. (2003a), 'Regulation of space in the contemporary postcolonial Pacific city: Port Moresby and Suva', *Asia Pacific Viewpoint*, 44, 3, 243–258.

Connell, J. (2003b), 'Losing ground? Tuvalu, the greenhouse effect and the garbage can', *Asia Pacific Viewpoint*, 44, 2, 89–108.

Connell, J. and Gibson, C. (2003), *Sound Tracks: Popular Music, Identity and Place*, London, Routledge.

Corbridge, S. (1993), *Debt and Development*, Oxford, Blackwell.

Corbridge, S. (2002), 'Third World debt', *in* Desai, V. and Potter, R.B. (eds), *The Companion to Development Studies*, London, Arnold.

Corbridge, S., Thrift, N. and Martin, R. (eds) (1994), *Money, Power and Space*, Oxford, Blackwell.

Cowen, M.P. and Shenton, R.W. (1996), *Doctrines of Development*, London, Routledge.

Cox, K.R. (ed.) (1997), *Spaces of Globalization: Reasserting the Power of the Local*, New York, London, Guilford Press.

Crang, M. (1998), *Cultural Geography*, London, Routledge.

Crang, M. (2002), 'Qualitative methods: the new orthodoxy?', *Progress in Human Geography*, 26, 647–655.

Crang, M. and Thrift, N.J. (eds) (2000), *Thinking Space*, London, Routledge.

Crang, P. (1999), 'Local–global', *in* Cloke, P.J., Crang, P. and Goodwin, M. (eds), *Introducing Human Geographies*, London, Arnold.

Crush, J. (ed.) (1995), *Power of Development*, London, Routledge.

Crystal, D. (1997), *Cambridge Encyclopaedia of Language*, Cambridge, Cambridge University Press.

Daniels, P.W. (2001), 'The geography of the economy', *in* Daniels, P.W., Bradshaw, M.J., Shaw, D.J.B. and Sidaway, J.D. (eds), *Human Geography: Issues for the 21st Century*, Harlow, Prentice Hall.

Daniels, P.W. and Lever, W. (eds) (1996), *The Global Economy in Transition*, Harlow, Longman.

Daniels, P.W., Bradshaw, M.J., Shaw, D.J.B. and Sidaway, J.D. (eds) (2001), *Human Geography: Issues for the 21st Century*, Harlow, Prentice Hall.

Dicken, P. (1998), *Global Shift: Transforming the World Economy* (3rd edn), London, Paul Chapman.

Dicken, P. (2000), 'Localization', *in* Johnston, R.J., Gregory, D., Taylor, P.J. and Watts, M. (eds), *The Dictionary of Human Geography*, London, Blackwell.

Dicken, P. (2003), *Global Shift: Reshaping the Global Economic Map in the 21st Century* (4th edn), London, Sage.

Dicken, P. (2004), 'Geographers and "globalization": (yet) another missed boat?', *Transactions of the Institute of British Geographers*, 29, 5–26.

Dicken, P., Kelly, P.F., Olds, K. and Yeung, H.W.C. (2001), 'Chains and networks, territories and scales: towards a relational framework for analysing the global economy', *Global Networks*, 1, 2, 89–112.

Dickenson, J.P., Gould, B., Clarke, C., Mather, S., Prothero, M., Siddle, D., Smith, C. and Thomas-Hope, E. (1996), *A Geography of the Third World* (2nd edn), London, Routledge.

Dunning, J.H. (1988), *Multinationals, Technology and Competitiveness*, London, Unwin Hyman.

Dwyer, C. (1999), 'Migrations and diaspora', *in* Cloke, P.J., Crang, P. and Goodwin, M. (eds), *Introducing Human Geographies*, London, Arnold.

Economist (2003), 'The revenge of geography', *The Economist*, 15 March, 19–23.

Ehrlich, P.R. (1971), *The Population Bomb*, Cutchogue, NY, Buccaneer Books.

Escobar, A. (1995), *Encountering Development: The Making and Unmaking of the Third World*, Princeton, NJ, Princeton University Press.

Esteva, G. (1992), 'Development', *in* Sachs, W. (ed.), *The Development Dictionary: A Guide to Knowledge as Power*, London, Zed Books.

Esteva, G. and Prakash, M.S. (1998), *Grassroots Post-modernism: Remaking the Soil of Cultures*, London, Zed Books.

Findlay, A. (1995), 'Population crises: the Malthusian specter?', *in* Johnston, R.J., Taylor, P.J. and Watts, M. (eds), *Geographies of Global Change: Remapping the World in the Late Twentieth Century*, Oxford, Blackwell.

Firth, S. (2000), 'The Pacific Islands and the globalization agenda', *The Contemporary Pacific*, 12, 1, 178–192.

Frank, A.G. (1969), *Latin America: Essays on the Development of Underdevelopment*, New York, Monthly Review Press.

Frèobel, F., Heinrichs, J. and Kreye, O. (1980), *The New International Division of Labour: Structural Unemployment in Industrialised Countries and Industrialisation in Developing Countries*, Cambridge, Paris, Cambridge University Press, Maison des sciences de l'homme.

Friedmann, J. (1986), 'The world city hypothesis', *Development and Change*, 17, 69–83.

Friedmann, J. (1996), 'Where we stand: a decade of World City research', *in* Knox, P.L. and Taylor, P.J. (eds), *World Cities in a World System*, Cambridge, Cambridge University Press.

Fukuyama, F. (1992), *The End of History and the Last Man*, London, Hamish Hamilton.

Geddes, P. (1915), *Cities in Evolution, an Introduction to the Town Planning Movement and to the Study of Civics*, London, Benn.

George, S. (2001), 'Corporate globalization', *in* Bircham, E. and Charlton, J. (eds), *Anti-capitalism: A Guide to the Movement*, London, Bookmark.

Gereffi, G. (1994), 'The organization of buyer-driven global commodity chains: how US retailers shape overseas production networks', *in* Gereffi, G. and Korzeniewicz, M. (eds), *Commodity Chains and Global Capitalism*, Westport, CT, and London, Praeger.

Gereffi, G. (1996), 'Global commodity chains: new forms of coordination and control among nations and firms in international industries', *Competition & Change*, 1, 427–439.

Gibson, L. (2004), 'Empty shells: demographic decline and opportunity in Niue', *in* Terry, J.P. and Murray, W.E. (eds), *Niue Island: Geographical Perspectives on the Rock of Polynesia*, Paris, INSULA, UNESCO.

Gibson-Graham, J.K. (2002), 'Beyond global vs. local: economic politics outside the binary frame', *in* Herod, A. and Wright, M.W. (eds), *Geographies of Power: Placing Scale*, Oxford, Blackwell.

Giddens, A. (1985), *A Contemporary Critique of Historical Materialism*, Cambridge, Polity Press.

Giddens, A. (1990), *The Consequences of Modernity*, Cambridge, Polity Press.

Giddens, A. (1991), *Modernity and Self-identity: Self and Society in the Late Modern Age*, Cambridge, Polity Press.

Giddens, A. (1999), *Runaway World: How Globalization is Reshaping our Lives*, London, Profile.

Gilbert, A. (1996), *The Mega-city in Latin America*, Tokyo, United Nations University Press.

Gill, S. (1995), 'Globalization, market civilization, and disciplinary neoliberalism', *Millenium*, 24, 3, 399–425.

Gilpin, R. (1987), *The Political Economy of International Relations*, Princeton, NJ, Princeton University Press.

Gilpin, R. (2001), *Global Political Economy: Understanding the International Economic Order*, Princeton, NJ, Princeton University Press.

Glasius, M., Kaldor, M. and Anheier, H. (eds) (2002), *Global Civil Society*, Oxford, Oxford University Press.

Goldsmith, E. and Mander, J. (eds) (2001), *The Case Against the Global Economy*, London, Earthscan.

Goodman, D. and Watts, M. (eds) (1997), *Globalizing Food*, London, Routledge.

Goodwin, M. (1999), 'Citizenship and governance', *in* Cloke, P.J., Crang, P. and Goodwin, M. (eds), *Introducing Human Geographies*, London, Arnold.

Gordon, D.M. (1988), 'The global economy: new edifice or crumbling foundations?', *New Left Review*, 168, 24–64.

Grassman, S. (1980), 'Long term trends in openness of international economies', *Oxford Economic Papers*, 32, 1, 122–133.

Gray, J. (1998), *False Dawn: The Delusions of Global Capitalism*, London, Granta Books.

Gregory, D. (1994), *Geographical Imaginations*, Blackwell, Oxford.

Gregory, D. (2004), *The Colonial Present: Afghanistan, Palestine, Iraq*, Oxford, Blackwell.

Greider, W. (1997), *One World, Ready or Not: The Manic Logic of Global Capitalism*, London, Allen Lane.

Gruffudd, P. (1999), 'Nationalism', *in* Cloke, P.J., Crang, P. and Goodwin, M. (eds), *Introducing Human Geographies*, London, Arnold.

Gwynne, R.N. (1985), *Industrialization and Urbanization in Latin America*, London, Croom Helm.

Gwynne, R.N. (1990), *New Horizons? Third World Industrialization in an International Framework*, Harlow, Essex, Longman.

Gwynne, R.N. (2004), 'Political economy, resource use, and Latin American environments', *Singapore Journal of Tropical Geography*, 24, 3, 247–260.

Gwynne, R.N. and Kay, C. (2000), 'Views from the periphery: future of neoliberalism in Latin America', *Third World Quarterly*, 21, 1, 121–156.

Gwynne, R.N. and Kay, C. (2004), *Latin America Transformed: Globalization and Modernity* (2nd edn), London, Arnold.

Gwynne, R.N., Klak, T. and Shaw, D.J.B. (2003), *Alternative Capitalisms: Geographies of Emerging Regions*, London, Arnold.

Hägerstrand, T. (1968), *Innovation Diffusion as a Spatial Process*, Chicago, IL, University of Chicago Press.

Hägerstrand, T. (1975), 'Space, time and human conditions', *in* Karlquist, A., and Lundquist, L. and Snickers, F. (eds) *Dynamic Location of Urban Space*, Farnborough, Saxon House.

Hall, P.G. (1984), *The World Cities* (3rd edn), London, Weidenfeld & Nicolson.

Hamnett, C. (2003), 'Contemporary human geography: fiddling while Rome burns', *Geoforum*, 34, 1–4.

Hardin, G.J. (1968), 'The tragedy of the commons', *Science*, 162, 1243–1248.

Hardt, M. and Negri, A. (2000), *Empire*, Cambridge, MA, and London, Harvard University Press.

Hardt, M. and Negri, A. (2001), 'What the protestors in Genoa want', *in* Negri, A. (ed.), *On Fire: The Battle of Genoa and the Anti-capitalist Movement*, London, One-Off Press.

Hart, G. (2001), 'Development critiques in the 1990s: culs de sac and promising paths', *Progress in Human Geography*, 25, 649–658.

Harvey, D. (1989), *The Condition of Postmodernity: An Enquiry into the Origins of Cultural Change*, Oxford, Blackwell.

Harvey, D. (1995), 'Globalization in question', *Rethinking Marxism*, 8, 4, 1–17.

Harvey, D. (1999), *The Limits to Capital* (new edn), London, Verso.

Harvey, D. (2000), *Spaces of Hope*, Edinburgh, Edinburgh University Press.

Harvey, D. (2001), *Spaces of Capital: Towards a Critical Geography*, Edinburgh, Edinburgh University Press.

Harvey, D. (2003), *The New Imperialism*, Oxford, Oxford University Press.

Hay, C. and Marsh, D. (eds) (2000), *Demystfying Globalization*, London, Macmillan.

Hayter, R., Barnes, T.J. and Bradshaw, M.J. (2003), 'Relocating resource peripheries to the core of economic geography's theorizing: rationale and agenda', *Area*, 35, 15–23.

Held, D. (ed.) (1991), *Political Theory Today*, Cambridge, Polity Press.

Held, D. (1995), *Democracy and the Global Order*, Cambridge, Polity Press.

Held, D. and McGrew, A.G. (2002), *Gloabalization/Anti-globalization*, Cambridge, Polity Press.

Held, D., McGrew, A.G., Goldblatt, D. and Perraton, J. (1999), *Global Transformations: Politics, Economics and Culture*, Cambridge, Polity Press.

Henderson, J. (2003), 'The future of democracy in Melanesia: what role for outside powers?', *Asia Pacific Viewpoint*, 44, 3, 225–242.

Herdt, G.H. (1997), *Same Sex, Different Cultures: Gays and Lesbians across Cultures*, Oxford, Westview Press.

Herman, E.S. and Chomsky, N. (1988), *Manufacturing Consent: The Political Economy of the Mass Media*, London, Vintage.

Herod, A. (2002), 'Global change in the world of organised labor', *in* Johnston, R.J., Taylor, P.J. and Watts, M. (eds), *Geographies of Global Change*, Oxford, Blackwell.

Herod, A. (2003), 'Scale: the local and the global', *in* Holloway, S., Rice, S. and Valentine, G. (eds), *Key Concepts in Geography*, London, Sage.

Hertz, N. (2001), *The Silent Takeover: Global Capitalism and the Death of Democracy*, London, Heinemann.

Hill, R.C. (1989), 'Comparing transnational production systems: the automobile in the USA and Japan', *International Journal of Urban and Regional Research*, 13, 3, 462–480.

Hillis, K. (1999), 'Cyberspace and cybercultures', *in* Cloke, P.J., Crang, P. and Goodwin, M. (eds), *Introducing Human Geographies*, London, Arnold.

Hines, C. (2000), *Localization: A Global Manifesto*, London, Earthscan.

Hirst, P.Q. (1997), *From Statism to Pluralism: Democracy, Civil Society and Global Politics*, London, UCL Press.

Hirst, P.Q. and Thompson, G. (1999), *Globalization in Question: The International Economy and the Possibilities of Governance* (2nd edn), Cambridge, Polity Press.

Houtart, F. (2001), 'Alternatives to the neoliberal model', *in* Hourtart, F. and Polet, F. (eds), *The Other Davos: The Globalization of Resistence to the World Economic System*, London, Zed Books.

Hughes, H. (2003), 'Aid has failed the Pacific', *Issue Analysis 33*, Sydney, The centre for independent studies.

Huntington, S.P. (1991), *The Third Wave: Democratization in the Late Twentieth Century*, Norman, London, University of Oklahoma Press.

Huntington, S.P. (1996), *The Clash of Civilizations and the Remaking of the World Order*, New York, Simon & Schuster.

Ilbery, B. (2001), 'Changing geographies of global food production', *in* Daniels, P.W., Bradshaw, M.J., Shaw, D.J.B. and Sidaway, J.D. (eds), *Human Geography: Issues for the 21st Century*, Harlow, Pearson Education.

Imry, R. (2004), 'Urban geography, relevance, and resistance to the "policy turn"', *Urban Geography*, 25, 8, 697–708.

Inglehart, R. (1990), *Culture Shift in Advanced Industrial Society*, Princeton, NJ, Princeton University Press.

Intergovernmental Panel on Climate Change (IPCC) (2001), *Climate Change 2001*, Cambridge, IPCC and Cambridge University Press.

International Federation of the Phonographic Industry (IFPI) (2004), 'Global music retail sales', available online at http://www.ifpi.org/site-content/publications/rin_order.html, accessed 12 November.

Jackson, P. (1989), *Maps of Meaning*, London, Unwin Hyman.

Jackson, P. (2002), 'Consumption in a globalizing world', *in* Johnston, R.J., Taylor, P.J. and Watts, M. (eds), *Geographies of Global Change*, Oxford, Blackwell.

Janelle, D.G. (1968), 'Central place development in a time-space framework', *Professional Geographer*, 20, 1, 5–10.

Janelle, D.G. (1969), 'Spatial reorganization: a model and concept', *Annals of the Association of American Geographers*, 59, 348–364.

Janelle, D.G. (1973), 'Measuring human extensibility in a shrinking world', *Journal of Geography*, 72, 8–15.

Jenkins, R.O. (1987), *Transnational Corporations and Uneven Development: The Internationalization of Capital and the Third World*, London, Methuen.

Johnston, R.J. (1986), *On Human Geography*, Oxford, Blackwell.

Johnston, R.J. (1996), *Nature, State and Economy: The Political Economy of Environmental Problems*, Chichester, John Wiley.

Johnston, R.J. (2000), 'Relevance', *in* Johnston, R.J., Gregory, D., Pratt, G. and Watts, M. (eds), *The Dictionary of Human Geography*, Oxford, Blackwell.

Johnston, R.J. and Sidaway, J.D. (2004), *Geography and Geographers: Anglo-American Human Geography since 1945* (6th edn), London, Arnold.

Johnston, R.J., Taylor, P.J. and Watts, M. (eds) (2002), *Geographies of Global Change: Remapping the World*, Oxford, Blackwell.

Johnston, R.J., Gregory, D., Pratt, G. and Watts, M. (eds) (2000), *The Dictionary of Human Geography*, Blackwell, Oxford.

Jones, R.L. (2000), 'Deforestation', *in* Thomas, D.S.G. and Goudie, A. (eds), *The Dictionary of Physical Geography*, Oxford, Blackwell.

Jubilee 2000 (2002), 'About Jubilee research', UK, www.jubilee2000uk.org.

Karl, T.L. (1997), *The Paradox of Plenty: Oil Booms and Petro-states*, Los Angeles, University of California Press.

Kay, C. (1989), *Latin American Theories of Development and Underdevelopment*, London, Routledge.

Kay, C. (2001), 'Reflections on rural violence in Latin America', *Third World Quarterly*, 22, 5.

Kay, C. (2002), 'Why East Asia overtook Latin America: agrarian reform, industrialisation and development', *Third World Quarterly*, 23, 6, 1072–1102.

Kelly, P.F. (1999), 'The geographies and politics of globalization', *Progress in Human Geography*, 23, 3, 379–400.

Kelsey, J. (1995), *The New Zealand Experiment: A World Model for Structural Adjustment?*, Auckland, NZ, Auckland University Press.

Keohane, R.O. (1995), ''Hobbes' dilemma and institutional change in world politics: sovereignty in international society', *in* Holm, H.H. and Sorensen (eds), *Whose World Order?*, Boulder, CO, Westview Press.

Khan, L.A. (1996), *The Extinction of the Nation State: A World Without Borders*, The Hague, Kluwer Law International.

Kindleberger, C. (1967), *Europe's Post-war Growth*, Oxford and New York, Oxford University Press.

Kitchin, R. and Dodge, M. (2002), 'The emerging geographies of cyberspace', *in* Johnston, R.J., Taylor, P.J. and Watts, M. (eds), *Geographies of Global Change: Remapping the World* (2nd edn), Oxford, Blackwell.

Klak, T. (1998), *Globalization and Neoliberalism: The Caribbean Context*, Lanham, Oxford, Rowman & Littlefield.

Klein, N. (2001), *No Logo: No Space, No Choice, No Jobs*, London, Flamingo.

Knox, P.L. (2002), 'World cities and organization of global space', *in* Johnston, R.J., Taylor, P.J. and Watts, M. (eds), *Geographies of Global Change*, Oxford, Blackwell.

Knox, P.L. and Agnew, J.A. (1998), *The Geography of the World-economy* (3rd edn), London, Edward Arnold.

Knox, P.L. and Taylor, P.J. (eds) (1995), *World Cities in a World Economy*, Cambridge, Cambridge University Press.

Korten, D.C. (1995), *When Corporations Rule the World*, West Hartford, Kumarian Press.

Krugman, P.R. (1996), *Pop Internationalism*, Cambridge, MA, and London, MIT Press.

Krugman, P.R. (1998), 'What's new about the "new" economic geography?', *Oxford Review of Economic Policy*, 14, 7–17.

Larner, W. and LeHeron, R.B. (2002), 'From economic globalisation to globalising economic processes: towards post-structural political economies', *Geoforum*, 33, 4, 415–419.

Lash, S. and Urry, J. (1994), *Economies of Signs and Space*, London, Sage.

Lee, H. (2004), 'Second generation Tongan transnationalism: hope for the future?', *Asia Pacific Viewpoint*, 45, 2, 235–254.

Lee, R. (2000a), 'Economic geography', *in* Johnston, R.J., Gregory, D., Pratt, G. and Watts, M. (eds), *The Dictionary of Human Geography*, Oxford, Blackwell.

Lee, R. (2000b), 'New international division of labour (NIDL)', *in* Johnston, R.J., Gregory, D., Pratt, G. and Watts, M. (eds), *The Dictionary of Human Geography* (4th edn), Oxford, Blackwell.

Lee, R. (2000c), 'Transational corporation', *in* Johnston, R.J., Gregory, D., Pratt, G. and Watts, M. (eds), *The Dictionary of Human Geography*, Oxford, Blackwell.

Lee, R. (2002), '"Nice maps, shame about the theory"? Thinking geographically about the economic', *Progress in Human Geography*, 26, 3, 333–355.

Lee, R. and Smith, D.M. (2004), *Geographies and Moralities: International Perspectives on Development, Justice and Place*, Malden, MA, Blackwell.

LeHeron, R.B. (1993), *Globalised Agriculture: Political Choice*, Oxford, Pergamon Press.

Levitt, P. (2001), *The Transnational Villagers*, Berkeley, London, University of California Press.

Lewis, A. (1981), 'The rate of growth of world trade, 1830–1973', *in* Grassman, S. and Lundberg, E. (eds), *The World Economic Order: Past and Prospects*, Basingstoke, Macmillan.

Lewis, W.A. (1955), *The Theory of Economic Growth*, London, Allen & Unwin.

Leyshon, A. (1995), 'Annihilating space? the speed up of communications', *in* Allen, J. and Hamnett, C. (eds), *A Shrinking World?*, Oxford, Oxford University Press.

Leyshon, A. (1996), 'Dissolving difference? Money, disembedding and the creation of global financial space', *in* Daniels, P.W. and Lever, W. (eds), *The Global Economy in Transition*, London, Longman.

Lin Sien, C. (ed.) (2003), *Southeast Asia Transformed: A Geography of Change*, Institute of Southeast Asian Studies, Singapore.

Lipeitz, A. (1987), *Mirages and Miracles: The Crises of Global Fordism*, London, Verso.

Love, J.F. (1995), *McDonalds: Behind the Arches*, New York, Bantam.

Lowe, P., Murdoch, J., Marsden, T., Munton, R. and Flynn, A. (1993), 'Regulating the new rural spaces: the uneven development of land', *Journal of Rural Studies*, 9, 205–222.

Luard, E. (1990), *The Globalization of Politics: The Changed Focus of Political Action in the Modern World*, London, Macmillan.

McCarthy, J. (2000), 'Social movements', *in* Johnston, R.J., Gregory, D., Pratt, G. and Watts, M. (eds), *The Dictionary of Human Geography*, Oxford, Blackwell.

McCormick, J. (1989), *Reclaiming Paradise: The Global Environmental Movement*, Bloomington, Indiana University Press.

McDowell, L. (1997), *Undoing Place?: A Geographical Reader*, London, Arnold.

McDowell, L. (1999), *Gender, Identity and Place: Understanding Feminist Geographies*, Cambridge, Polity Press.

McEwan, C. (2001), 'Geography, culture, and global change', *in* Daniels, P.W., Bradshaw, M.J., Shaw, D.J.B. and Sidaway, J.D. (eds), *Human Geography: Issues for the 21st Century*, London, Pearson Education.

McGrew, A.G. (2000), 'Sustainable globalization?', *in* Allen, T. and Thomas, A. (eds), *Poverty and Development: Into the 21st Century*, Oxford, Oxford University Press.

McGrew, A.G. and Lewis, P.G. (1992), *Global politics: Globalization and the Nation-state*, Cambridge, Polity Press.

McIntyre, M. and Soulsby, J. (2004), 'Land use and land degradation on Niue', *in* Terry, J.P. and Murray, W.E. (eds), *Niue Island: Geographical Perspectives on the Rock of Polynesia*, Paris, INSULA, UNESCO.

McKenna, M.K.L. and Murray, W.E. (2002), 'Jungle law in the orchard: comparing globalization in the New Zealand and Chilean apple industries', *Economic Geography*, 78, 4, 495–514.

McLuhan, M. (1962), *The Gutenberg Galaxy: The Making of Typographic Man*, Toronto, University of Toronto Press.

McLuhan, M. (1964), *Understanding Media: The Extensions of Man* (1st edn), New York, McGraw-Hill.

McMichael, P. (2004), *Development and Social Change: A Global Perspective* (3rd edn), Thousand Oaks, CA, and London, Pine Forge.

Malecki, E.J. (1991), *Technology and Economic Development*, New York, Longman.

Mann, M. (1988), *States, War and Capitalism: Studies in Political Sociology*, Oxford, Blackwell.

Mansvelt, J. (2005), *Geographies of Consumption*, London, Sage.

Martin, R. (1994), 'Stateless monies, global financial integration and national ecoomic autonomy: the end of geography?', *in* Corbridge, S., Martin, R. and Thrift, N.J. (eds), *Money, Power and Space*, Oxford, Blackwell.

Martin, R. (2004), 'Geography: making a difference in a globalising world', *Transaction of the Institute of British Geographers*, 29, 2, 147–150.

Massey, D.B. (1984), *Spatial Divisions of Labour: Social Structures and the Geography of Production* (1st edn), London, Macmillan.

Massey, D.B. (1991), 'A global sense of place', *Marxism Today*, June, 24–29.

Massey, D.B. (1994), *Space, Place and Gender*, Cambridge, Polity Press.

Massey, D.B. (1995), *Spatial Divisions of Labour: Social Structures and the Geography of Production* (2nd edn), London, Macmillan.

Massey, D.B. (2001), 'Geography on the agenda', *Progress in Human Geography*, 25, 5–18.

Meadows, D.H. (1972), *The Limits to Growth: A Report for the Club of Rome's Project on the Predicament of Mankind*, London, Earth Island.

Mercier, K. (2003), 'The anti-globalisation movement in London', *Institute of Geography*, Victoria University of Wellington, Masters of Development Studies.

Meyer, W.R. and Turner, B.L. (1995), 'The Earth transformed: trends, trajectories and patterns', *in* Johnston, R.J., Taylor, P.J. and Watts, M. (eds), *Geographies of Global Change*, Oxford, Blackwell.

Mittelman, J.H. (2000), *The Globalization Syndrome: Transformation and Resistance*, Princeton, NJ, Princeton University Press.

Morgan, S. (1984), *Sisterhood is Global*, London, Feminist Press.

Murray, W.E. (1998), 'The globalisation of fruit, neoliberalism and the question of sustainability – lessons from Chile', *European Journal of Development Studies*, 10, 1, 201–227.

Murray, W.E. (1999), 'Local responses to global change in the Chilean fruit complex', *European Review of Latin American and Caribbean Studies*, 66, 19–38.

Murray, W.E. (2000), 'Neoliberal globalisation, "exotic" agro-exports and local change in the Pacific Islands: a study of the Fijian kava sector', *Singapore Journal of Tropical Geography*, 21, 3, 355–375.

Murray, W.E. (2001), 'The second wave of globalisation and agrarian change in the Pacific Islands', *Journal of Rural Studies*, 17, 2, 135–148.

Murray, W.E. (2002a), 'From dependency to reform and back again: the Chilean peasantry in the twentieth century', *Journal of Peasant Studies*, 29, 3/4, 190–227.

Murray, W.E. (2002b), 'Sustaining agro-exports in Niue: the inevitable failure of free-market restructuring', *Journal of Pacific Studies*, 24, 2, 211–228.

Murray, W.E. (2004a), 'Mercosur', *in* Forsyth, T.J. (ed.), *Encyclopedia of International Development*, London, Routledge.

Murray, W.E. (2004b), 'Neocolonialism', *in* Forsyth, T.J. (ed.), *Encyclopedia of International Development*, London, Routledge.

Murray, W.E. and Silva, E. (2004), 'The political economy of sustainable development', *in* Gwynne, R.N. and Kay, C. (eds), *Latin America Transformed: Globalization and Modernity*, London, Arnold.

Murray, W.E. and Storey, D. (2003), *Post-colonial Transformations and Political Conflict in Oceania*, Melbourne, Blackwell, Special edition of *Asia Pacific Viewpoint*, 44/3.

Murray, W.E. and Terry, J.P. (2004), 'Niue's place in the Pacific', *in* Terry, J.P. and Murray, W.E. (eds), *Niue Island: Geography on the Rock of Polynesia*, Paris, INSULA, UNESCO.

Nagar, R., Lawson, V., McDowell, L. and Hanson, S. (2002), 'Locating globalization: feminist (re)readings of the subjects and spaces of globalization', *Economic Geography*, 78, 3, 257–284.

Nicole, R. (2000), *The Word, the Pen and the Pistol: Orientalism in the Pacific*, New York, State University of New York Press.

Nunn, P.D. (2003), 'Revising ideas about environmental determinism: human–environment relations in the Pacific Islands', *Asia Pacific Viewpoint*, 44, 1, 63–72.

O'Brien, R. (1992), *Global Financial Integration: The End of Geography*, London, Royal Institute of International Affairs, Pinter Publishers.

Ohmae, K. (1990), *The Borderless World*, London, Collins.

Ohmae, K. (1995), *The End of the Nation State*, New York, Free Press.

O'Riordan, T. (1981), *Environmentalism* (2nd edn), London, Pion.

Ó' Tuathail, G. (1996), *Critical Geopolitics*, Minneapolis, Minneapolis University Press.

Ó' Tuathail, G. (2002), 'Post Cold-War geopolitics: contrasting superpowers in a world of global dangers', *in* Johnston, R.J., Taylor, P.J. and Watts, M. (eds), *Geographies of Global Change*, Oxford, Blackwell.

Overton, J.D. (2000), 'Vakavanua, vakamatanitu: discourses of development in Fiji', *Asia Pacific Viewpoint*, 40, 2, 121–134.

Overton, J.D. (2003), 'Understanding coups', *Asia Pacific Viewpoint*, 44, 3, 351–356.

Overton, J.D. and Scheyvens, R. (eds) (1999), *Strategies for Sustainable Development: Experiences from the Pacific*, London, Zed Books.

Pacione, M. (1999), *Applied Geography: Principles and Practice*, London, Routledge.

Painter, J. (1995), *Politics, Geography and Political Geography: A Critical Perspective*, London, New York, Arnold, Wiley.

Painter, J. (2000), 'Localization', *in* Johnston, R.J., Gregory, D., Pratt, G. and Watts, M. (eds), *The Dictionary of Human Geography*, Oxford, Blackwell.

Painter, J. and Philo, C. (1995), 'Spaces of citizenship', *Political Geography*, 14, 2, 107–120.

Peck, J. (2001), 'Neoliberalizing states', *Progress in Human Geography*, 25, 3, 445–455.

Peck, J. and Tickell, A. (1994), 'Jungle law breaks out: neoliberalism and global–local disorder', *Area*, 26, 4, 317–326.

Peck, J. and Tickell, A. (2002), 'Neoliberalizing space', *Antipode*, 34, 380–404.

Peck, J. and Yeung, H.W.C. (eds) (2003), *Remaking the Global Economy: Economic Geographical Perspectives*, London, Sage.

Peet, R. (1991), *Global Capitalism: Theories of Societal Development*, London, Routledge.

Peet, R. and Watts, M. (1993), 'Development theory and environment in an age of market triumphalism', *Economic Geography*, 69, 227–253.

Perlmutter, H.V. (1991), 'On the rocky road to the first global civilisation', *Human Relations*, 44, 9, 902–906.

Perraton, J. (2001), 'The global economy – myths and realities', *Cambridge Journal of Economics*, 25, 669–684.

Petras, J. (1993), 'Cultural imperialism in the late 20th century', *Journal of Contemporary Asia*, 23, 2, 139–148.

Petras, J. (1999), *The Left Strikes Back: Class Conflict in Latin America in the Age of Neoliberalism*, Boulder, CO, Westview Press.

Petras, J. and Vettmeyer, H. (2001), *Globalization Unmasked*, London, Zed Books.

Philo, C. (ed.) (1991), *New Words, New Worlds: Reconceptualising Social and Cultural Geography*, Aberystwyth, Cambrian Printers.

Pickering, K.T. and Owen, L.A. (1997), *An Introduction to Global Environmental Issues* (2nd edn), London, Routledge.

Pieterse, J.N. (1995), 'Globalization as hybridisation', *in* Featherston, S., Nash, S. and Robertson, R. (eds), *Global Modernities*, London, Sage.

Pieterse, J.N. (2001), 'Globalization and collective action', *in* Hamel, P., Lustiger-Thaler, H., Pietrse, J.N. and Roseneil, S. (eds), *Globalization and Social Movements*, Basingstoke, Palgrave.

Pieterse, J.N. and Parekh, B. (eds) (1995), *The Decolonisation of Imagination: Cultural Knowledge and Power*, London, Zed Books.

Pollard, J. (2001), 'The global financial system: worlds of monies', *in* Daniels, P.W., Bradshaw, M.J., Shaw, D.J.B. and Sidaway, J.D. (eds), *Human Geography: Issues for the 21st Century*, Harlow, Prentice Hall.

Porter, M.E. (1990), *The Competitive Advantage of Nations*, London, Macmillan.

Potter, D. (2000), 'Democratization, good governance, and development', *in* Allen, T. and Thomas, A. (eds), *Poverty and Development into the 21st Century*, Oxford, Oxford University Press.

Potter, L. and Cooke, F.M. (eds) (2004), *Negotiating Modernity and Globalisation in Rural and Marine Environments*, Special edition of *Asia Pacific Viewpoint*, Oxford, Blackwell.

Potter, R.B. (1993), 'Little England and little geography: reflections on Third World teaching and research', *Area*, 25, 291–294.

Potter, R.B., Binns, T., Elliott, J.A. and Smith, D. (1999), *Geographies of Development* (1st edn), Harlow, Prentice Hall.

Potter, R.B., Binns, T., Elliot, J.A. and Smith, D. (2004), *Geographies of Development* (2nd edn), Harlow, Prentice Hall.

Power, M. (2003), *Rethinking Development Geographies*, London, Routledge.

Power, M. and Sidaway, J.D. (2004), 'The degeneration of tropical geography', *Annals of the Association of American Geographers*, 94, 3, 585–601.

Prebisch, R. (1950), *The Economic Development of Latin America and its Principal Problems*, New York, United Nations.

Prebisch, R. (1964), *Towards a New Trade Policy for Development*, New York, UNCTAD.

Preston, P.W. (1996), *Development Theory*, Oxford, Blackwell.

Radcliffe, S.A. (2004), 'Civil society, grassroots politics and livelihoods', *in* Gwynne, R.N. and Kay, C. (eds), *Latin America Transformed: Globalization and Modernity*, London, Arnold.

Rapaport, M. (ed.) (1999), *Pacific Islands: Environment and Society*, Honolulu, University of Hawai'i Press.

Rennie-Short, J. (1993), *An Introduction to Political Geography* (2nd edn), London, Routledge.

Rennie-Short, J. (2004), 'Black holes and loose connections in a global urban network', *The Professional Geographer*, 56, 2, 295–302.

RIAA (2004), 'Anti-piracy', http://www.riaa.com/issues/piracy/default.asp.

Rigg, J. (2002), 'Of miracles and crises: (re-)interpretations of growth and decline in East and Southeast Asia', *Asia Pacific Viewpoint*, 43, 2, 137–156.

Rigg, J. (2003), *Southeast Asia: The Human Landscape of Modernization and Development*, London, Routledge.

Ritzer, G. (1993), *The McDonaldizaton of Society*, London, Pine Forge/Sage.

Roberts, D. (2001), *Guinness World Records: British Hit Singles and Albums*, Enfield, Guinness.

Roberts, J.T. and Hite, A. (eds) (2000), *From Modernization to Globalization: Perspectives on Development and Social Change*, Malden, MA, and Oxford, Blackwell.

Roberts, S.M. (2002), 'Global regulation and trans-state organization', *in* Johnston, R.J., Taylor, P.J. and Watts, M. (eds), *Geographies of Global Change*, Oxford, Blackwell.

Robertson, R. (1992), *Globalization: Social Theory and Global Culture*, London, Sage.

Robertson, R. (2003), *The Three Waves of Globalization*, London, Zed Books.

Rosenau, J.N. (1980), *The Study of Global Interdependence: Essays on the Transnationalization of World Affairs*, London, Pinter.

Rosenau, J.N. (1990), *Turbulence in World Politics: A Theory of Change and Continuity*, New York, London, Harvester Wheatsheaf.

Rostow, W. (1960), *The Stages of Economic Growth: A Non-Communist Manifesto*, Cambridge, Cambridge University Press.

Routledge, P. (1999), 'Survival and resistence', *in* Cloke, P.J., Crang, P. and Goodwin, M. (eds), *Introducing Human Geographies*, London, Arnold.

Routledge, P. (2002), 'Resisting and reshaping destructive development: social movements and globalizing networks', *in* Johnston, R.J., Taylor, P.J. and Watts, M. (eds), *Geographies of Global Change: Remapping the World* (2nd edn), Oxford, Blackwell.

Rugman, A. (2000), *The End of Globalization*, London, Random House.

Ruigrok, W. and van Tulder, R. (1995), *The Logic of International Restructuring*, London, Routledge.

Rupert, M. (2000), *Ideologies of Globalization*, London, Routledge.

Sassen, S. (2000), *Cities in a World Economy* (2nd edn), Thousand Oaks, CA, and London, Pine Forge Press.

Sassen, S. (2001), *The Global City: London, New York, Tokyo*, Princeton, NJ, Princeton University Press.

Sauer, C. (1925), 'The morphology of landscape', *University of California Publications in Geography*, 2, 19–54.

Scheyvens, R. (2002), *Tourism for Development: Empowering Communities*, Harlow, Prentice Hall.

Schirato, T. and Webb, J. (2003), *Understanding Globalization*, London, Sage.

Scholte, J.A. (2000), *Globalization: A Critical Introduction*, Basingstoke, Palgrave.

Scott, A.J. (1988), *New Industrial Space*, London, Pion Press.

Shurmer-Smith, P. and Hannam, K. (1994), *Worlds of Desire, Realms of Power: A Cultural Geography*, London, Edward Arnold.

Sidaway, J.D. (2000), 'Postcolonial geographies: an exploratory essay', *Progress in Human Geography*, 24, 591–612.

Sidaway, J.D. (2001a), 'Geopolitical traditions', *in* Daniels, P.W., Bradshaw, M.J., Shaw, D.J.B. and Sidaway, J.D. (eds), *Human Geography: Issues for the 21st Century*, London, Longman.

Sidaway, J.D. (2001b), 'The place of the nation-state', *in* Daniels, P.W., Bradshaw, M.J., Shaw, D.J.B. and Sidaway, J.D. (eds), *Human Geography: Issues for the 21st Century*, London, Longman.

Sidaway, J.D., Bunnell, T. and Yeoh, B.S.A. (2003), 'Editors' introduction: Geography and postcolonialism', *Singapore Journal of Tropical Geography*, 24, 3, 269–272.

Silva, E. (2004), 'Authoritarianism, democracy and development', *in* Gwynne, R.N. and Kay, C. (eds), *Latin America Transformed: Globalization and Modernity*, London, Edward Arnold.

Sklair, L. (2001), *The Transnational Capitalist Class*, Oxford, Blackwell.

Smith, D.M. (1994), *Geography and Social Justice*, Oxford, Blackwell.

Smith, D.M. (2000), *Moral Geographies: Ethics in a World of Difference*, Edinburgh, Edinburgh University Press.

Smith, G. (2000a), 'Nation-state', *in* Johnston, R.J., Gregory, D., Pratt, G. and Watts, M. (eds), *The Dictionary of Human Geography*, Oxford, Blackwell.

Smith, G. (2000b), 'Geopolitics', *in* Johnston, R.J., Gregory, D., Pratt, G. and Watts, M. (eds), *The Dictionary of Human Geography*, Oxford, Blackwell.

Smith, N. (1984), *Uneven Development*, Oxford, Blackwell.

Smith, N. (2000), 'Global Seattle', *Environment and Planning D*, 18, 1, 1–5.

Smith, N. (2003), *American Empire: Roosevelt's Geographer and the Prelude to Globalization*, Berkeley, University of California Press.

Soja, E.W. (1996), *Thirdspace: Journeys to Los Angeles and Other Real and-imagined Places*, Cambridge, MA, and Oxford, Blackwell.

Spybey, T. (1996), *Globalization and World Society*, Cambridge, Polity Press.

Sriskandarajah, D. (2003), 'Inequality and conflict in Fiji: from purgatory to hell?', *Asia Pacific Viewpoint*, 44, 3, 305–324.

Starr, A. (2000), *Naming the Enemy: Anti-corporate Movements Confront Globalization*, London, Zed Books.

Steger, M.B. (2002), *Globalism*, Maryland, Rowman and Littlefield.

Stiglitz, J.E. (2002), *Globalization and its Discontents*, London, Penguin.

Stoddard, D.R. (1987), 'To claim the high ground: geography for the end of the century', *Transactions of the Institute of British Geographers*, NS, 12, 327–336.

Storey, D. and Murray, W.E. (2001), 'Dilemmas of development in Oceania: the political economy of the Tongan agro-export sector', *Geographical Journal*, 167, 4, 291–304.

Storey, D., Bulloch, H. and Overton, J.D. (2005), 'The poverty consensus: some limitations of the "popular agenda"', *Progress in Development Studies*, 5, 1, 30–44.

Strange, S. (1996), *The Retreat of the State: The Diffusion of Power in the World Economy*, Cambridge, Cambridge University Press.

Stutz, F.P. and De Souza, A.R. (1998), *The World Economy: Resources, Location, Trade, and Development* (3rd edn), Upper Saddle River, NJ, Prentice Hall.

Sunkel, O. (ed.) (1993), *Development from Within: Towards a Neo-structuralist Approach for Latin America*, Boulder, CO, and London, Lynne Rienner.

Swyngedouw, E. (1997), 'Neither global or local: "glocalization" and the politics of scale', *in* Cox, K.R. (ed.), *Spaces of Globalization: Reasserting the Power of the Local*, New York, London, Guilford Press.

Taylor, G. (2002), 'We haven't gone away', *Observer Worldview* extra online www.observer.co.uk, 21 July.

Taylor, P.J. and Flint, C. (1989), *Political Geography: World Economy, Nation-state and Locality* (2nd edn), Harlow, Longman Scientific and Technical.

Taylor, P.J. and Flint, C. (2000), *Political Geography: World-Economy, Nation-state and Locality* (4th edn), Harlow, Prentice Hall.

Taylor, P.J., Watts, M. and Johnston, R.J. (2002), 'Geography/globalization', *in* Johnston, R.J., Taylor, P.J. and Watts, M. (eds), *Geographies of Global Change*, Oxford, Blackwell.

Thompson, C.J. and Tambyah, S.K. (1999), 'Trying to be cosmopolitan', *Journal of Consumer Research*, 26, 3, 214–241.

Thompson, G. and Allen, J. (1997), 'Think global, then think again: economic globalization in context', *Area*, 29, 3.

Thrift, N.J. (2000), 'Local–global dialectic', *in* Johnston, R.J., Gregory, D., Pratt, G. and Watts, M. (eds), *The Dictionary of Human Geography*, Oxford, Blackwell.

Thrift, N.J. (2002), 'A hyperactive world', *in* Johnston, R.J., Taylor, P.J. and Watts, M. (eds), *Geographies of Global Change*, Oxford, Blackwell.

Todaro, M.P. (1997), *Economic Development* (6th edn), London, Longman.

Toffler, A. (1970), *Future Shock*, London, The Bodley Head.

Tomlinson, J. (1991), *Cultural Imperialism: A Critical Introduction*, London, Pinter.

Tomlinson, J. (1999), *Globalization and Culture*, Cambridge, Polity Press.

Tyler-Millar, G. (2002), *Living in the Environment* (12th edn), London and New York, Brookes-Cole.

UNCTAD (2002), *The Least Developed Countries Report 2002*, New York, United Nations.

UNCTAD (2004), *World Investment Report*, New York, United Nations.

UNDP (various years), *Human Development Report*, Oxford, Oxford University Press.

Urry, J. (2003), *Global Complexity*, Cambridge, Polity Press.

Vernon, R. (1977), *Storm Over the Multinationals: The Real Issues*, London, Macmillan.

Wallerstein, I.M. (1980), *The Modern World-system II*, New York, Academic Press.

Waters, M. (2001), *Globalization* (2nd edn), London, Routledge.

Weber, M., Roth, G., Wittich, C. and Fischoff, E. (1978), *Economy and Society: An Outline of Interpretive Sociology*, Berkeley, London, University of California Press.

Weiss, L. (1998), *The Myth of the Powerless State: Governing the Economy in a Global Era*, Cambridge, Polity Press.

Whatmore, S. (2002), 'From farming to agri-business: global agri-food networks', *in* Johnston, R.J., Taylor, P.J. and Watts, M. (eds), *Geographies of Global Change*, Oxford, Blackwell.

Whatmore, S. and Thorne, J. (1997), 'Nourishing networks: alternative geographies of food', *in* Goodman, D. and Watts, M. (eds), *Globalizing Food*, London, Routledge.

Wieringa, S. (ed.) (1995), *Subversive Women: Women's Movements in Africa, Asia, Latin America and the Caribbean*, London, Zed Books.

Williamson, J. and Milner, C. (1991), *The World Economy: A Textbook in International Economics*, New York, London, Harvester Wheatsheaf.

World Bank (various years), *World Development Report*, Washington, DC, World Bank.

Yeung, H.W.C. (2002), 'The limits to globalization theory: a geographic perspective on global economic change', *Economic Geography*, 78, 3, 285–306.

Index